Nonlinear Problems in Future Particle Accelerators

Proceedings of the Workshop on

Nonlinear Problems in Future Particle Accelerators

Capri, Italy   19-25 April 1990

Editors

W. Scandale

G. Turchetti

World Scientific

Singapore • New Jersey • London • Hong Kong

Published by

World Scientific Publishing Co. Pte. Ltd.
P O Box 128, Farrer Road, Singapore 9128
USA office: 687 Hartwell Street, Teaneck, NJ 07666
UK office: 73 Lynton Mead, Totteridge, London N20 8DH

Library of Congress Cataloging-in-Publication data is available.

NONLINEAR PROBLEMS IN FUTURE PARTICLE ACCELERATORS

ISBN 981-02-0231-8

Printed in Singapore by JBW Printers & Binders Pte. Ltd.

FOREWORD

The design of future particle accelerators, based on superconducting magnets and cavities, requires an extensive analysis of the unavoidable nonlinear effects and the definition of effective correction strategies. As a consequence the underlying theoretical framework drastically changes from the linear analysis, on which the classical accelerator theory is essentially based, to the modern analysis of nonlinear dynamical systems.

The basic ideas were developed at the end of the last century by outstanding mathematicians like Poincaré and Cantor, who proved that even simple nonlinear systems could exhibit very unstable chaotic motion and provided the geometrical framework for their description. Long before the computers became available the use of discrete time systems was suggested and the basic mathematical theory quite extensively developed.

A nonlinear particle accelerator has many common features with a planetary system and the stability analysis raises the same difficulties and leads to the same orders of magnitudes, typically one billion orbits, far beyond the possibility of numerical simulations.

As a consequence the new accelerators are a challenge to the theory of dynamical systems, with a vast possibility of changing the model, unknown in celestial mechanics.

In this respect the interaction of people coming from two quite different areas, namely accelerator physics and nonlinear dynamics seemed to be quite promising and was the basic motivation for organizing this workshop. The answer was quite successful and physicists from all the major laboratories, where new projects are being developed or ready to start, have provided an overview of the major existing problems, which range from the definition of the relevant parameters of nonlinearity, the reliability and interpretation of long time simulation, the correction strategies of multipolar errors. On the other side the mathematical physicists outlined the major role of the map approach and the normal form theory to understand the region of moderate nonlinearities where the cahoticity is very limited, and possible strategies to understand the diffusion in extensively chaotic regions. The way of developing a mathematical theory of stochastic perturbations and the foundations of discrete dynamics were outlined.

In spite of the difference in language the barriers were easily overcome and several round tables were organized to discuss two of the major problems which emerged; the long time simulations and the stochastic and modulated diffusion. While it appeared that the nonlinear theory in weakly perturbed region (1/2 of the dynamical aperture) is well established, the last two problems are completely open and possibly will be the subject of a future workshop.

Acknowledgements

The editors would like to thank Prof. V. Vaccaro of the University of Naples and Prof. L. Palumbo of the Laboratori Nazionali di Frascati for their constant help and crucial collaboration in organizing the Workshop.

The financial support of the following institutions is specially acknowledged:

INFN (Istituto Nazionale di Fisica Nucleare) Presidenza
INFN Sezione di Bologna
INFN Sczione di Napoli
CERN
Dipartimento di Scienze Fisiche, Università de Napoli
Regione Campania

CONTENTS

INVITED PAPERS

THE LHC : GENERAL PARAMETERS AND MAGNETIC ERRORS

J. Gareyte, CERN, Geneva, Switzerland

Abstract :

The main features of the LHC are described, with emphasis on the choices made to maximize energy and luminosity, and to reduce cost. The different sources of non linear errors in the field of the magnets are reviewed, ant their consequences on particle dynamics and machine operation are outlined.

I Introduction

The LHC[1] is a high magnetic field twin aperture collider that CERN proposes to install on top of LEP in the LEP tunnel (fig. 1).

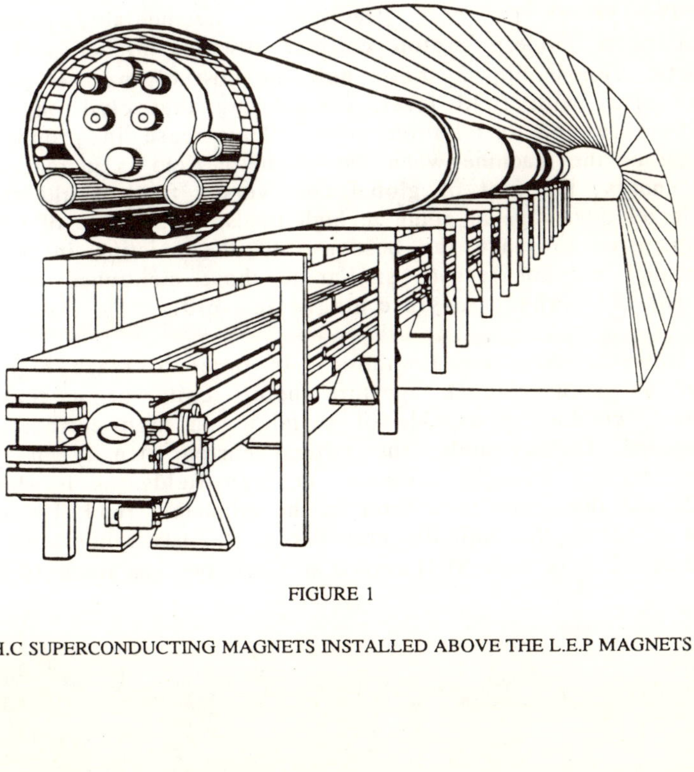

FIGURE 1

L.H.C SUPERCONDUCTING MAGNETS INSTALLED ABOVE THE L.E.P MAGNETS

It will produce collisions between two counterrotating beams of protons at a maximum energy of about 8 TeV per beam, with a very high luminosity in the range of a few $10^{34} cm^{-2}s^{-1}$. It can also be used to collide heavier species, like for instance lead nuclei, in this case providing enormous amounts of energy in the center of mass system of the collisions, although with a lower luminosity around $10^{28} cm^{-2}s^{-1}$. And since the world largest electron storage ring, LEP, will be sitting about 1m below the LHC, it will be possible to organize collisions between electrons and protons at a modest additional cost.

The superconducting bending magnet is by far the most important single item in the project, and deserves special attention in this report. In fact, it is because it would be technologically difficult and economically prohibitive to produce superconducting magnets of the same field quality as classical ones that we, accelerator physicists, are present at this workshop seeking the help of other disciplines. The price to pay for getting compact, high field machines is a non linear, hostile environment for the particles.

II <u>The LHC, a high field collider</u>

The search for cost effectiveness irremediably pushes the design of a modern proton collider towards higher and higher magnetic fields[2]. Fig. 2 shows how the cost of magnets per Tesla meter varies with field level. Considering that civil engineering, infrastructure and installation costs all decrease in proportion to the size of the machine when the magnetic field is increased for a fixed energy, the plot of global cost versus field would show an even larger advantage in going to high fields. In the case of CERN, it is imperative to make the best use of the existing infrastructure and tunnel and for reasons similar to those mentioned above this calls for the highest possible magnetic field.

However there are limitations. In order to reach high fields within a given magnet aperture using coils of a reasonable thickness, conductors capable of sustaining a high current density are needed. Unfortunately, the critical current density in type II superconductors strongly decreases at high fields, as is shown in Fig. 3, and this leads to a limit in the attainable field. Using NbTi cooled at 4.2 K, the limit lies around 6 to 7 Tesla, while with Nb_3S_n cooled at 4.2 K or with NbTi cooled at 1.8 K one can reach 10 Tesla.

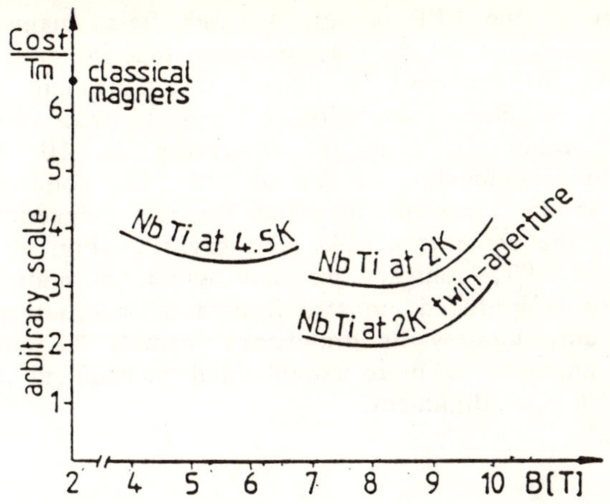

FIGURE 2

COST OF MAGNET + CRYOSTATS FOR 50 mm diam COIL APERTURE DIPOLES

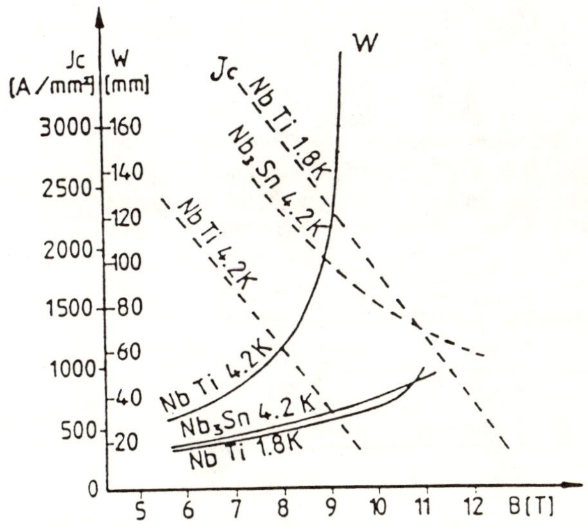

FIGURE 3

CURRENT DENSITY IN COMMERCIAL SUPERCONDUCTORS (Jc) AND AN EXAMPLE OF THE NECESSARY COIL THICKNESS IN DIPOLE MAGNETS (W)

Another limitation arises in the case of the LHC from the fixed cross section of the LEP tunnel. At high field, magnet dimensions tend to increase, one of the reasons being that the iron shield which surrounds the magnet must be placed at a sufficient distance from the coils to minimize saturation effects. It was found that two independent magnetic channels operating at 10 T would be impossible to accommodate on top of LEP. The solution was the so called "two in one" concept, in which the two magnetic channels are embedded in the same iron yoke and cryostat (Fig. 4). This allows the design of 10T, compact magnets which in addition have the advantage of a reduced construction and installation cost. This design has unfortunately its drawbacks, namely a mutual influence of the two channels, a more complicated mechanical structure, and difficulties for the alignment.

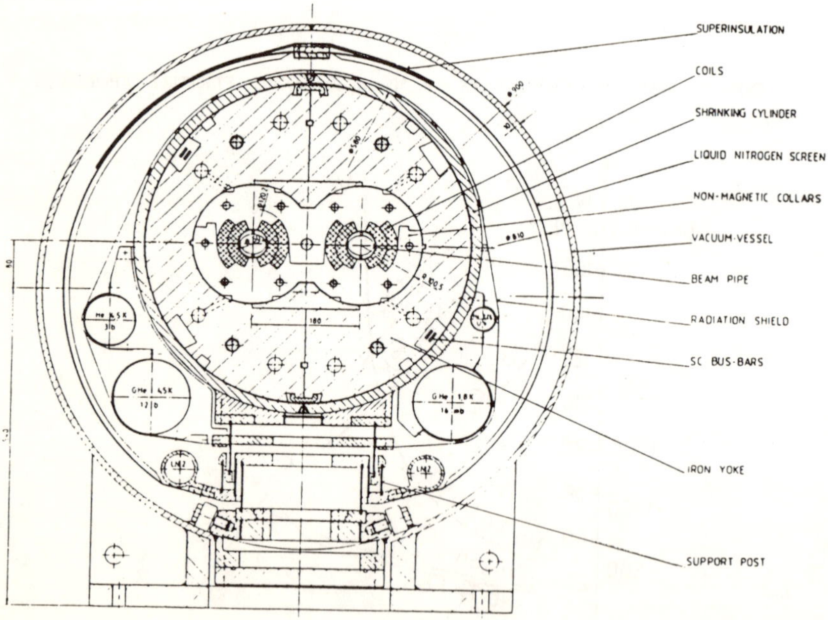

FIGURE 4

LHC DIPOLE : STANDARD CROSS-SECTION

III A high luminosity collider

There is a wide consensus in the high energy physics community about the necessity of providing particle collisions in the near future at an energy scale of about 1 TeV at the parton level. This aim can be realized with protons provided the energy is larger than 10 TeV, which is well within reach of the LHC. However, as the energy E is increased so that parton collisions can probe matter properties at smaller and smaller distances, the cross section of the interesting reactions tend to decrease like $1/E^2$. For this reason very high luminosities are required to assure reasonable event rates.

The luminosity of a collider is given by :

$$L = \frac{N^2 k f \gamma}{\pi \varepsilon^* \beta^*}$$

where N is the number of particles per bunch, k the number of bunches, f the revolution frequency, γ the energy divided by the rest energy, β^* the betatron function at the interaction point and

$$\varepsilon^* = \frac{4\sigma^2 \gamma}{\beta}$$

the normalized emittance, with σ the rms beam radius (the beam cross section is supposed to be round).

The main limitation to the luminosity comes from the electromagnetic beam-beam interaction. This produces a change of the particle tune, which is for the central particle in a beam :

$$\xi = \frac{r_o N}{\pi \varepsilon^*}$$

where r_0 is the classical particle radius. It is know from experience with the CERN SP$\bar{\text{P}}$S collider that the total tune spread across the beam particle distribution must not increase beyond 0.02 to preserve reasonable beam lifetimes. In the LHC, as we will see later, some unavoidable tune spread will be produced by various non linearities in the guiding fields. If we share equally the allowed tune spread between these effects and the beam-beam effect, we arrive at the following condition :

$$n\xi \leq 0.01$$

where n is the number of interaction regions.

For 3 interaction regions in the LHC one has $\xi \leq 0.0033$ and one obtains the maximum usable beam density :

$$\frac{N}{\varepsilon *} \leq 6.16\ 10^{15}$$

When this beam-beam imposed limit is reached, one can only increase the luminosity by increasing at the same time the beam intensity and the emittance.

The beam intensity is limited by collective effects, by the amount of synchrotron radiation power emitted (which has to be absorbed at cryogenic temperatures and hence at high cost) and by the difficulty of handling beams with hundreds of megajoules of strored energy. The emittance is limited by the dynamic aperture of the machine. In a linear machine the limit to the oscillation amplitude of the particles is the radius of the vacuum chamber in which the beam circulates (in the LHC about 20 mm). In presence of non linear restoring forces particles may become unstable and subsequently hit the aperture for starting amplitudes considerably smaller than the chamber radius. The dynamic aperture corresponds to the maximum amplitude of the particles which are just stable for a certain duration (from half and hour at injection energy to about 10 hours at collision energy, that is for 2.10^7 and 4.10^8 revolutions respectively). We learned at this workshop that this is a physicist's definition of the "basin of infinity".

IV Structure of the machine

LHC being in the same tunnel as LEP will also have 8 arcs and 8 long straight sections. The two proton beams, horizontally separated by 180 mm in the arcs, alternate from the outside to the inside in the middle of each of the 8 long straight sections, where in principle they could interact (Fig. 5).

The LHC lattice of FODO type is constituted by :

- 8 arcs, each of them containing 50 half cells,

- 8 insertions, each of them containing one long straight section and two dispersion suppressors of a type that allows trajectories of identical length for the hadrons in LHC and for the leptons in LEP.

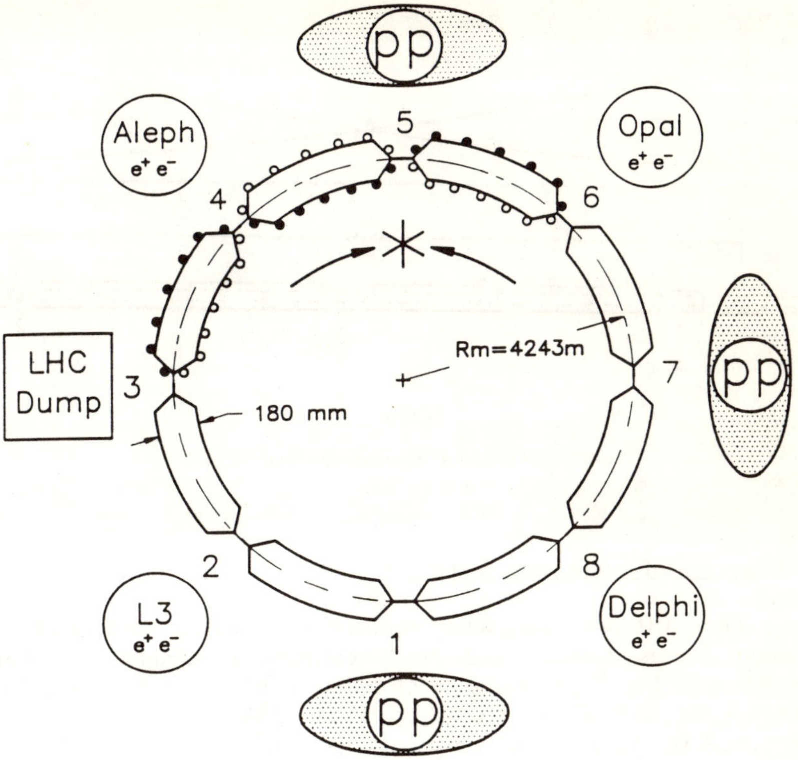

FIGURE 5

PRINCIPLE LAYOUT OF THE LHC

One half of the regular are cell (Fig. 6) consists of four dipoles D about 10 m long, a focussing (QF) and a defocussing (QD) quadrupole. On each beam stand individual correctors which are lumped near the main quadrupoles (a combined tuning quadrupole QT and octupole 0, a combined sextupole S, decapole D and closed orbit correction dipole COD, and a beam observation monitor BOM) and in the middle of the half cell (a combined sextupole, octupole and decapole).

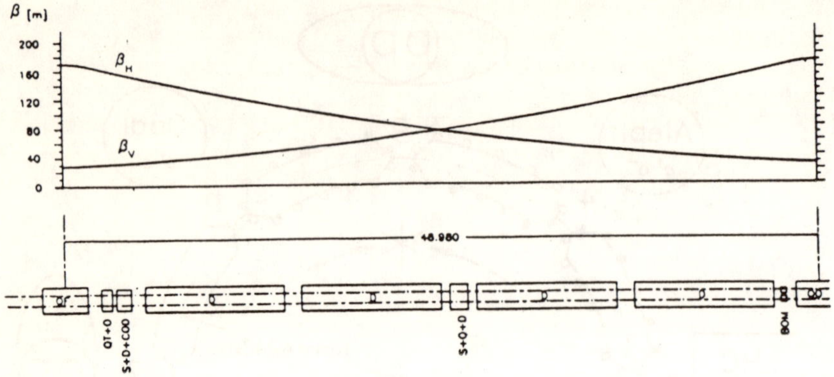

FIGURE 6

LAYOUT OF THE STANDARD LHC HALF-CELL

V Sources of magnetic errors

The different magnetic multipoles can be generated by current sheets flowing outside a cylindrical vacuum chamber[3] as shown in Fig. 7. A multipole of order m is generated by a current density of the form $I(\Phi)=I_0\cos(m\Phi)$. "Skew" multipoles are obtained in the same way by rotating the distribution $I(\Phi)=I_0 \sin(m\Phi)$.

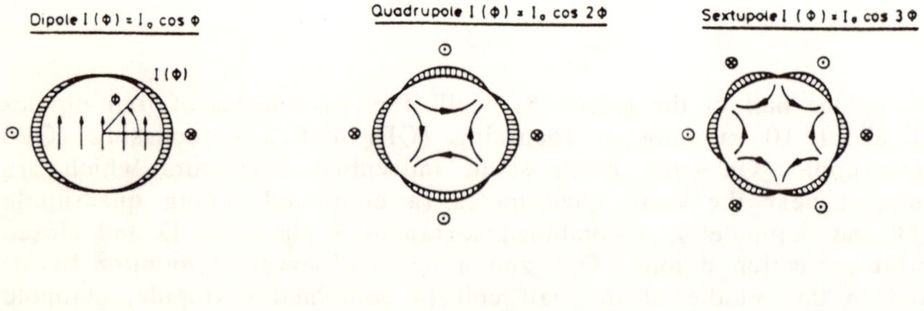

FIGURE 7

GENERATION OF MULTIPOLE FIELDS

In order to build an accelerator the minimum one needs is dipoles to bend the particle trajectories and quadrupoles to focus them. In addition, in order to achieve high performance it is necessary to add sextupoles to correct the chromatic aberrations which arise from the variation of the focal length of quadrupoles with the particle momentum, and skew quadrupoles to cancel the horizontal-vertical coupling effects generated by unavoidable tilts of the main quadrupoles. In proton machines octupoles are also necessary to control the tune spread in the beam and in this way damp instabilities due to collective effects.

Therefore the particle motion in accelerators or storage rings is by necessity fundamentally non linear. In addition unavoidable errors in the magnetic fields introduce further important non linearities, and in superconducting machines these are generally dominant.

It is customary to expand the magnetic field in the main bending dipoles in the following way :

$$B_x + iB_y = Bo \sum_{i=1}^{\infty} (b_n + ia_n) \left(\frac{Z}{R_r} \right)^{n-1}$$

where Bo is the wanted vertical bending field and B_x and B_y are the actual components of the field in the horizontal and vertical planes respectively. The coefficients b_n correspond to the normal and a_n to the skew multipole components. The reference radius Rr is usually chosen close to the half aperture of the magnet coils. For LHC as well as SSC studies Rr = 1 cm is assumed and an's and bn's are expressed in units of 10^{-4}. The above formula corresponds to the European notation where n = 1 describes the dipole, whereas in the US the dipole is n = 0.

The first, systematic source of errors comes from the fact that the ideal cos (Φ) current distribution is approximated by current shells of constant thickness as shown in Fig. 8. This generates unwanted normal multipoles of odd order. The lowest order multipoles are minimized by a suitable choice of the angle subtented by the coils. For instance with two layers as in Fig. 8 one can cancel out both the sextupole (n = 3) and the decapole (n = 5) terms. By inserting non-conducting wedges one can minimize higher order multipoles. In the LHC magnet, residual n = 7 and n = 9 terms are large enough to pose problems, as we will see later on, and ways are sought to reduce their values.

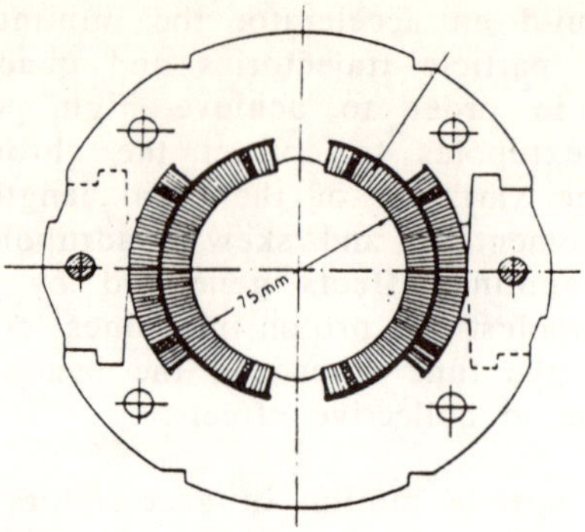

FIGURE 8

A TWO LAYER DIPOLE COIL (HERA TYPE)

Mechanical tolerances on the dimensions and positions of the coils are responsible for additional errors, although precisions as high as 0.02 mm can be reached in practice. At high field the enormous magnetic forces which tend to compress the coils (about 400 Tons/meter for the LHC) are potential sources of deformation. Adequate prestress has to be applied in order to diminish their effect.

At low field, by far the most important source of errors is the persistent currents. These are eddy currents which establish themselves as soon as a magnetic field is applied to the superconducting wires. They flow in opposite directions on each side of the wires so as to oppose the applied magnetic field inside the superconductor. (Fig. 9).

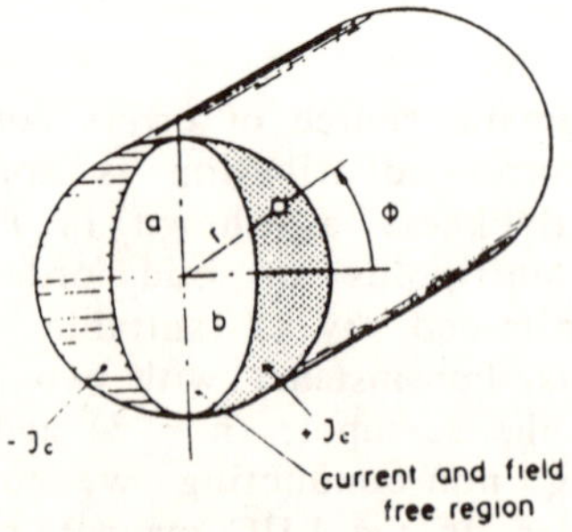

FIGURE 9

PERSISTENT CURRENTS IN A SUPERCONDUCTING FILAMENT

Their net effect at the center of the coil is the generation of odd order (for a dipole magnet) and even order (for a quadrupole) high multipoles. In the dipoles the strongest component is sextupole, while in the quadrupoles the dodecapole is dominant. Their strength depend on the magnetization history and vary slowly (time constants of about half an hour) when the main field is held constant, like for instance during the injection process. For these reasons it is very difficult to compensate this effect accurately.

The only way to diminish the effect of the persistent currents is to use conductors of extremely small diameter. The Hera magnet coils are wound with wires made of 15μm diameter superconducting filaments, and it seems technologically possible now to go down to diameters of 5μm or even 2.5μm for the LHC and the SSC. The limit is set by the tunnelling of the Cooper pairs in between filaments, which could possibly be prevented by a suitable doping of the copper matrix in which the filaments are embedded.

VI Effect of magnetic errors and their corrections

Table I gives the values of the error coefficients which are expected for the LHC bending magnets. The systematic errors (those which are the same in all magnets of a production line) originate mainly from the persistent currents (b_3 and b_5) and the coil design (b_7 and b_9). The random errors are due to geometrical variations from magnet to magnet.

Systematic errors drive structure resonances which are excited by harmonics of the machine periodicity (8 for the LHC) and in addition produce a tune variation with the betatron amplitude and the momentum of the particles.

Structure resonances can be avoided by adequately choosing the integer part of the tune, but the tune variation remains and constitutes the dominant effect of systematic errors. In order to assure a sufficient beam lifetime all resonances up to a rather high order have to be avoided for all beam particles. This is only possible if the tune dispersion in the beam is kept below a certain limit, around 0.005.

Sextupole, octupole and decapole correctors are inserted in the lattice (see Fig. 6) to reduce the tune dispersion caused by the systematic errors. However, at large amplitudes and large momentum deviations the high order multipoles b_7 and b_9 dominate and their effect cannot be corrected.

Random errors mainly drive non linear resonances, and experience shows that even high order ones (up to 7th or 8th order) can produce a slow increase of particle amplitudes, leading to the subsequent loss of the particles. Since the largest error is the sextupole, it is possible after measuring it in each single magnet to install the magnets in a determined sequence so as to minimize the excitation of the more dangerous resonances. Such "sorting" schemes have been found interesting, although their efficiency is limited by the presence of the higher order terms.

The detrimental effect of random errors reveals itself by an increase of the "Smear", a value easy to calculate from computer simulation of the particle trajectories. The Smear is defined as the variance of the square of the amplitude of a particle measured turn after turn ; in a linear machine the amplitude is constant and the value of the Smear is zero. It is assûmed that the motion will be "sufficiently linear" if the Smear stays below 0.1. The "linear aperture" of a machine is defined as the region of the vacuum chamber where the tune shift is smaller than 0.005 and the Smear smaller than 0.1.

The optimization of a machine is made using a computer "tracking" programme. Tracking over a few hundred turns is sufficient to obtain the <u>short term dynamic aperture</u> (where the particles stay in the machine) as well as the <u>linear aperture</u> defined above. One can also make a few selected long term trackings (over say 10^6 turns) or evaluate the Lyapunov exponents to determine a "chaotic boundary" inside which the motion appears regular. The result of these studies is a cascade of successive "apertures" (as in Fig. 10) which are useful to judge wether the magnet quality is sufficient or not.

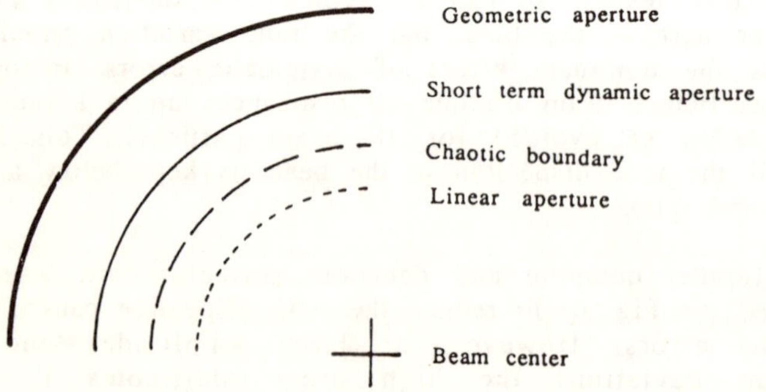

FIGURE 10

DIFFERENT CONCEPTS OF "MACHINE APERTURES"

TABLE I

ESTIMATED ERRORS FOR THE LHC $\left(\begin{cases} coil\phi = 5cm \\ fil.\phi = 5\mu m \end{cases}\right)$ Injection at 450 GeV

Multipole	Systematic	Random (rms)
a_2 b_2	± 0.6 ± 1.6	1.7 1.2
a_3 b_3	± 0.1 - 3.7	0.5 1.7
a_4 b_4	± 0.03 ± 0.05	0.2 0.15
a_5 b_5	± 0.03 0.45	0.07 0.22
a_7 b_7	± 0.01 0.13	0.04 0.02
a_9 b_9	± 0.001 0.024	0.002 0.005

Références :

1. G. Brianti, K. Hubner
The large Hadron Collider in the LEP Tunnel CERN 87-05

2. R. Perin
State of the art in high field superconducting magnets for particle accelerators.
XIV th International Conference on high energy Accelerators, Tsukuba, Japan, August 22-26-1989
Particle Accelerators, V26, N1-4 (1990) P. 625

3. K. H. Mess, P. Schmüser
Sperconductivity in Particle Accelerators
CAS school, Hamburg, Fed. Rep. Germany
30 May-3 June 1988 CERN 89-04

BEAM STABILITY ANALYSIS VIA NORMAL FORMS AND NON PERTURBATIVE METHODS

GIORGIO TURCHETTI †

Dipartimento di Fisica, Università di Bologna, ITALY
INFN – Sezione di Bologna, ITALY

ABSTRACT

The analysis of the nonlinear effects on the stability of the betatronic motion is crucial for the design of new hadron accelerators based on superconducting magnets. The tracking techniques prove to be insufficient in order to define the relevant nonlinear parameters and carry out correction strategies. A magnetic lattice is conveniently described by a symplectic map in the nonlinear case just as in the linear one. The normal forms are, for the case of nonresonant linear tunes, the analogue of the Courant-Sneider theory and allow to compute both the invariant tune and the geometric deformation of the orbits. The normal forms are given by asymptotic series because of the presence of small divisors, associated to true or virtual resonances. The remainders growth as n^{nn} has been estimated allowing to obtain upper bounds to the Arnold diffusion and lower bounds to the lifetimes. The case of quasi resonant or resonant linear tunes has also been treated by the so called resonant normal forms and interpolating hamiltonians, wich describe the geometric structure of the islands and allow to compute the tune shift beyond. Information on the short term dynamical aperture is obtained by a careful analysis of the normal forms. Finally non perturbative approaches have been considered: in one case by using the Padé approximants, which are ideally suited to analytically continue the piecewise geometric series of the normal forms, in the other by exploiting a variational method to construct of the integrals of motion. After the first application to a beam-beam map the methods based on normal forms have been succesfully used to compute the tune shift and smear for the LHC lattice, to analyze the 1988 SPS experiment and to compute the corrections of the multipolar errors in LHC.

† Work supported in part with a MPI 60% grant and a Worldlab grant

1. Introduction

The modern analysis of nonlinear systems goes back to Poincaré [1] who proved that generic systems are non integrable and introduced the hamiltonian maps by taking sections of an orbit with a smooth transverse manifold (1898). After the works of Birkhoff [2] (1920) on the construction of the formal series to conjugate hamiltonian flows or maps with their integrable counterpart, the relevant progress was due to Siegel [3] , who first proved the convergence of the perturbation series arising in a holomorphic small divisors problem (1942). Similar results on the robustness of individual strongly non resonant tori for hamiltonian systems were obtained by Kolmogorov [4] , Arnold [5] and Moser [6] (1954, 1962), whose theory provided the foundations of the nonlinear stability analysis of magnetic surfaces in Tokamaks and of beam trajectories in particle accelerators. Finally a careful analysis of the remainders of perturbation series allowed Nekhoroshev [7] to prove the stability of hamiltonian flows (1977) in open sets for finite but exponentially long times. In between these milestones, which are found about every 20 years, there is a lot of quite significant work carried out especially after the sixties, when the interest of astronomers and of physicists was awaken. Very significant are the numerical works by Gustavson [8] Contopoulos [9] and Hénon [10,11] who showed that the behaviour predicted by the KAM theory was indeed observable and physically relevant.

In a physical context the use of perturbation theory in hamiltonian dynamics was promoted by Dragt [12,13] , who showed how to compute hamiltonian flows using the Lie series and by Galgani and coworkers [14–17] in Milan, who showed how to construct the integrals of motion and the normal forms for the anharmonic oscillators and other nonlinear systems, providing exponential bounds on the remainders. A similar program was developed in Bologna by myself and coworkers [18–23] for the symplectic maps and significant applications to the beam dynamics were carried out in collaboration with W. Scandale at CERN [24–31] .

By a very careful analysis of the quadratic map the mechanism of divergence of the Birkhoff series was explained [18,19] . The proposed algorithmic scheme for the *normal forms* allowed very high orders to be reached [20] and a first application to the beam-beam effect [24] was proposed (1983). The extension of the method to symplectic maps of arbitrary dimensions was carried out [21,22] and the interpolating hamiltonian to treat the resonant and quasi resonant frequencies [23] was introduced (1987). The use of normal forms to treat the nonlinearities of the magnetic lattice was considered later (1986,1987) by analyzing the LHC lattice [26,27] and the 1988 SPS experiment [29] . More recently (1989) bounds on the remainders and Nekhoroshev–like stability estimates were obtained [32,33] , a very effective correction method based on normal forms was proposed [31] and non perturbative approaches were examined [34] . Independently (1988) a very convenient algorithmic scheme for polynomials manipulations was proposed by Berz [35] and used to compute the normal forms by Forrest [36] .

Here we shall review the main ideas and techniques developed for symplectic maps emphasizing their relevance for the beam dynamics.

In section 2 we motivate the description of a magnetic lattice by a symplectic map. In section 3 we introduce the normal forms. In section 4 we specify the relevant parameters of nonlinearity. In section 5 we introduce the Birkhoff series. In section 6 the small divisors and resonances are reviewed. In section 7 the growth of the perturbation series is analyzed. In section 8 the leading singularities of the normal forms are derived. In section 9 the resonant normal forms and interpolating hamiltonians are described. In section 10 the integrals of motion and a variational methods to compute them are examined. In section 11 the Nekhoroshev estimates are reviewed. In section 12 the analyticity results on the majorant series are summarized. In section 13 the dynamical aperture and diffusion are examined. In section 14 the results on the LHC lattice and SPS experiments are presented. In section 15 the correction strategy for multipolar errors and the results for LHC are outlined.

2. Description of a magnetic lattice

Consider a magnetic lattice formed by E magnetic elements we denote with $\mathcal{M}_1, \ldots,$ $\mathcal{M}_E$. Letting x, y be the horizontal and vertical axes perpendicular to the orbit, one can write the nonlinear Hill equation to investigate the motion. To simplify the presentation we shall consider only the vertical motion assuming the coupling with the horizontal motion to be negligible. The Hill equation for the single particle motion reads [37]

$$\frac{d^2y}{ds^2} + k_1(s)y = \sum_{\ell \geq 2} \frac{k_\ell(s)}{\ell!} y^\ell \tag{2.1}$$

The multipolar coefficients of the magnets $k_1(s)$, $k_2(s), \ldots$ are basically discontinuous functions since the edge effects are negligible. The linear equation can be solved according to

$$y(s) = A\sqrt{\beta(s)} \sin(\psi(s) + \delta) \tag{2.2}$$

where $\beta(s)$ is periodic, A and δ are constants and

$$\psi(s) = \int_0^s \frac{d\sigma}{\beta(\sigma)} \tag{2.3}$$

The Fourier spectrum of $\beta(s)$ will decay linearly and the application of canonical perturbation theory [38] in order to treat the non linear terms faces serious difficulties. Indeed the Fourier spectrum cannot be truncated at low orders without introducing substantial errors and the analytic resummation of infinite series is necessarily limited to the first two orders by its algorithmic complexity. For this reason all the tracking codes are based on the transfer map approach: on each magnetic element $\mathcal{M}_j$ which extends to the arc $[s_{j-1}, s_j]$ the equation (2.1) has constant coefficients $k_1, k_2, \ldots$ and the solution is a map $\mathbf{M}_j$ which tranforms the point $y(s_{j-1})$, $p(s_{j-1})$ in phase space into the point $y(s_j), p(s_j)$ where $p \equiv dy/ds$. The overall transfer map $\mathbf{F}(0)$ corresponding to the Poincaré section of the orbit taken at $s = 0$ between the first magnetic element $\mathcal{M}_1$ and the last one $\mathcal{M}_E$, is given by the ordered composition of the transfer maps for the individual magnetic elements

$$\mathbf{F}(0) = \mathbf{M}_E \circ \mathbf{M}_{E-1} \circ \ldots \circ \mathbf{M}_1 \tag{2.4}$$

We notice that $\mathbf{F}(0) = \mathbf{F}(s_E)$ where $s_E = 2\pi R$ is the length of the ring. If we change the point where the Poincaré section is taken from $s = 0$ to s_k the new one turn map $\mathbf{F}(s_k)$ is given by

$$\mathbf{F}(s_k) = \mathbf{M}_k \circ \mathbf{M}_{k-1} \circ \ldots \circ \mathbf{M}_1 \circ \mathbf{M}_E \circ \ldots \circ \mathbf{M}_{k+2} \circ \mathbf{M}_{k+1} \tag{2.5}$$

The one turn maps on adjacent sections are related by

$$\mathbf{F}(s_k) = \mathbf{M}_k \circ \mathbf{F}(s_{k-1}) \circ \mathbf{M}_k^{-1} \tag{2.6}$$

We separate each transfer map and one turn map into linear and non linear terms according to $\mathbf{M}_k = \mathbf{L}_k + \mathbf{N}_k$ and $\mathbf{M}(s_j) = \mathbf{L}(s_j) + \mathbf{N}(s_j)$.
Letting $\mathbf{x}(s) \equiv (y(s), p(s))$ the solution of the linear Hill equation (2.1) reads

$$\mathbf{x}(s) = \mathbf{T}(s)\mathbf{R}(\psi(s))\mathbf{T}^{-1}(0)\mathbf{x}(0) \qquad \mathbf{T}(s) = \begin{pmatrix} \beta^{1/2} & 0 \\ \frac{1}{2}\dot{\beta}\beta^{-1/2} & \beta^{-1/2} \end{pmatrix} \tag{2.7}$$

Since $\mathbf{T}(s)$ is periodic with period $2\pi R$ from (2.7) we have

$$\mathbf{x}(s + 2\pi R) = \mathbf{T}(s)\mathbf{R}(2\pi\nu)\mathbf{T}(s)^{-1}\mathbf{x}(s) \tag{2.8}$$

where $\nu = (2\pi)^{-1}\psi(2\pi R)$ is the *tune* and $\mathbf{T}(s)$ is the transformation which maps the orbits at any Poincaré section s into circles. As a consequence in the language of trasfer maps the invariant tune and the geometric deformation factors are obtained by determining the similarity transformation which maps $\mathbf{L}(s_k)$ into a rotation

$$\mathbf{L}(s_k) = \mathbf{T}(s_k)\mathbf{R}(2\pi\nu)\mathbf{T}(s_k)^{-1} \tag{2.9}$$

From (2.6), which for linear maps reads $\mathbf{L}(s_k) = \mathbf{L}_k\mathbf{L}(s_{k-1})\mathbf{L}_k^{-1}$, we obtain the recurrence

$$\mathbf{T}(s_k) = \mathbf{L}_k\, \mathbf{T}(s_{k-1}) \tag{2.10}$$

The matrix $\mathbf{T}(s_k)$ is defined up to a rotation we choose to be the identity.
In the nonlinear case the procedure is identical. We look for a nonlinear transformation $\boldsymbol{\Phi}(s_k)$ which reduces the map $\mathbf{F}(s_k)$ to a rotation $\mathbf{U} = \mathbf{R}(\Omega(||\mathbf{x}||))\mathbf{x}$. The transformed orbits are still circles but the rotation angle depends on its radius.

$$\mathbf{F}(s_k) = \boldsymbol{\Phi}(s_k) \circ \mathbf{U} \circ \boldsymbol{\Phi}(s_k)^{-1} \tag{2.11}$$

As in the linear case it is sufficient to achieve the transformation (2.11) only at one section point since $\boldsymbol{\Phi}(s_k) = \mathbf{M}_k \circ \boldsymbol{\Phi}(s_{k-1})$; the transformations $\boldsymbol{\Phi}$ are also defined up to a rotation $\mathbf{R}(\Xi(||\mathbf{x}||))$.

For convenience from now on we shall refer to the $s = 0$ section, use linearly normalized coordinates $\hat{\mathbf{x}} = (\hat{y}, \hat{p})$

$$\hat{\mathbf{x}} = \mathbf{T}^{-1}\mathbf{x} \qquad \mathbf{T} \equiv \mathbf{T}(0) \tag{2.12}$$

and factorize the linear rotation in the transformed nonlinear map $\hat{\mathbf{F}}$

$$\hat{\mathbf{F}} = \mathbf{R}(2\pi\nu)\left[\mathbf{I} + \hat{\mathbf{N}}\right] \qquad \hat{\mathbf{N}} = \mathbf{R}(-2\pi\nu)\mathbf{T}^{-1} \circ \mathbf{N} \circ \mathbf{T} \tag{2.13}$$

We shall also use the complex coordinates denoting with z the linear normal coordinates and with ς the nonlinear normal coordinates where

$$z = \hat{y} - i\hat{p}, \qquad \varsigma = \eta - ip_\eta \tag{2.14}$$

The two components of the map will then be complex conjugate and read

$$z' = F(z, z^*) = e^{i\omega}\left(z + \sum_{n\geq 2}[F]_n(z, z^*)\right) \tag{2.15}$$

where $[F]_n$ denote homogeneous polynomials of order n. The trasformation

$$z = \Phi(\varsigma, \varsigma^*) = \varsigma + \sum_{n\geq 2}[\Phi]_n(\varsigma, \varsigma^*) \tag{2.16}$$

puts the map into the nonlinear normal form $U(\varsigma, \varsigma^*)$ according to the conjugation equation (2.11) which reads

$$F \circ \Phi = \Phi \circ U \tag{2.17}$$

We can choose the normal form U to be a rotation, just as in the linear case, when ω is not resonant.

$$U = e^{i\Omega(\varsigma\varsigma^*)}\varsigma, \qquad \Omega = \omega + \sum_{n\geq 1}\Omega_{2n}(\varsigma\varsigma^*)^n \tag{2.18}$$

However when ω is resonant U must be a function capable of describing new topological structures. This *resonant normal form* has no analogue in the linear case.

3. Normal forms

The normal forms are specified by an *invariance group.* We define a map U to be a *normal form with respect to* α if it commutes with the group of rotations of an angle α namely if

$$\Delta_\alpha U \equiv U(e^{i\alpha}\varsigma, e^{-i\alpha}\varsigma^*) - e^{i\alpha}U(\varsigma,\varsigma^*) = 0 \tag{3.1}$$

Notice that if $\alpha/2\pi$ is *irrational* then the group is the rotation group. This implies $U(\varsigma,\varsigma^*) = \varsigma u(\varsigma\varsigma^*)$ and if we impose that U is *symplectic* namely that its jacobian is equal to 1, then it takes the form (2.18). We observe that its invariant curves are circles since $|\varsigma| = |U(\varsigma,\varsigma^*)|$.

If $\alpha/2\pi = p/q$ $(p,q \in \mathbf{Z})$ is a rational number, then the group generated by $e^{i\alpha}$is a discrete subgroup of the rotation group and the normal form reads

$$U(\varsigma,\varsigma^*) = e^{i\omega}\left[\varsigma + \sum_{n,l}(\varsigma\varsigma^*)^n(u_{n,l}\,\varsigma^{lq+1} + u_{n,-l}\varsigma^{*lq-1})\right] \tag{3.2}$$

This form is called resonant normal form and the invariants of U are not trivially obtained.

We remark that in general α need not to be equal to the linear frequency ω. This is mandatory whenever the tune is resonant namely $\nu = \omega/2\pi$ is a rational number. However if the tune is non resonant one can compute a resonant normal form by choosing $\alpha/2\pi$ equal to a rational approximation to the tune itself. As we shall see this is even strongly recommended when the tune is very close to a rational number p/q with q small.

Interpolating hamiltonians When U is a resonant normal form the construction of the invariants and the iteration of the map are not trivial. In this case, which has no linear analogue, it is necessary to introduce an *interpolating hamiltonian,* whose *flow interpolates* the orbits $\varsigma_n = U^{\circ n}(\varsigma,\varsigma^*)$ of the map.

Letting $H = 2i\mathcal{H}(\varsigma,\varsigma^*)$ denote the interpolating hamiltonian and $S(t;\varsigma,\varsigma^*)$ denote its flow where the interpolation condition reads

$$S(n;\varsigma,\varsigma^*) = U^{\circ n}(\varsigma,\varsigma^*) \qquad n \in \mathbf{Z} \tag{3.3}$$

In the non resonant case $S(t;\varsigma,\varsigma^*) = e^{it\Omega(\varsigma\varsigma^*)}\varsigma$ and

$$H = 2i\mathcal{H}(\varsigma\varsigma^*) \qquad \mathcal{H}(\rho) = \frac{1}{2}\int_0^\rho \Omega(\rho')d\rho' \tag{3.4}$$

We notice that setting $\varsigma = \sqrt{2r}e^{i\theta}$, the hamiltonian $\mathcal{H}(r,\theta)$ is a real interpolating hamiltonian for the normal coordinates (η,p_η) after a canonical transformation to action and angle variables (r,θ) (the factor $2i$ relating H to it is due to the use of complex coordinates). In the resonant case H is time dependent but becomes time independent in a *rotating frame.* We denote with $\hat{\varsigma}$ the coordinates in the rotating frame and with $\hat{H}$ the corresponding hamiltonian. Setting

$$\alpha \equiv \omega_R = 2\pi\frac{p}{q} \qquad \omega = \omega_R + 2\pi\epsilon \tag{3.5}$$

We have

$$\varsigma = e^{i\omega_R t}\hat{\varsigma}, \qquad H(\varsigma,\varsigma^*;t) = i\omega_R\,\varsigma\varsigma^* + \hat{H}\left(e^{-i\omega_R t}\varsigma, e^{i\omega_R t}\varsigma^*\right) \tag{3.6}$$

The reason for going into a moving frame is due to the fact that the resonant normal form allows to describe a chain of q islands. The orbit *jumps* from one island to another (jumps exactly p islands each time): in the rotating frame it is brought back to the same initial island on which it will continue to move forever. The hamiltonian $\hat{H}(\hat{\varsigma},\hat{\varsigma}^*)$ in the moving frame is time independent.
Since both $\hat{\mathcal{H}}$ and $\mathcal{H}$ are in normal form it turns out that $\hat{H}$ is an integral of motion for the orbit of the map not only in the moving frame but also in the fixed frame. Indeed letting ς_n denote a point of the orbit after n iterations we have

$$\hat{H}(\varsigma_n,{\varsigma^*}_n) \equiv \hat{H}\left(e^{i\omega_R n}\hat{\varsigma}_n, e^{-i\omega_R n}\hat{\varsigma^*}_n\right) = \hat{H}(\hat{\varsigma}_n,\hat{\varsigma^*}_n) = \text{const.} \tag{3.7}$$

The interpolating hamiltonian allows to describe the islands associated to the q resonance. The period of the orbits can be computed and allows to reproduce the tune shift and its locking at the islands; even the secondary tune for the rotations around the center of the islands (elliptic fixed points) can be evaluated and the geometric effects of orbit distorsions determined through the Fourier spectrum of the orbits of the interpolating hamiltonian.

Invariants of motions We conclude by observing that in the non resonant case the invariant of motion is the radius of the orbit in the normal coordinates. Denoting with

$$\varsigma = \Psi(z,z^*) \equiv \Phi^{-1}(z,z^*) \tag{3.8}$$

the inverse function, the invariant is given by

$$I(z,z^*) = \varsigma\varsigma^* = \Psi(z,z^*)\Psi^*(\varsigma,z^*) \tag{3.9}$$

In the resonant case the invariant reads

$$I(z,z^*) = \hat{\mathcal{H}}(\Psi(z,z^*),\Psi^*(z,z^*)) \tag{3.10}$$

where $\hat{\mathcal{H}} = \hat{H}/(2i)$. We observe that in both cases any function of $I(z,z^*)$ is also an invariant.

4. Parameters of nonlinearity

In order to determine the effects of the nonlinearity the *tune shift*, the *smear* (deviation of the orbits from circularity) and the *dynamic aperture* are used.
The tune shift $\delta\nu$ is an invariant quantity which does not depend on the particular Poincaré section of the orbit we are considering. For the nonresonant normal forms we have $\delta\nu = \Omega - \nu$. For the resonant normal forms letting $\varsigma = \rho e^{i\theta}$, we have to compute the period of an orbit specified by $\hat{\mathcal{H}}(\rho e^{i\theta}, \rho e^{-i\theta}) = E$. From $\rho = \rho(E,\theta)$, which can be computed by a Newton's method or convergent series expansion we obtain

$$\delta\nu = \frac{1}{T} \qquad T = \int_0^{2\pi} \frac{\rho d\theta}{\frac{\partial \hat{\mathcal{H}}}{\partial \rho}(\rho e^{i\theta}, \rho e^{-i\theta})} \tag{4.1}$$

The smear is not invariant, just as the β function in the linear case, and can be defined as the mean square deviation of the distance r from origin. Letting

$$r = |z(\rho,\theta)| = |\Phi(\rho e^{i\theta}, \rho e^{-i\theta})| \tag{4.2}$$

where ρ is an invariant for the non resonant case and $\rho = \rho(E,\theta)$ for the resonant case, we define the smear as the mean square deviation of r^2 namely

$$\sigma = \frac{\langle (r^2 - \langle r^2 \rangle)^2 \rangle^{1/2}}{\langle r^2 \rangle} \tag{4.3}$$

In order to compute σ is convenient to refer to the Fourier spectrum $A_k(\rho)$ of $\Phi(\rho e^{i\theta}, \rho e^{-i\theta})$. Letting

$$[\Phi]_n(\varsigma, \varsigma^*) = \sum_{j=0}^{n} a_{n,j} \varsigma^j \varsigma^{*n-j} \tag{4.4}$$

we write

$$\begin{aligned} &\Phi(\rho e^{i\theta}, \rho e^{-i\theta}) = \sum_k A_k(\rho) e^{ik\theta} \\ &A_k(\rho) = \rho^{|k|} \sum_{n=0}^{\infty} \rho^{2n} b_{k,n} \qquad\qquad b_{k,n} = a_{|k|+2n,\ n+\frac{k+|k|}{2}} \end{aligned} \tag{4.5}$$

The spectrum $A(k)$ provides information on the presence of resonances and allows moreover to compute rapidly and efficiently the smear according to

$$\langle r^2 \rangle = \sum_k |A_k|^2, \qquad \langle r^4 \rangle = \sum_{k,j,l} A_k\, A_j\, A_l^*\, A_{k+j-l}^* \tag{4.6}$$

The extension of these formulae to the coupled x, y motion namely for maps of $\mathbf{R}^4$ is straightforward.

The dynamic aperture for maps of R^2 is the radius of the largest disc contained in the basin boundary of infinity namely the complement of the stability boundary of the origin. For maps of R^4 or higher dimensional such a definition cannot be used because of the slow diffusions through the tori. In this case a mathematically founded definition involves the observation time (or duration of the experiment) and the radius of the stability region (physical aperture i.e. radius of the vacuum tube).

Definition: Let $S_R : \{|z| \leq R\}$ be a disc of radius R in phase space and let $T \in \mathbf{Z}$ be a given number of iterations. The dynamic aperture is the radius $A(R,T)$ of the largest sphere contained in the a domain D such that

$$S_A \subset D, \qquad \bigcup_{0 \leq j \leq T} F^{\circ j}(D) \subset S_R \tag{4.7}$$

It can be noticed that $\lim_{T\to\infty} A(R,T) = 0$ for maps in dimension higher or equal to 4.

Usually choosing R equal to the physical aperture and t small of the order of $10^3 \sim 10^4$ one refers to $A(R,T)$ as the short term dynamical aperture, while for $T = 10^8 \sim 10^9$ corresponding to the duration on a run, $A(R,T)$ is called long term dynamical aperture.

A more satisfactory definition implies the introduction of a measure invariant under the map and depends on a parameter η which gives the fraction of the initial particles which we require to remain confined. As a consequence $A(R,t;\eta)$ can be defined as follows;

Definition Given R, T and $0 < \eta < 1$ we define the dynamic aperture A with probability η is the radius of the largest sphere S_A such that letting μ be an invariant measure normalized to A namely $\mu(A) = 1$ one has

$$\mu(\underbrace{S_R \cap F(S_R \cap F(S_R \cap \ldots S_R \cap F(S_R \cap F(A))\ldots)))}_{T \text{ times}} \geq \eta \tag{4.8}$$

An upper bound $\overline{A}$ to A is given by

$$\mu(S_R \cap F^{\circ\, T}(S_{\overline{A}})) \geq \eta \tag{4.9}$$

Even though this definition is more satisfactory it is much more difficult to give analytic estimates. On the contrary via the formulation of the Nekhoroshev theorem for symplectic maps allows to give lower bounds to the dynamic aperture or to the stability time at fixed dynamical aperture.

The typical result is formulated by looking for $A(R,T) = R/2$ which provides $T = T(R)$ denominated stability time. The theorem gives $\overline{T}(R) \leq T(R)$ or by inverting $\overline{R}(T) \leq R(T)$ as we shall see in the next section.

5. Birkhoff series for the normal forms

For the nonlinear polynomial maps the functional equations conjugating a map with its normal forms in the non resonant and resonant case do not have analytic solutions defined in open sets of phase space. As a consequence the perturbative solutions provide asymptotic series whose properties need to be carefully examined. We shall describe them in the non resonant case and review the divergence mechanisms and the rigorous results on the remainder estimates. We first observe that when a map F is defined by the composition of a large number M of polynomial maps, it is convenient to compute its truncation $[F]_{\leq N}$ by composing and truncating at order N the individual maps according to the following algorithm

$$\mathcal{F}_j = [M_j \circ \mathcal{F}_{j-1}]_{\leq N} \qquad j = 1, \dots, N \qquad \mathcal{F}_0 = \mathcal{I} \quad [F]_{\leq N} = \mathcal{F}_N \tag{5.1}$$

The map $[F]_{\leq N}$ is a symplectic truncation of F and the normal forms computed by perturbation at order N on F or $[F]_{\leq N}$ are exactly the same. The conjugation equation reads

$$F \circ [\Phi]_{\leq N} - [\Phi]_{\leq N} \circ [U]_{\leq N} = R_N \tag{5.2}$$

where R_N is a remainder of order $N+1$. Letting $[Q]_n$ be the homogeneous component of order n in the l.h.s. of equation (5.2) once we subtract the order n unknowns $[U]_n$ and $\Delta[\Phi]_n$, the recurrent equations are obtained by projecting (5.5) into the space of normal form polynomials and its complement. Letting Π_α be the projector into the space of polynomials in normal form with respect to α we have

$$\begin{cases} \Delta_\omega \Pi_\alpha [\Phi]_n + [U]_n = \Pi_\alpha\, [Q]_n \\ \Delta_\omega (1 - \Pi_\alpha)[\Phi]_n = (1 - \Pi_\alpha)[Q]_n. \end{cases} \tag{5.3}$$

The kernel of Δ_ω is $\Pi_\omega \mathcal{E}$ where $\mathcal{E}$ is the space of polynomials. As a consequence when $\alpha = \omega$ we have $\Delta_\omega \Pi_\alpha = 0$ and the equations (5.6) determine $[U]_n$ and $(1-\Pi_\omega)[\Phi]_n$. The normal form component $\Pi_\omega[\Phi]_n$ is partially determined requiring that the jacobian of Φ is one. The remaining arbitrariness corresponds to the freedom of changing Φ and U into $\Phi \circ \Gamma$ and $\Gamma^{-1} \circ U \circ \Gamma$ where Γ is any map in normal form with unit jacobian.
If ω is nonresonant and α is resonant namely $\alpha \neq \omega$ then $\Pi_\alpha \mathcal{E} \supset \Pi_\omega \mathcal{E}$ and consequently Δ_ω^{-1} exists in $(1 - \Pi_\alpha)\mathcal{E}$. (we recall that if $\omega/2\pi = p/q + \epsilon$ with $\epsilon \ll 1$ the choice is $\alpha/2\pi = p/q$ is recommended). In this case the condition of unit jacobian for Φ must be first used at any order n in order to fix $\Pi_\alpha[\Phi]_n$ (up to the arbitrariness related to the composition with a map Γ in normal form and unitary jacobian); then the first of equations (5.3) determines $[U]_n$ and the second determines $(1-\Pi_\alpha)[\Phi]_n$. We wish to stress once more that the normal form is not unique but defined up to a gauge Γ of normal form maps with unit jacobian, which in the nonresonant case reads $\Gamma = e^{i\Xi(\varsigma\varsigma^*)}\varsigma$.

Denoting with $\Psi = [\Phi]^{-1}_{\leq N}$ the inverse transformation we introduce the remainder $\hat{R}_n(t)$ corresponding to (5.5) iterated an integer number t of times

$$\Psi \circ F^t \circ [\Phi]_{\leq N} = e^{it[\Omega]_{\leq N}}\zeta + \hat{R}_N(t). \tag{5.4}$$

6. Small divisors and resonances

The non existence of uniform solutions (defined in open sets) of the conjugation equation is due to the topology of the orbits. Indeed the nonresonant normal forms can only describe orbits diffeomorphic to circles, whereas the resonant normal forms for a given q can describe in addition only chains of islands whose number is a multiple of q. They cannot describe other chains of islands which form a dense set nor the homoclinic tangles of the stable and unstable manifolds emerging from the hyperbolic fixed points.

For the nonresonant normal forms the divergence of the series is caused by the propagation of the small divisors corresponding to the sequence of leading resonances which converge to the linear tune.

We shall examine this mechanism in the simplest but significant case of the quadratic map $z' = e^{i\omega}[z + i(z + z^*)^2/4]$ assuming that $\nu \in [0, 1/2]$. A leading rational approximation to the linear tune is a true resonance, producing a chain of islands, whenever it approximates the linear tune from *below* except for a very small interval in a neighbourhhood of 1/3. A leading rational approximation to the linear tune from *above* is a *virtual* resonance since it would occur at an imaginary distance from the origin; as a consequence no islands are observed but the divergence effect on the series of the related divisor is the same as for a true resonance.

We recall that the divisors arise from the solution of the homologic equations (5.3): Using (4.4) the solution of the second equation in (5.3) with $\alpha = \omega$ reads

$$a_{n,k} = \frac{Q_{n,k}}{e^{i(2k-n)\omega} - e^{i\omega}} \qquad \text{for} \quad 2k - n \neq 1 \tag{6.1}$$

where $Q_{n,k}$ are the coefficients of $[Q]_n$. The real part of the undetermined coefficients $a_{2n+1,n}$ is fixed by requiring that the jacobian of the transformation is 1. The imaginary part remains undetermined and its choice corresponds to fixing the gauge Ξ (the standard choice is $\mathrm{Im}\, a_{2n+1,n} = 0$).

For any integer q letting p be the integer closest to $q\nu$

$$p = \{q\nu\} \equiv \begin{cases} [q\nu] & \text{if} \quad q\nu - [q\nu] \leq \frac{1}{2} \\ [q\nu] + 1 & \text{if} \quad q\nu - [q\nu] > \frac{1}{2} \end{cases} \tag{6.2}$$

and noticing that this definition implies $|\nu - p/q| \le 1/2q$ we can write for q large

$$e^{iq\omega} - 1 = e^{i\pi(q\nu - p)}\, 2\,i \sin \pi(q\nu - p) \simeq\ 2\pi\, i\, q \left(\nu - \frac{p}{q}\right) \tag{6.3}$$

The modulus of the divisor is bounded just by π however there are sequences q_j of integers for which the following estimate holds $|\nu - p_j/q_j| \le 1/q_j^2$ where p_j is still defined as above. For these sequences we have

$$|e^{iq_j\omega} - 1| \le \frac{2\pi}{q_j} \tag{6.4}$$

The inverses in this case grow at least as q_j. We observe that these sequences are obtained with the closest integer continued fraction algorithm ($0 \le \nu < 1$)

$$\nu = \cfrac{1}{s_1 + \cfrac{1}{s_2 + \cfrac{1}{\ddots + \cfrac{1}{s_n + r_n}}}} \qquad s_j \in \mathbf{Z} \tag{6.5}$$

where

$$r_{n-1} = \frac{1}{s_n + r_n} \qquad s_n = \left\{\frac{1}{r_{n-1}}\right\} \quad r_n = \frac{1}{r_{n-1}} - s_n \tag{6.6}$$

and

$$\begin{cases} p_{n+1} = s_{n+1}p_n + p_{n-1} \\ q_{n+1} = s_{n+1}q_n + q_{n-1} \end{cases} \tag{6.7}$$

initialized by $p_{-1} = 1,\ q_{-1} = 0,\ p_0 = 0, q_0 = 1$. The entries of the continued fraction s_k are not all positive; as a consequence no bounding properties are obtained. Upper and lower bounds are obtained from the ordinary continued fraction, defined by the above algorithm where in (6.6) s_n is obtained by replacing the closest integer operator {} with the integer part [ ] operator. The entries are then all positive and it can be proved that

$$\frac{p_2}{q_2} \le \ldots \le \frac{p_{2n}}{q_{2n}} \le \nu \le \frac{p_{2n+1}}{q_{2n+1}} \le \ldots \le \frac{p_1}{q_1} \tag{6.8}$$

Real and virtual resonances In order to understand how the small divisors affect the perturbation series we observe that if we perform a transformation to the normal form up to order 3, assuming $\Omega_2 \neq 0$ the map can be written as

$$\varsigma' = e^{i[\omega + \Omega_2\varsigma\varsigma^*]}\varsigma + R_4(\varsigma, \varsigma^*) \tag{6.9}$$

where R_4 is a remanider of order 4 and Ω_2 is the second order approximation to the non linear tune explicitly given by

$$\Omega_2 = -\frac{1}{16}\left[3\cot\frac{\omega}{2} + \cot\frac{3\omega}{2}\right] \tag{6.10}$$

We notice that $\Omega_2 < 0$ for $0 < \nu < 1/2$ and $\nu \notin [0.291, 0.334]$.
Letting p/q be a rational approximation to the linear tune and letting

$$\nu \equiv \frac{\omega}{2\pi} = p/q + \epsilon \tag{6.11}$$

the nonlinear tune obtained by neglecting the perturbation ($R_4 = 0$) crosses the resonance p/q at a distance $\rho = |\varsigma|$ from the origin such that

$$\omega + \rho^2\Omega_2 = 2\pi\frac{p}{q} \qquad \Longrightarrow \qquad \rho^2 = \frac{2\pi\epsilon}{-\Omega_2} \tag{6.12}$$

The crossing of the resonance occurs for a real value of ρ whenever the product $\epsilon(-\Omega_2)$ is positive. Excluding a small interval around 1/3 (more precisely for $\nu \notin [0.291, 0.334]$, where the following statements have to be reversed) whenever p/q is below the linear tune ($\epsilon > 0$) the corresponding resonance is crossed at a real value of ρ. Conversely if p/q is above the linear tune, the resonance is virtual because it is crosses for a purely imaginary value of ρ.
The even approximants of the ordinary continued fraction are real resonances ($\epsilon_{2j} = \nu - p_{2j}/q_{2j} > 0$) the odd approximants virtual resonances ($\epsilon_{2j+1} = \nu - p_{2j+1}/q_{2j+1} < 0$). The problem of determining the relevant resonances is essentially solved by the ordinary continued fraction in the case of one frequency * but remains completely open in the case of several frequencies, namely for maps of $\mathbf{R}^{2d}$ with $d \geq 2$, because algorithms like the continued fraction are not available.

---

* One should observe that some of the ordinary continued fraction approximations are marginal since they do not appear in the nearest integer continued fraction. This occurs when the bound $1/q_j^2$ to $|\epsilon_j|$ is almost saturated; in this case the next approximant differs only by one unit $q_{j+1} = q_j + 1$. In this case the approximant p_j/q_j is not relevant. The approximants produced by the nearest integer continued fraction do not exhibit this phenomenon and seem to be all relevant. We consider the following instructive example: letting $0 \leq \eta \ll 1$ we have

$$\nu = \frac{1}{q+\eta} \Longrightarrow \frac{p_1}{q_1} = \frac{1}{q} \quad \epsilon_1 \leq \frac{\eta}{q^2} \qquad\qquad \nu = \frac{1}{q-\eta} \Longrightarrow \frac{p_1}{q_1} = \frac{1}{q-1} \quad \epsilon_1 \leq \frac{1}{(q-1)^2} \qquad \frac{p_2}{q_2} = \frac{1}{q} \quad -\epsilon_2 \leq \frac{\eta}{q^2}$$

7. Leading terms analysis of the Birkhoff series

The way in which the leading divisors propagate into the coefficients of the normal forms has been analyzed in the case of the quadratic map. Any leading approximation p/q such that $|\epsilon| \leq 1/q^2$, first contributes to the coefficient $a_{q-1,0}$ of Φ at perturbation order $n = q - 1$. At next orders it propagates in such a way that for $n = q + 2m - 1$ the leading coefficient $a_{n,m}$ and $a_{n,q+m}$ are of order ϵ^{-m-1}. A similar mechanism holds when $n - q$ is even, namely $n = q + 2m$; the separation into even and odd subsequences is a natural consequence of the structure of the perturbative expansions.
The leading coefficients satisfy the following recurrence

$$a_{q+2m+1,\ m+1} = \frac{q+m+1}{q+m} \frac{-\Omega_2}{2\pi\epsilon} a_{q+2m-1,\ m}$$

$$a_{q+2m+1,\ q+m+1} = \frac{m+1}{m} \frac{-\Omega_2}{2\pi\epsilon} a_{q+2m-1,\ q+m} \tag{7.1}$$

initialized by $a_{q-1,0} = (q\epsilon)^{-1}c$. The nature of the recurrence becomes even more transparent on the coefficints $b_{n,k}$ of the series defining the Fourier components $A_k(\rho)$ of Φ defined by (4.5) .
In this case one has

$$b_{-q+1,\ m+1} = \frac{q+m+1}{q+m} \frac{-\Omega_2}{2\pi\epsilon} b_{-q+1,\ m}$$

$$b_{q+1,\ m+1} = \frac{m+2}{m+1} \frac{-\Omega_2}{2\pi\epsilon} b_{q+1,\ m} \tag{7.2}$$

For q large these sequences define almost geometric series whose radius of convergence can be read from previous equations. Moreover by a summation procedure we obtain a rational function whose leading poles are located at $\overline{\rho}^2 = \frac{2\pi\epsilon}{-\Omega_2}$. This pole occurs for a physical value of the radius ρ in the case of a true resonance $\epsilon > 0$, for a purely imaginary value of the radius in the case of a virtual resonance $\epsilon < 0$ in complete agreement with our previous analysis.
It is evident that for any sign of ϵ the series will diverge for $\rho = |\overline{\rho}|$. However in one case the divergence is removable by analytic continuation, in the second case the divergence is not removable. Indeed the analytic continuation of the series will locate a polar singularity on the real positive axis. This point will be discussed later.

The difficulty of the problem is due to the presence of infinitely many leading rational approximations p/q with $|\epsilon| < q^{-2}$ which compete.To each resonance we associate

$$\rho_j^2 = \frac{2\pi\epsilon_j}{-\Omega_2} \tag{7.3}$$

where the sequence ρ_j alternates on the real and imaginary axis. A resonance determines the behaviour of the Birkhoff series once its contribution has become larger with respect to the previous resonance. This produces a *retardation effect*: the dominance of a resonance p_j/q_j, which appears through a constant ratio $|\rho_j|^2$ in the Birkhoff series, does not begin at order $n = q_j$ but at considerably higher order.

In order to illustrate it we consider a sequence $\{f_j\}$ where the j-th resonance starts to act at order $n \coloneqq q_j$ and another sequence $\{g_j\}$ which exibits the *retardation effect.*

$$f_{n-1} = \frac{1}{\rho_j} f_{n-2} \qquad \text{for} \quad q_j \le n < q_{j+1}$$

$$g_{n-1} = \frac{1}{\rho_j} g_{n-2} \qquad \text{for } n_j \le n < n_{j+1} \qquad n_j = \min n : \left\{ |g_{n-2}| < \left| \frac{1}{\rho_j} \right|^{n-q_j+2} \right\} \tag{7.4}$$

initialized by $f_0 = g_0 = 1$. The even and odd subsequences of $|f_n|$ and $|g_n|$ give essentially upper and lower bounds to the even and odd subsequences of $||[\Phi]_{2n}||$ and $||[\Phi]_{2n+1}||$ of the conjugation function ($||\ ||$ is the L^2 norm with respect to the angle θ). The coefficients Ω_{2n} of the nonlinear tune are also bounded by the same subsequences.

The analysis of the sequences (7.4) simplifies considerably when ν is a quadratic irrational namely $\nu = (-s + \sqrt{4+s^2})/2$ corresponding to a continued fraction with equal entries $s_j = s$ so that $p_j = q_{j-1}$. In this case for $j \to \infty$ we have $q_{j-1}/q_j \to \nu$ and $|\epsilon_j| q_j^2 \to \alpha$ where α is a real constant.

It is easy to show that both sequences grow as $n!$ and the corresponding series diverge. Indeed from equations (7.3) and (7.4) we see that $|f_n|\ |2\pi\alpha/\Omega_2|^{n/2}$ is bounded above by $n!$ and below by $\nu^n n!$.

It can also be proven that for j large one has

$$n_j \sim \frac{1-\nu}{\log(1/\nu)}\ q_j \log q_j \tag{7.5}$$

As a consequence the subsequence $|g_{n_j}|$ grows as $\exp(n_j \log n_j)$.

8. Analytic continuation and asymptotic estimates

Using the leading terms approximation we can determine the leading singularities of the conjugation function. The Fourier coefficients $A_{1\pm q_j}$ corresponding to the main resonances have poles at $\rho = \pm\rho_j$ with residues exponentially small with q_j and proportional to the width of the resonance. The poles alternate on the real and imaginary axis corresponding to real and virtual resonances.
Taking the subleading singularities into account a whole sequence of poles is found for any Fourier coefficient with slightly smaller residues. This picture is only the main framework of a more complex structure as the presence of *islands within the islands ad infinitum* would suggest.

The use of the sequences $\{f_j\}$ and $\{g_j\}$ to approximate the leading cofficients of Φ lead essentially to very similar results, the only change being $f_l \to g_l$ and $q_j \to n_j$. First of all we observe that

$$f_{q_j-2+l} = f_{q_j-2}\left(\frac{1}{\rho_j}\right)^l \qquad f_{q_j-2} = \left(\frac{1}{\rho_0}\right)^{q_1-2}\left(\frac{1}{\rho_1}\right)^{q_2-q_1}\cdots\left(\frac{1}{\rho_{j-1}}\right)^{q_j-q_{j-1}} \tag{8.1}$$

for $1 \le l < q_{j+1} - q_j$.
The leading terms in the $a_{n,k}$ sequence of the perturbative expansion of Φ are given by

$$a_{q_j-1+2l,\,l} = c_j^-\, f_{q_j-2+(2l+1)} \qquad a_{q_j-1+2l,\,q_j+l} = c_j^+\, f_{q_j-2+(2l+1)} \tag{8.2}$$

for $1 \le l \le \ell_j \equiv [q_{j+1} - q_j - 1]/2$. Using (8.2) we obtain

$$\begin{aligned} \Phi \underset{asympt}{=} & \sum_{j=1}^{\infty}\sum_{l=0}^{\ell_j} [a_{q_j-1+2l,\,l}\zeta^l\zeta^{*q_j-1+l} + a_{q_j-1+2l,\,q_j+l}\zeta^{q_j+l}\zeta^{*l-1}] \\ = & \sum_{j=1}^{\infty} [c_j^-\, e^{i(-q_j+1)\theta} + c_j^+\, e^{i(q_j+1)\theta}]\, \rho^{q_j-2} f_{q_j-2}\left(\frac{\rho}{\rho_j}\right)\sum_{l=0}^{\ell_j}\left(\frac{\rho}{\rho_j}\right)^{2l} \end{aligned} \tag{8.3}$$

Replacing the finite sums up to ℓ_j with a sum up to ∞ we obtain the required continuation of Φ. The new terms we include when $l > \ell_j$ are in any case smaller with respect to the neglected ones having the same order.

$$\Phi \underset{cont}{=} \sum_{j=1}^{\infty} [c_j^-\, e^{i(-q_j+1)\theta} + c_j^+\, e^{i(q_j+1)\theta}]\, \rho^{q_j-2} f_{q_j-2}\frac{\left(\frac{\rho}{\rho_j}\right)}{1-\left(\frac{\rho}{\rho_j}\right)^2} \tag{8.4}$$

If the sequence $\{g_j\}$, is used the result is still (8.3) where f_{q_j-2} becomes g_{n_j-2}. The retardation effect changes only the residues of the poles.

The residues γ_j of the poles of (8.3) are bounded by

$$|\gamma_j| = |f_{q_j-2}\, \rho_j^{q_j-2}| = \left| \left(\frac{\rho_j}{\rho_0}\right)^{q_1-2} \left(\frac{\rho_j}{\rho_1}\right)^{q_2-q_1} \cdots \left(\frac{\rho_j}{\rho_{j-1}}\right)^{q_j-q_{j-1}} \right| \leq \nu^{q_j-2} \qquad (8.5)$$

Writing $|\rho_j| = b/q_j$ with $b = |2\pi\alpha/\Omega_2|^{1/2}$ the residue appears to be exponentially small with $|\rho_j|$. Being interested only in asymptotic results we can replace ν^{q_j-2} with ν^{q_j} in (8.5) and write the width of the islands associated to the resonance q_j as $\Delta_j \propto \epsilon_j^{(q_j-2)/4} \sim q_j^{-q_j/2}$, see (9.8).
Taking the retardation effect into account we see that the residues are bounded by the resonance widths. Indeed we have $|\gamma_j| \leq \nu^{n_j} \leq \exp[-(1-\nu)q_j \log q_j]$ so that $|\gamma_j| \leq \Delta_j^{2(1-\nu)} < \Delta_j$ since $\nu < 1/2$.

The contribution of the subdominant terms can also be examined. To simplify the analysis we refer to the $k = 0$ component. We first observe that the subleading contributions of a resonance q_j are given by $a_{q_j-1+2l,l+m} \simeq (1/\rho_j)^{2(l-m)+1}$ for $m \leq l$. As a consequence using (4.5) for $k = 0$ where $b_{0,n} = a_{2n,n}$ and setting $b_{0,q_j-1+m} = c_j f_{q_j-1+2m}$ for $0 \leq m \leq \ell_j$ we obtain

$$\begin{aligned} A_0(\rho^2) \underset{asympt}{=} & \sum_{j=1}^{\infty} \sum_{m=0}^{\ell_j} b_{0,q_j-1+m}\rho^{2(q_j-1+m)} \\ = & \sum_{j=1}^{\infty} c_j\, f_{q_j-2}\, \rho^{2(q_j-1)} \left(\frac{1}{\rho_j}\right) \sum_{m=0}^{\ell_j} \left(\frac{\rho}{\rho_j}\right)^{2m} \end{aligned} \qquad (8.6)$$

After the continuation also A_0 turns out to be the sum of an infinite sequence of poles according to

$$A_0(\rho^2) \underset{cont}{=} \sum_{j=1}^{\infty} c_j f_{q_j-2} \frac{1}{\rho_j} \frac{\rho^{2(q_j-1)}}{1 - \left(\frac{\rho}{\rho_j}\right)^2} \qquad (8.7)$$

Taking (8.4) into account the residues for the poles at ρ_j for the Fourier coefficients are bounded by $|\gamma_j| \leq |\rho_j|^{q_j-1}\ \nu^{q_j-2} \leq |\rho_j^{q_j-1}|$ and decrease as Δ_j^2. Using the sequence $\{g_l\}$, which takes the retardation effect into account, we have to change q_j into n_j in (8.7) and the residues decrease as Δ_j raised to a constant times $\log(1/\rho_j)$. As one should expect the residues of these secondary poles are much smaller than the previous ones.

Even though the present picture is only the consequence of an asymptotic analysis it is numerically confirmed typically for $j \leq 3$. This singularity structure justifies the use of Padé approximants to analytically continue the Fourier coefficients. They consist in replacing a Taylor series truncated at order $2N-1$ or $2N$ by the ratios $[N-1/N]$ or $[N/N]$ of polynomials whose degrees are specified by the square brackets. Numerical results, see the report by G. Servizi in this volume, fully confirm that the divergence due to a virtual resonance is overcome since the Padé locate the poles on the imaginary ρ axis. In the case of a true resonance the poles are located on the real ρ axis and wild oscillations are observed both in the tune and the orbits given by the conjugation function. When the residue is very small the oscillations extend on a small region comparable with the resonance width and the tune or conjugation function are correctly continued further on.

9. Resonances and interpolating hamiltonians

When the tune is exactly resonant or close to a resonance the nonresonant normal forms fail to reproduce the orbits of the map.
In this case the resonant normal forms and the interpolating hamiltonian are mandatory. We refer to [23] where the theory has been developed and to [39] where a careful analysis of the low order resonances has been carried out; here we limit ourselves to summarize the main results.

Let the frequency ω be close to a resonant frequency $\omega_R = 2\pi\ p/q$ according to $\omega = \omega_R + 2\pi\epsilon$. In this case the resonant normal form with respect to $\nu_R = p/q$ reads

$$\varsigma' = e^{i(\omega_R+2\pi\epsilon)}\left[e^{i[\Omega_2\varsigma\varsigma^*+\ldots+\Omega_{2m}(\varsigma\varsigma^*)^m]}\varsigma + i\alpha\varsigma^{*q-1}\right] + O(|\varsigma|^q) \tag{9.1}$$

where $m = \frac{q-2}{2}$ if q is even, $m = \frac{q-3}{2}$ if q is odd. We denote with $H = 2i\mathcal{H}$ the interpolating hamiltonian in the rotating frame, see (3.6), defined by

$$U = e^{i\omega_R} e^{D_{\hat{H}}}\varsigma \qquad D_{\hat{H}} = \frac{\partial \hat{H}}{\partial \varsigma^*}\frac{\partial}{\partial \varsigma} - \frac{\partial \hat{H}}{\partial \varsigma}\frac{\partial}{\partial \varsigma^*} \tag{9.2}$$

where $D_{\hat{H}}$ is the Lie derivative, and notice that $\hat{\mathcal{H}} = \hat{H}/(2i)$ is a real function. If we use polar canonical coordinates $\varsigma = \sqrt{2r}e^{i\vartheta}$ then $\hat{\mathcal{H}}(r,\theta)$ is just the hamiltonian for the action–angle real canonical coordinates (r,θ) (the factor $2i$ arises from the use of complex coordinates treated as canonically conjugate). The interpolating hamiltonian up to order q reads

$$\hat{H} = i\left[2\pi\epsilon\varsigma\varsigma^* + \frac{\Omega_2}{2}(\varsigma\varsigma^*)^2 + \ldots \frac{\Omega_{2m}}{m+1}(\varsigma\varsigma^*)^{2m} + \frac{\alpha}{q}(\varsigma^q + \varsigma^{*q})\right] \tag{9.3}$$

Here we shall rather use ordinary canonical coordinates $\varsigma = \rho e^{i\theta}$ as we did in the last section so that the interpolating hamiltonian will read

$$\begin{aligned} \hat{\mathcal{H}}(\rho,\theta) &= h(\rho) + \frac{\alpha}{q}(\rho)^q \cos q\theta + \ldots \\ h(\rho) &= \frac{2\pi\epsilon}{2}\rho^2 - \frac{\Omega_2}{4}\rho^4 + \ldots + \frac{\Omega_{2m}}{2(m+1)}\rho^{2(m+1)} \end{aligned} \tag{9.4}$$

The hamiltonian $\hat{\mathcal{H}}$ has an elliptic fixed point at the origin and assuming $\epsilon > 0$, $\Omega_2 < 0$, has q elliptic fixed points at $\varsigma_{2k} = \rho_+ e^{i\theta_{2k}}$ with $k = 1, \ldots, q$ and q hyperbolic fixed points at $\varsigma_{2k+1} = \rho_- e^{i\theta_{2k+1}}$ with $k = 1, \ldots, q$ where $\vartheta_k = \pi k/q$ and $\rho_\pm$ are solutions of

$$h'(\rho_\pm) \pm \alpha(\rho_\pm)^{q-1} = 0 \tag{9.5}$$

if $\alpha > 0$. If $\alpha < 0$ the elliptic and hyperbolic fixed points are interchanged. The solutions $\rho_\pm$ can be written as series in ϵ and one has

$$\rho_\pm^2 = \frac{2\pi\epsilon}{-\Omega_2} + O(\epsilon^2) \qquad \rho_+^2 - \rho_-^2 = O(\epsilon^{q/2-1}) \tag{9.6}$$

for $q \geq 6$. For $q < 6$ one has to replace $O(\epsilon^2)$ with $O(\epsilon^{q/2-1})$ in the above equation. If the $|\Omega_k|$ do not increase too fast we can approximate h with the hamiltonian of a pendulum

$$\hat{\mathcal{H}} = h(\rho_0) + \frac{h''(\rho_0)}{2}(\rho')^2 + \frac{\alpha}{q}\rho_0^q \cos q\theta \tag{9.7}$$

The separatrices have a maximal distance given by

$$\Delta = 2\left|\frac{\alpha}{qh''(\rho_0)}\right|^{1/2}(\rho_0)^{q/2} \simeq \epsilon^{(q-2)/4} \tag{9.8}$$

since one has $h''(\rho_0) = -4\pi\epsilon$.

We consider finally the explicit form of the hamiltonians at the exact resonances of order 3 and 4 [39]:

resonance 3 ($\nu = p/3$)

$$\hat{\mathcal{H}} = h_0^1 \varrho^3 \cos 3\vartheta + h_2^0 \varrho^4 \tag{9.9}$$

resonance 4 ($\nu = p/4$)

$$\hat{\mathcal{H}} = (h_0^1 \cos 4\vartheta + h_2^0)\varrho^4 + \varrho^6(h_3^0 + h_1^1 \cos 4\vartheta) \tag{9.10}$$

The resonance 3 is unstable if $h_0^1 \neq 0$. The resonance 4 can be stable or unstable depending on the values of h_0^1, h_2^0, h_3^0, h_1^1. Hénon's map in this case exhibits a singular behaviour. Since $h_0^1 = -h_2^0$, if we use perturbative theory only up to order 3 (corresponding to order 4 in the interpolating hamiltonian) then the origin appears to be unstable while for $N \geq 4$ it is stable. When we approach the origin, the orbits do not look like circles, but rather like hyperbolae, even though they join smoothly at a finite distance from the origin. This singular phase space structure disappears if we add a small perturbation to the map. The resonances of order higher than 5 are always stable.
The origin of any polynomial map can be made unstable with respect to a given resonance if the coefficients of the tune below the resonance order vanish. This is precisely what is done for the multipole error corrections. Indeed the parameters

of the correction magnets are choosen in order to make the tune shift coefficients vanish at second or fourth order. It is evident that continuing the correction strategy up to very high orders would not be convenient because the origin would become unstable with respect to higher and higher order resonances.
We conclude by observing that the series of the resonant normal forms are are affected by divisors D_n which are periodic in n with period q and are therefore all bounded. No plateaus corresponding to resonances are observed. The ratios $||[\Phi]_{2n-2}/[\Phi]_{2n}||$ have a monotonic behaviour growing linearly, at most with periodic oscillations superimposed.

10. First integrals and variational method

When the linear frequency is non resonant one can write linear equations for the invariants or first integrals of motion of the map. From the conjugation equation we can write

$$e^{i\Xi}\Psi \circ F = e^{i\Omega} e^{i\Xi} \Psi \tag{10.1}$$

where Ξ. s the arbitrary gauge. We then obtain

$$|\Psi\Psi^*| \circ F = |\Psi\Psi| \tag{10.2}$$

namely the integral of motion $I \equiv \Psi\Psi^*$ satisfies a linear equation with boundary conditions

$$I \circ F(\varsigma, \varsigma^*) = I(\varsigma, \varsigma^*) \qquad I = \varsigma\varsigma^* + O(|\varsigma|^3) \tag{10.3}$$

which can be directly solved by perturbation according to

$$\Delta[I]_n = (1 - \Pi) \sum_{k=2}^{n-1} [I_k \circ F]_n \tag{10.4}$$

where Π denotes the normal form projector. The invariant I provides only the geometry of the orbits by its level lines, not the dynamics. The perturbative expansion faces the same convergence problems as the the previous nonlinear conjugation equation.

The linearity of (10.3) allows a variational formulation of the problem. A variational analysis was suggested by Remiddi and will be presented by Bazzani in this volume. The interest of the method is that it provides unique approximate solutions when a finite dimensional trial space is considered. Let

$$\mathcal{E}(I) = \int_{\mathcal{E}} |I \circ F(\varsigma, \varsigma^*) - I(\varsigma, \varsigma*)|^2 d\mu(\varsigma) \tag{10.5}$$

be a quadratic functional of I, where $\mu(\varsigma)$ is the Lebesgue measure and $\mathcal{E}$ a region where I is defined. Typically $\mathcal{E}$ can be either a disc $|\varsigma| \le R$ or an anulus, $R_1 \le |\varsigma| \le R_2$. If $\mathcal{E}$ is a disc, in the limit of vanishing radius the variational solution agrees with the perturbative result. For R large the improvement can be very important such as to allow resonant structures to be reproduced.

11 Nekhoroshev stability estimates

The heuristic arguments developed so far on the divergence of the Birkhoff series can be made rigorous using the method of majorant series. A theorem was recently proved which shows that for symplectic maps in dimension $2d$ with d arbitray both the norms of Φ_N and Ω_N grow as $N^{a(N-2)}$ and that the remainder $\hat{R}_N(t)$ defined by (5.8) is bounded by $t^2\, N^{a(N+1)}$, where $a = 3+\eta$ and η is the diophantine exponent of ω namely

$$|e^{iq\omega} - 1|^{-1} \le (4q|\epsilon_q|)^{-1} \le \gamma q^{\eta} \tag{11.1}$$

If, for a fixed r, we look for the truncation order which makes the remainder as small as possible we find that $N_{\rm opt} = e^{-1}(r_*/r)^{\frac{1}{3+\eta}}$ where $r_* = 2/(49\gamma)$ provided that the norms of the coefficients of the map satisfy

$$||\,[F]_2||_1 \le 3^{-\eta}\gamma^{-1}, \qquad ||\,[F]_n||_1 \le 1 \text{ for } n > 2 \tag{11.2}$$

where $||\ \ ||_r$ denotes the norm on the polydisc of radius r

$$||F||_r = \sum_{n=2} |||[F]_n||_1 r^n \qquad |||[F]_n||_1 = \sum_{j=0}^{k} |F_{k,j}| \tag{11.3}$$

The optimal time for which the orbit remains bounded can be estimated. The result is expressed by the following theorem [32,33]

Theorem

If $F(z, z^*)$ is a symplectic map of $\mathbf{R}^{2d}$, with linear part $e^{i\omega}z$ where ω are non resonant frequencies satisfying a diophantine condition (11.1) and such that (11.3) are verified, then assuming $||z^{(0)}|| \le r/2$ and letting $z^{(t)} = F(z^{(t-1)}, z^{*(t-1)})$ one has $||z^{(t)}|| \le (1 + \frac{1}{24})r$ for all $t \le T$ such that

$$T = \frac{7}{8}\sqrt{\frac{r}{6}}\exp\left[\frac{3+\eta}{2e}\left(\frac{r_*}{r}\right)^{\frac{1}{3+\eta}}\right] \qquad r_* = \frac{2}{49\gamma} \tag{11.4}$$

provided that $\qquad r \le r_*(3e)^{-3-\eta}$.

The estimate is not optimal because the remainder after t iterations was estimated to be t^2 times the one step remainder (this causes the $\sqrt{r}$ factor and the presence of a $1/2$ in the exponential) and because the factor $3+\eta$ comes from an initialization, the factorial growth due to the recurrence giving rise to $1+\eta$. As a consequence an optimal estimate reads

$$T = C\, r \exp\left[\frac{1+\eta}{e}\left(\frac{r_*}{r}\right)^{\frac{1}{1+\eta}}\right] \tag{11.5}$$

where $C \geq 7/48$ is a numerical constant.

We wish to sketch how the factorial growth of the coefficients of the series and of the remainder is obtained in the case of the integrals of motion where the derivation is immediate because the functional equation is linear. The recurrence for the perturbative solution is given by (10.4). Using the norms (11.3), the diophantine condition (11.1) and the inequality (3.1) of reference [33] we obtain

$$\begin{aligned} \|[I]_n\|_1 &\leq \gamma n^\eta \sum_{s=2}^{n-1} \|[I]_s\| \sum_{m_1+\ldots+m_s=n} \|[F]_{m_1}\|\cdots\|[F]_{m_s}\| \\ &\leq \gamma n^\eta \sum_{s=2}^{n-1} \|[I]_s\| \sum_{m_1+\ldots+m_s=n} 1 = \gamma n^\eta \sum_{s=2}^{n-1} \|[I]_s\| \binom{n-1}{s-1} \end{aligned} \tag{11.6}$$

where formaula (A.3) of Appendix 1 of ref. [33] has been used.Letting $\mu_2 = 1$ and

$$\|[I]_n\| \leq \xi_n \mu_n \qquad \xi_n \leq \xi_{n-1} \gamma\, n^\eta \qquad \mu_n = \sum_{s=2}^{n-1} \mu_s \binom{n-1}{s-1} \tag{11.7}$$

It follows easily by induction that $(n-1)! \leq \mu_n \leq 2^n n!$ so that

$$\|[I]_n\|_1 \leq (2\gamma)^n n^{n(1+\eta)} \tag{11.8}$$

As consequence since the remainder is given by $R_N = \sum_{n>N} [[I]_{\leq N} \circ F]_n$ we have the following estimate

$$\|[R_N]_n\|_1 = \|\, [[I]_{\leq N} \circ F]_n\| \leq \sum_{s=2}^{N} \|[I]_s\| \sum_{m_1+\ldots+m_s} 1 = \sum_{s=2}^{N} \|[I]\|_s \binom{n-1}{s-1} \leq \|[I]_N\|_1 2^n \tag{11.9}$$

Using these results we obtain the following estimate in a disc of radius $r \leq 1/4$

$$\|R_N\|_r \leq \|[I]_N\|_1 r \sum_{n=N}^{\infty} r^n 2^n = \|[I]_N\| \frac{(2r)^N}{1-2r} \leq 2(4\gamma r)^N\, N^{N(1+\eta)} \tag{11.10}$$

For a fixed r the remainder is minimum for $N = e^{-1}(r_*/r)^{1/(1+\eta)}$ with $r_* = 1/4\gamma$ and the minimum is given by

$$R_N^{\min} = \exp\left[\frac{1+\eta}{e}\left(\frac{r_*}{r}\right)^{\frac{1}{1+\eta}}\right] \qquad r^* = \frac{1}{4\gamma} \tag{11.11}$$

The final estimate, which leads to (11.5), is obtained by showing that after t iteration the remainder is t times the above remainder. This part is more tricky and has been developed in [40].

As an exercize we estimate the stability time for a typical SPS experiment with 8 strong sextupoles. For convenience we choose a strongly non resonant tune such as the golden mean $\nu = (\sqrt{5}-1)/2$. The sextupole strength is $k_3 = (\partial B/\partial x)(B_0\rho)^{-1} = 0.15$ with an average $\beta = 100$ meters. With normalized and rescaled coordinates the map becomes the quadratic Hénon map where

$$y = \frac{Y}{\beta k_3} \qquad p = \frac{P}{k_3\beta^2} \tag{11.12}$$

Letting $\rho = \sqrt{Y^2+P^2}$ and $r = \sqrt{y^2+p^2}$, setting $\rho_* = 2/(49\gamma) = 0.08$ we have $T = 0.01(\rho/\rho_*)\exp(0.73\sqrt{\rho_*/\rho})$ where $\eta = 1, \gamma \simeq 0.5$ have been used. As a consequence $r_* = \rho_*/\beta k_3 \sim 5$ mm and T, which is 1/8 the actual number of turns in the true machine, becomes

$$\begin{aligned} &T \simeq 1 &&\text{for} \quad \rho = \rho_* 10^{-2} && r = 10^{-2} r_* = 0.05 \text{ mm} \\ &T \simeq 10^6 &&\text{for} \quad \rho = \rho_* 10^{-3} && r = 10^{-3} r_* = 0.005 \text{ mm} \\ &T \simeq 10^{26} &&\text{for} \quad \rho = \rho_* 10^{-4} && r = 10^{-4} r_* = 0.0005 \text{ mm} \end{aligned} \tag{11.13}$$

We are off about 4 orders of magnitude with respect to the expected result. These estimates which provide a bound on the Arnold diffusion could be improved, but we doubt that the expected lifetimes could be reproduced. This is suggested also by tracking experiments and different mechanisms such as modulational diffusion must be explored.

12. Analyticity results from the majorant series

The majorant series for the conjugation function and the normal form, excluding the divisor contribution which is absent in the resonant case, satisfy the equation

$$\mu_n = \sum_{s=2}^{n-1} \mu_s \sum_{\substack{m_1+\ldots+m_s=n \\ m_j \geq 1}} \mu_{m_1} \cdots \mu_{m_s} \qquad 3 \leq n, \qquad (12.1)$$

initialized by $\mu_1 = \mu_2 = 1$. The growth of the μ_n is $n!$. We observe that the divergent character due to nonlinearity is captured by the sequence μ_n defined above; it is interesting to observe that its generating function $g(x) = \sum_{n \geq 1} \mu_n x^n$ satisfies the functional equation

$$3g(x) = g(g(x)) + x^2 + 2x \qquad g(0) = 0,\ g'(0) = \frac{1}{2}\ g''(0) = 1 \qquad x \in \mathbf{C} \quad (12.2)$$

and that $g(x)$ is analytic in a neighborhood of $x = -1$ with a fractal natural boundary which includes the origin [41]. It has been found that the fractal near the origin has a parabolic behaviour ($y = x^2$) and that the measure of points (being normalized to 1 on the whole fractal) on an arc fom the origin to (x, x^2) behaves as $\exp(-1/x)$, see the contribution by Xie in this volume.

The majorant sequence for the invariants of motion (11.8) has a generating function which satisfies the functional equation

$$2f(x) - f\left(\frac{x}{1-x}\right) = x^2 \qquad (12.3)$$

In this case the singularities are arbitrarily close to the real axis and the discontinuity at x still has an exponential decrease [40]. As a consequence the heuristic analysis of section 8 and the majorant series all lead to the same growth factor and to a similar analyticity structure.

13. Dynamic aperture and diffusion

The problem of determining the dynamic aperture is by far the most difficult. Some information can be obtained from the loss of simplecticity of the normal forms, especially the resonant case. Other approaches workable in the 2 dimensional case are given by the inner envelope of the stable and unstable manifolds of the hyperbolic fixed points of the first main resonance. Finally the dynamic aperture can be viewed as the real section of the Julia set of the complexified map.

In the first case we determine the boundary of the domain $\mathcal{D}_\eta{}^{\mathrm{Jac}}$, image by Φ of the domain $\hat{\mathcal{D}}_\eta{}^{\mathrm{Jac}}$, where Φ itself has a significant deviation from simplecticity. To be more explicit $\hat{\mathcal{D}}_\eta{}^{\mathrm{Jac}}$ is given by:

$$\left| \frac{\partial(\Phi, \Phi^*)}{\partial(\varsigma, \varsigma^*)} - 1. \right| \geq \eta \tag{13.1}$$

Such a domain will depend on the perturbation order N and the linear tune ν as well. If N is lower than the first or the second main resonances of ν, according to the use of non resonant or resonant normal forms, then the region $\hat{\mathcal{D}}_\eta{}^{\mathrm{jac}}$ is comparable with the basin of attraction of infinity.
This is not surprising because in this case the only obstacle to the existence of invariant curves are the hyperbolic points and the related hyperbolic tangle through which the points escape to infinity. Of course when ν has a close dominant resonance of order less than N then $\hat{\mathcal{D}}_\eta{}^{\mathrm{Jac}}$ will touch the resonance region, which could be much inside the dynamic aperture. The result on the dynamic aperture would become more accurate as we increase N but the measure of the intervals in the tune axis where it can be applied decreases with N.
Very similar results are obtained by considering the region $\hat{\mathcal{D}}_\eta{}^{\mathrm{Rem}}$ where the remainder defined by (5.2) is above a given value $|R_N| \geq \eta$. The domain $\hat{\mathcal{D}}_\eta{}^{\mathrm{Inv}}$ where Φ ceases to be invertible is also comparable with $\hat{\mathcal{D}}_\eta{}^{\mathrm{Jac}}$ and $\hat{\mathcal{D}}_\eta{}^{\mathrm{Rem}}$ (letting $z = re^{i\vartheta}$ and $\varsigma = \rho e^{i\theta}$ the condition is $d\vartheta/d\theta = 0$).

The second method [42] consists in determining the hyperbolic fixed points of the main resonance via the interpolating hamiltonian and computing the inner envelope of their stable and unstable manifolds. To this end the use of hyperbolic normal forms can be useful [43]. The results obtained by this method for frequencies close to 0 and 1/3 are quite accurate.

The third approach [44] is based on complex periodic orbits. For the quadratic map the newtonian or second order map in the y variable is written

$$\Delta^2 y_n \equiv y_{n+1} - 2\cos\omega\ y_n + y_{n-1} = \sin\omega\ y_n^2 \tag{13.2}$$

and a solution $y_n = \Phi(e^{in\omega}\varsigma)$ is searched. The series will converge within a disc of radius R and the real section of the torus $x = \Phi(\rho e^{i\theta})$, $p = [\cos\omega\Phi(\rho e^{i\theta}) - \Phi(\rho e^{i(\theta-\omega)})] / \sin\omega]$ agrees qualitatively with the dynamic aperture. See also the report by Marmi in this volume.

Modulation diffusion When a four dimensional problem is considered the Arnold diffusion which is bounded by our Nekhoroshev stability estimates seems to be insufficient to explain the observed beam lifetimes. The very slow frequency modulation of the magnetic field in the quadrupoles due to the current ripples seems to be a good candidate to explain it. A very simple model can be proposed to analyze it and this corresponds to the equation for a FODO cell with a strong sextupole

$$\ddot{y} + \omega^2(s) = y^2 \delta_P(s) \qquad \omega^2(s) = \omega_0^2[1 + \epsilon \sin \Omega s] \tag{13.3}$$

where $\delta_P(s)$ is the periodic δ function of period 2π The solution can be written

$$\begin{pmatrix} y_{n+1} \\ p_{n+1} \end{pmatrix} = \mathcal{M}_n \begin{pmatrix} y_n \\ p_n \end{pmatrix} \tag{13.4}$$

where $\mathcal{M}_n = \mathbf{TMT}^{-1}$ is the solution of the Mathieu equation having the representation (2.7) where at first order in ϵ

$$\beta = \frac{1}{\omega_0}\left[1 - \frac{\epsilon}{4}\sin(2\pi n\Omega)\right]^2 \qquad \psi = 2\pi\omega_0\left[1 - \frac{2\epsilon\omega_0^2 \sin \pi\Omega}{\pi\Omega(\Omega^2 - 4\omega_0^2)} \sin 2\pi\Omega\left(n + \frac{1}{2}\right)\right] \tag{13.5}$$

A model further simplified, given by a quadratic map with modulated rotation $\beta = 1/\omega_0$ and $\psi = 2\pi\omega_0[1 - \epsilon/2 \, \sin 2\pi\Omega(n + \frac{1}{2})]$, has been considered by Malavasi, and reported in this volume.

14- Analysis of LHC lattice

The algorithms for the normal forms have been developed and numerically implemented first for the 2-dimensional maps [20], then in the resonant case [39] and later for the 4-dimensional maps in the non resonant case [45]. A general program for any type of normal forms in arbitrary dimensions is being developed.

The LHC tune shifs and smears

The first application considered has been the evaluation of the tune shift and smear for a LHC (Large Hadron Collider project at CERN) [30] for the horizontal betatronic motion [26]. Later the same analysis has been repeated for the complete x, y betatronic motion and excellent agreement was found with the tracking results by FASTTRACK[46]. Up to 20% from the dynamic aperture the computation of the tune shifts $\delta\nu_x, \delta\nu_y$ and smears σ_x, σ_y is faster and more accurate than by tracking.

The 1988 SPS experiment

The experiment carried out at the SPS in 1988, in order to evaluate the influence of the non linear fields on beam dynamics, was also analyzed using the normal forms[29]. In the experiment eight strong sextupoles were powered with opposite polarities in order to minimize the effect of a third order resonance. The experiment provided three sets of data: the first two correspond to a current of 140 A. and different working points, the third to a current of 250 A. In the three cases the results are in excellent agreement with the experimental data and the tracking simulations by SIXTRACK. In the first case a convergent behaviour is observed as the amplitude is increased, whereas in the second and third case the presence of a virtual resonance of order 5 restricts to low orders the use of the series. However the short term dynamic aperture was estimated by considering the boundary of the region where Φ ceases to be invertible and a good agreement with the experiment was found.

15. Correction of multipolar errors

The correction of the multipolar errors is a crucial point in the design of the new large hadronic machines such as LHC, SSC, ELOISATRON. The normal form approach is well founded and appears to be more effective than the techniques based on canonical perturbation theory [47]. The procedure consists in minimizing the tune shift, which is an invariant, rather than the smear which varies with the chosen reference section.

As a first example we consider a cubic map where the quadratic term is fixed and the cubic one is varied and can be thought as the corrector (sextupole error and octupole correctors)

$$z' = e^{i2\pi\nu}\left[z + \frac{i}{4}(z+z^*)^2 + i\alpha(z+z^*)^3\right] \tag{15.1}$$

The second order expression of the tune shift is (see also (6.10))

$$\delta\nu = -\frac{1}{32\pi}\left[3\cot(\pi\nu) + \cot(3\pi\nu)\right] + \frac{3}{2\pi}\alpha \tag{15.2}$$

At this order the tune shift can be made equal to zero and one has

$$\alpha = \frac{1}{48}\left[3\cot(\pi\nu) + \cot(3\pi\nu)\right] \tag{15.3}$$

In a real lattice the tunes are obtained at second order by differentiating with respect to $\rho_1 = \varsigma_1\varsigma_1^*$ and $\rho_2 = \varsigma_2\varsigma_2^*$ a function $\mathcal{H} = \nu_1\rho_1 + \nu_2\rho_2 + h_{20}\rho_1^2 + h_{21}\rho_1\rho_2 + h_{22}\rho_2^2 + \ldots$ and the tune shifts read

$$\begin{aligned} \delta\nu_x &= 2h_{20}\rho_1 + h_{21}\rho_2 + \ldots \\ \delta\nu_y &= h_{21}\rho_1 + 2h_{22}\rho_2 + \ldots \end{aligned} \tag{15.4}$$

The dependence of the multipolar coefficients in second and fourth order is the following

$$\text{second order } h_{2,i} \propto \begin{cases} k_2^2 & \text{sextupole field} \\ k_3 & \text{octupole field} \end{cases}$$

$$\text{fourth order } h_{3,i} \propto \begin{cases} k_5 & \text{dodecapole field} \\ k_2 k_4 & \text{decapole+sextupole field} \\ k_3(k_2)^2 & \text{octupole+ sextupole field} \\ (k_3)^2 & \text{octupole field} \\ (k_2)^4 & \text{sextupole field} \end{cases} \qquad (15.5)$$

Since the $h_{n,i}$ are multilinear in the integrated gradients k_j, the corresponding coefficients are obtained by solving a system of linear equations obtained by computing the $h_{n,i}$ with suitable values 0 or 1 to the k_i.
Let us introduce the functions $\chi_\ell(\mathbf{k})$ where

$$\chi_{2\ell}(\mathbf{k}) = R^{-\ell-2} \int_0^R d\rho_1 \int_0^R d\rho_2 \left\{ ([\delta\nu_x]_\ell)^2 + ([\delta\nu_y]_\ell)^2 \right\} \qquad \ell = 1, 2, \ldots, m \quad (15.6)$$

where $[\quad]_\ell$ denotes a homogeneous polynomial in ρ of degree ℓ. The correction strategy consists in minimizing the tune shift in the polydisc $0 \le \rho_1 \le R$, $0 \le \rho_2 \le R$ in two different ways:

i) order by order: minimizing the functions $\chi_2, \chi_4, \ldots, \chi_{2m}$

ii) globally: minimizing $\chi = \chi_2 + \chi_4 + \ldots + \chi_{2m}$.

We observe that $[\delta\nu]_2$ is linear in k_3 and quadratic in k_2, $[\delta\nu]_4$ is linear both in k_3 and k_4 and in general $[\delta\nu]_{2m}$ is linear in k_{2m-1} and k_{2m}. As a consequence the order by order minimization implies the solution of a non linear problem only when the second order is considered. One has to face the solution of 3 linear equations for the octupolar correctors and a cubic equation for the sextupolar corrector. Imposing some symmetry conditions it is possible to have a number of equations equal to the number of unknown and to make the tune shift exactly zero.
This correction srategy was developed in [39] and successfully applied to the correction of the LHC errors in [31].

REFERENCES

[1] H. Poincaré *Les méthodes nouvelles de la méchanique céleste* Gauthier-Villars, Paris (1898)
[2] G. D. Birkhoff Acta Mathematica **43**, 1 (1927)
[3] C.L. Siegel, Annals of Mathematics **43**, 607 (1942)
[4] A.N. Kolmogorov Dokl. Akad. Nauk. SSSR **98**, 527 (1954)
[5] V.I. Arnold Russ. Math. Surv. **18**, 9 (1963)
[6] J. Moser Nachr Akad. Wiss. Gottingen Math. Phys. **K1.1**, 1 (1962)
[7] N.N. Nekhoroshev Russ. Math. Surv. **18**, 85 (1977)
[8] Gustavson The Astronomical Journal **71**, 670 (1966)
[9] Contopoulos The Astronomical Journal **68**, 1 (1963)
[10] M. Hénon Quarterly of Appl. Math. **27**,291 (1969)
M. Hénon, C. Heiles The Astronomical Journal **69**, 73 (1964)
[11] J. Roels, M. Hénon Bull. Astr. Soc. **2**, 267 (1967)
[12] A. Dragt, J. M. Finn J. Math. Phys **17** pag. 2215 (1976)
A. Dragt *Nonlinear Dynamics and the beam-beam interaction* AIP Conf. Proc. **57**, 143 (1979)
[13] A. Dragt, E. Forrest J. M. P. **24**, 2734 (1983)
[14] L. Galgani, A. Giorgilli J. Cel. Mechanics **17**, 267 (1978)
A. Giorgilli Computer Physics Comm. **16**, 331 (1979)
[15] G. Benettin, L. Galgani, A. Giorgilli J. Cel. Mech. **37**, 1 (1985)
[16] G. Benettin, G. Gallavotti J. Stat. Phys. **44**, 293 (1986)
[17] A. Giorgilli, A. Delshams, E. Fontich, L. Galgani, C. Simó J. Diff. Eqs **77**, 167 (1989)
[18] G. Benettin, A. Giorgilli, G. Servizi, G. Turchetti Phys. Letters **A95**, 11 (1983).
[19] G. Servizi, G. Turchetti Il Nuovo Cimento **B95**, 121 (1986)
[20] G. Servizi, G. Turchetti Comp. Phys. Comm. **32**, 201(1984).
[21] A. Bazzani, G. Turchetti "Normal forms for symplectic map" from *Nonlinear Dynamics and stochastic processes II* Ed. by Livi, World Scientific pag. 25 (1988).
[22] A. Bazzani Celestial Mechanics **42**, 107 (1987)
[23] G. Turchetti "Perturbative methods for hamiltonian maps", from *Methods and applications of nonlinear dynamics* Ed. W. A. Saenz. World Scientific Publ. pag. 95 (1988).
[24] M. Pusterla, G. Servizi, G. Turchetti, IEEE Transactions on nuclear sciences **NS30**, 2642 (1983).
[25] A. Bazzani, A. Pisent, G. Turchetti *Normal forms for hamiltonian maps and future applications to accelerators* CERN report SPS 87/38(AMS) (1987)
[26] A. Bazzani, G. Turchetti, P. Mazzanti, G. Servizi Il Nuovo Cimento **B102**, 51 (1988)

[27] A. Bazzani, B. Leemann, W. Scandale, G. Servizi, G. Turchetti "Description of nonlinear beam dynamics in the CERN large hadron collider by using normal form algorithms" from *European Particle Accelerators Conference (EPAC 1988)* Ed. S. Tazzari, World scientific (1989)

[28] A. Bazzani, G. Turchetti *Second advanced ICFA beam dynamics workshop* Ed. by J. Hagel, E.Keil CERN yellow reports 88-04 pag 72 (1988)

[29] A. Bazzani, G. Turchetti "Normal form results for the 1988 dynamical aperture experiment" CERN SPS report 89-25 (AMS) (1989)

[30] W. Scandale CERN-SPS report 87-28 (AMS) (1987)

[31] W. Scandale, F. Schmidt, E. Todesco CERN SPS report to be published in Particle Accelerators (1990)

[32] G. Turchetti "Nekhoroshev stability estimates for symplectic maps and physical applications" from *Number theory in Physics* ed. P. Moussa, Springer (1989)

[33] A Bazzani, S. Marmi, G. Turchetti Journal of Celestial Mechanics **47**, 333 (1990)

[34] A. Bazzani this volume

[35] M. Berz Particle Accel. **24**, 109 (1989)

[36] E. Forrest, M. Berz, LBL Accelerator & Fusion Research Division report 25609 UC-414 (1988)

[37] E. Courant, H. Snyder Annals of Physics (N.Y.) **3**, 1 (1958)

[38] R. D. Routh "Single particles dynamics and nonlinear resonances in circular accelerators" from *Non linear dynamics aspects of particle accelerators* Ed. by Jowett,Month and Turner, Springer-Verlag pag. 37 (1985)

[39] E. Todesco "Fome normali risonanti per mappe simplettiche polinomiali e correzioni multipolari per LHC" University of Bologna Master thesis (1990)

[40] A. Bazzani, S. Marmi *First integrals and stability estimates for isochronous symplectic maps* Dip. di Fisica Bologna preprint (1990)

[41] D. Bessis, S. Marmi, G. Turchetti Rendiconti di Matematica **VII-9**, 645 (1989)

[42] M. Giovannozzi *Analisi di stabilità per mappe hamiltoniane e loro applicazioni in fisica degli acceleratori* Univ. of Bologna Master thesis (1990)

[43] G. L. Da Silva Ritter et Al. Physica D **29**, 181 (1987)

[44] M. Giovannozzi, S. Marmi Rendiconti di Matematica **VII-9**, 457 (1989)

[45] A. Bazzani, G. Servizi, G. Turchetti *POLYN a program for polynomial manipulations* and *BIRK4 a program for the normal forms* Dip. di Fisica Bologna preprints (1988)

[46] F. Schmidt Ph.D thesis DESY HERA report 88-02 (1988)

[47] D. Neuffer "Lumped multipole correction for the CERN Large Hadron Collider" CERN report 87-014 (AHF) (1987)

SINGLE RESONANCE THEORY WITH MAPS

ÉTIENNE FOREST
Accelerator and Fusion Research Division
Lawrence Berkeley Laboratory
1 Cyclotron Road
Berkeley, California 94720, U.S.A.

and

JOHN IRWIN
Stanford Linear Accelerator Center
P. O. Box 4349
Stanford, California 94309, U.S.A.

ABSTRACT

We describe how one can perform a single resonance calculation using the single resonance map and the co-moving map. Again we emphasize that all the concepts of this paper have been implemented for the most complicated situations using the Differential Algebra Package of Berz. In theory, it is possible and desirable to do similar calculations on fitted maps (à la Warnock) with fitted canonical transformations, unfortunately the tools based on fitted maps are not yet as versatile as DA-based tools.

Warning
The word ***convergence*** *should not be taken in its purest mathematical sense throughout this paper. All the series produced in this paper are probably asymptotic. By convergence we mean to say that the asymptotic series is still well behaved as we increase the order of the perturbation to its chosen maximum value.*

1. Introduction

In previous papers[1,2], we have argued that all the usual concept of single particle dynamics in a periodic Hamiltonian can be obtained from finite time maps. We have argued that both *thinking* and *computing* using finite time maps is a more efficient approach to accelerator physics than the usual Hamiltonian theory for two obvious reasons:

i) Circular rings are made of very discontinuous Hamiltonians in the time-like variable "s".

ii) We are interested in the motion at a finite number of surface of section (or locations in the "s" variable).

Hence it is most efficient to define the ring mathematically as an ordered set of maps between the locations of interest.

From a computational standpoint, this view has been reinforced by the development of the DA-Package by Berz[3]. Given a tracking code, one can automatically get the exact coefficients of the Taylor series map around any orbit. This map can then be analyzed with Lie methods using a program also based on the DA-package of Berz. We have described this in detail in reference[4]. In particular, we hinted at the possibility of studying a "single-resonance" map. In this note, we will describe how one gets from a 2N phase space the N pseudo-invariants in a system dominated by one resonance. As usual, the claim is that all calculations can be done to arbitrary order with DA-based software and that the results can be compared with the actual tracking code which produced the map.

2. The Single-Resonance Map

We start with the usual Taylor series expansion:

$$z^f_i = T^1_{ij} z_j + T^2_{ij_1j_2} z_{j_1} z_{j_2} + \ldots + T^{no-1}_{ij} z_{j_1} \ldots z_{j_{no-1}} + \ldots$$
$$\mathbf{z} = (\mathbf{q}, \mathbf{p}) \tag{1}$$

We know that a symplectic map **M** acting on functions of phase space can be found from Eq. 1. For example, it is possible to write **M** as a factored product (Dragt-Finn)[5]:

$$\mathbf{M} = \exp(:f_2:)\,\exp(:f_3:)\,\ldots.\exp(:f_{no}:) \tag{2}$$

such that

$$\mathbf{z^f} = \mathbf{M}\,\mathbf{z} = T^1_{ij} z_j + T^2_{ij_1j_2} z_{j_1} z_{j_2} + \ldots + T^{no-1}_{ij} z_{j_1} \ldots z_{j_{no-1}} + \ldots O(\|z\|^{no}) \tag{3}$$

This factorization is not really useful except for the fact that it assumes no special properties of the map **M**. Here however, we will assume that **M** is dominated by one resonance and that between the location of this resonance in action and the origin there appear to be no resonances (as depicted on Fig. 1). Let us call this region of phase space B . Within B, one can attempt the following transformation:

$$\mathbf{M} = \mathbf{A}^{-1}\mathbf{R}\mathbf{A} + \ldots O(\|z\|^{no}) \text{ in the Taylor series}$$

where $\mathbf{R} = \exp(:-\boldsymbol{\mu}\bullet\mathbf{J} + D(\mathbf{J}):)$

and $\mathbf{A} = \exp(:F_{no}:)\,\ldots\exp(:F_3:)\,\mathbf{A_2}\ldots.$ (4)

The computation of **A** from a DA-map has been described in detail in Reference 4. It should be said that **A** is bound to diverge somewhere inside the domain B. If we restrict ourselves to cases where **A** diverges at ∂B, we can introduce the transformation $\mathbf{A_r}$ such that:

$$\mathbf{M}= \mathbf{A_r}^{-1}\mathbf{N}\mathbf{A_r}$$

where

$$\mathbf{N}= \exp(:-\mu\bullet\mathbf{J}:)\ \exp(:T_3:)\ ....\exp(:T_{no}:)$$

$$T_i= D_i(\mathbf{J}) + \sum_{\mathbf{j}-\mathbf{k}=\lambda\mathbf{m}} A^i_{\mathbf{j},\mathbf{k}}|\mathbf{j},\mathbf{k}> \quad ;\ \lambda\in Z \tag{5}$$

$$|\mathbf{j},\mathbf{k}> = (h^+{}_1)^{j_1}(h^-{}_1)^{k_1} \ldots (h^+{}_N)^{j_N}(h^-{}_N)^{k_N}\ ,$$
$$\text{with}\quad h^{\pm}{}_j= q_j \pm i\, p_j = \sqrt{2J_j}\, \exp(\mp i\phi_j)\ . \tag{6}$$

The vector **m** is the defining vector of the resonance:

$$\mathbf{m}\bullet\mu(\mathbf{J})= p \qquad\qquad p=\text{integer} \tag{7}$$

Notice that the map **N** retains all the resonant terms.To the extent that the divergences in Eq. 4 are due to the small denominators in **A** which are produced by the $|\mathbf{j},\mathbf{k}>$ terms with $\mathbf{j}-\mathbf{k}=\lambda\mathbf{m}$, the map **N** is therefore well behaved.

In addition the single-resonance map **N** is "closed". By this we simply mean that any nonlinear manipulation on **N**, such as reordering its factorization, will always produce a single-resonance map. This is true because the set

$$S_{\mathbf{m}}= \{\ |\mathbf{j},\mathbf{k}>\ |\ \mathbf{j}-\mathbf{k}=\lambda\mathbf{m}\ ;\ \lambda\in Z\} \tag{8}$$

is closed under Poisson bracket.

3. The Orthogonal Invariants of N

The particular form of **N** allows for the immediate computation of N-1 invariants. Consider an arbitrary linear combination of the actions J_i:

$$I_{\mathbf{a}}= \mathbf{a}\cdot\mathbf{J} \tag{9}$$

Let us compute the Poisson bracket of $I_{\mathbf{a}}$ with $|\mathbf{j},\mathbf{k}>$:

$$[\mathbf{a}\cdot\mathbf{J},|\mathbf{j},\mathbf{k}>] = \{i(\mathbf{j}-\mathbf{k})\cdot\mathbf{a}\}\ |\mathbf{j},\mathbf{k}> =i\lambda\mathbf{m}\cdot\mathbf{a}\ |\mathbf{j},\mathbf{k}> \qquad i=(-1)^{1/2} \tag{10}$$

From Eq. 10 and the equation for **N**, we conclude that if $\mathbf{m}\bullet\mathbf{a}=0$ then $I_{\mathbf{a}}$ is an invariant of the map **N**. Since there are N-1 vectors orthogonal to **m**, **N** has N-1 invariants of the type $I_{\mathbf{a}}$. The corresponding invariants $K_{\mathbf{a}}$ of the map **M** are given by the usual prescription:

$$K_{\mathbf{a}} = \mathbf{A_r}^{-1}\ I_{\mathbf{a}} \tag{11}$$

4. The Co-Moving Map N_c

As it is done in Hamiltonian theory, we can now introduce a co-moving phase ϕ_c :

$$\phi_c = m \cdot \phi \tag{12}$$

In fact using the new N-1 invariants, we introduce a new set of angles and actions:

$$\Phi_{i+1} = \mathbf{a}^i \cdot \phi \quad K_{i+1} = \mathbf{a}^i \cdot \mathbf{J} \qquad \text{with} \quad i=1,N-1 \ , \mathbf{m} \cdot \mathbf{a}^i = 0, \text{ and } \|\mathbf{a}^i\|=1$$

$$\Phi_1 = \phi_c = m \cdot \phi \qquad K_1 = \mathbf{m} \cdot \mathbf{J}/\|\mathbf{m}\|^2 \tag{13}$$

Using the co-moving action K_1,we construct the co-moving map N_c:

$$\begin{aligned} \mathbf{N_c} &= \exp(:p2\pi\ K_1:)\ \mathbf{N} \\ &= \exp(:p2\pi\ K_1 - \boldsymbol{\mu} \cdot \mathbf{J}:)\ \exp(:T_3:)\ \ldots.\exp(:T_{no}:) \end{aligned} \tag{14}$$

The map has $\mathbf{N_c}$ has two fundamental properties:

i) It has the same invariants in B as the map **N**.
ii) The map $\mathbf{N_c}$ can be written with a single exponent which is still invariant in a region extending beyond the domain B.

Proof of i)

We must evaluate the canonical transformation **B** such that:

$$\mathbf{N} = \mathbf{B}^{-1}\mathbf{R}\mathbf{B} \text{ where } \mathbf{R} = \exp(:-\boldsymbol{\mu} \cdot \mathbf{J} + D(\mathbf{J}):) \tag{15}$$

Lemma

$$\exp(:p2\pi\ K_1:)\ |\mathbf{j},\mathbf{k}\rangle = |\mathbf{j},\mathbf{k}\rangle \text{ if } |\mathbf{j},\mathbf{k}\rangle \in S_{\mathbf{m}} \tag{16}$$

Using Eq. 10 and the definition of K_1:

$$\exp(:p2\pi\ K_1:)\ |\mathbf{j},\mathbf{k}\rangle = \exp(\ i\ p2\pi\ \mathbf{m} \cdot (\mathbf{j}-\mathbf{k})/\|\mathbf{m}\|^2\ )\ |\mathbf{j},\mathbf{k}\rangle \tag{17}$$

but since $|\mathbf{j},\mathbf{k}\rangle \in S_{\mathbf{m}}$ there exists an integer λ such that $\mathbf{j}-\mathbf{k}=\lambda\mathbf{m}$. Hence Eq. 17 simplifies to

$$\exp(\ i\ p2\pi\ \mathbf{m} \cdot (\mathbf{j}-\mathbf{k})/\|\mathbf{m}\|^2\ )\ |\mathbf{j},\mathbf{k}\rangle = \exp(\ i\ p2\pi\ \lambda)\ |\mathbf{j},\mathbf{k}\rangle = |\mathbf{j},\mathbf{k}\rangle \tag{18}$$

This proves the lemma.

Going back to Eq. 15 and multiplying it by $\exp(:p2\pi\ K_1:)$, we get :

$$\exp(:p2\pi\, K_1:)\, \mathbf{N} = \exp(:p2\pi\, K_1:)\mathbf{B^{-1}RB}$$

$$\Rightarrow \quad \mathbf{N_c} = \exp(:p2\pi\, K_1:)\, \mathbf{B^{-1}RB}$$

$$\Rightarrow \quad \mathbf{N_c} = \exp(:p2\pi\, K_1:)\, \mathbf{B^{-1}}\exp(:-p2\pi\, K_1:)\, \exp(:p2\pi\, K_1:)\, \mathbf{RB} \tag{19}$$

However, using closure of S_m, we conclude that the map $\mathbf{B^{-1}}$ contains Lie exponents in S_m only, hence, by application of the lemma, we obtain the final result:

$$\mathrm{N} = \mathbf{B^{-1}R_cB}$$

$$\text{where } \mathbf{R_c} = \exp(:p2\pi\, K_1:)\, \mathbf{R} = \exp(:p2\pi\, K_1 - \boldsymbol{\mu}\cdot\mathbf{J} + \mathrm{D}(\mathbf{J}):) \tag{20}$$

The N pseudo-invariants around the origin are given by

$$\mathrm{I_i} = \mathbf{B^{-1}}\mathrm{K_i} \tag{21}$$

For i=2,N, we have simply $I_i = K_i$. For i=1, the invariant diverges again at the boundary ∂B.

Heuristic discussion of ii)

Ignoring the divergence of I_1, let us write $\mathbf{N_c}$ as a single exponent:

$$\begin{aligned}\mathbf{N_c} = \mathbf{B^{-1}R_cB} &= \exp(\mathbf{B^{-1}}:p2\pi\, K_1 - \boldsymbol{\mu}\cdot\mathbf{J} + \mathrm{D}(\mathbf{J}):\mathbf{B})\\ &= \exp(:\mathbf{B^{-1}}\{p2\pi\, K_1 - \boldsymbol{\mu}\cdot\mathbf{J} + \mathrm{D}(\mathbf{J})\}:)\\ &= \exp(:-\mathrm{H}(\mathbf{J};\boldsymbol{\phi}):)\end{aligned} \tag{22}$$

The pseudo-Hamiltonian $\mathrm{H}(\mathbf{J};\boldsymbol{\phi})$ is in the set S_m. To see why $\mathrm{H}(\mathbf{J};\boldsymbol{\phi})$ might be a good invariant beyond ∂B, we need to express the map $\mathbf{N_c}$ of Eq. 14 using the co-moving actions and angles:

$$p2\pi\, K_1 - \boldsymbol{\mu}\cdot\mathbf{J} = (p2\pi - \mathbf{m}\cdot\boldsymbol{\mu})\, K_1 - \sum_{i=2,N} \boldsymbol{\mu}\cdot\mathbf{a^i}\, K_i \tag{22}$$

Presumably, the zeroth order tune μ is not very far from resonance condition:

$$(p2\pi - \mathbf{m}\cdot\boldsymbol{\mu}) = \varepsilon << \mathbf{m}\cdot\boldsymbol{\mu} \tag{23}$$

Hence, the operator $\mathbf{N_c}$ of Eq. 14 can be rewritten as

$$\mathbf{N_c} = \exp(:\varepsilon K_1 - \sum_{i=2,N} \boldsymbol{\mu}\cdot\mathbf{a^i}\, K_i:)\, \exp(:T_3:)\, \ldots.\exp(:T_{no}:) \tag{24}$$

If we try to rewrite Eq. 24 as a single exponent using the Campbell-Baker-Hausdorff (CBH) formula, its convergence will be controlled by the parameter ε. In the original map **N**, the parameter in front of K_1 was $-\mathbf{m}\cdot\boldsymbol{\mu}$; a parameter assumed to be in the range of $p2\pi$ and therefore restricting the convergence of the CBH formula within B.

However, there is a more geometrical reason to expect $H(\mathbf{J};\boldsymbol{\phi})$ to be a good invariant beyond ∂B. Geometrically, the factor $\exp(:p2\pi\ K_1:)$, in Eq. 14, takes a point propelled forward by **N** and brings it back in the neighborhood of the initial condition. At the resonance, the angles $\boldsymbol{\phi}$ do not occupy the entire open set $(0,2\pi)^N$, but are congregated into clumps or islands. The map $\exp(:p2\pi\ K_1:)$ prevents a phase space point from leaving a given island. Hence the tune of the map $\mathbf{N_c}$ drifts from nearly zero at the origin to zero at the island and away from zero beyond the island. This type of motion is topologically compatible with the motion produced by a time-independent Hamiltonian. Therefore, the small denominator divergences in **B** are cancelled by the terms proportional to the small numerator ε in Eq. 22.

5. Acknowledgements

Roger Servrancks and Uli Wienands of TRIUMF by their encouragements and requests provided the driving force behind this work. Also Bob Warnock of SLAC, whose fitted maps had some problems near resonances, contributed in a similar manner. We felt that we should understand the one map resonance in regular perturbation theory before jumping into the world of fitted maps.

Finally, we would like to acknowledge the work of Turchetti et al., they have work on techniques and tools mathematically equivalent to ours. Unfortunately, there has been a scandalous lack of cross-referencing with the Antlantic acting as an unexplicable barrier. This state of affair has work to the detriment of both groups (See References 6 and 7).

6. References

1. E. Forest, A Hamiltonian-Free Description of Single Particle Dynamics for Hopeless Complex Systems, *J. Math. Phys.* (1990), pp. 1133-1144.
2. E. Forest, SSC-111, on the same topic as reference 1.
3. M. Berz, Differential Algebraic Description of Beam Dynamics to Very High Orders, *Part. Accel.* **24** (1989), pp. 109.
4. E. Forest, J. Irwin and M. Berz, Normal Form Methods for Complicated Periodic Systems: A Complete Solution Using Differential Algebra and Lie Operators, *Part. Accel.* **24** (1989), pp. 91-107.
5. A.J. Dragt and John M. Finn, Lie Series and Invariant Functions for Analytic Symplectic Maps, *J. Math. Phys.* **17** (1976), pp. 2215-2227.
6. A. Bazzani, M. Giovannozzi, G. Servizi, G. Turchetti and E. Todesco, Normal Forms for Area Preserving Maps, Resonances and Dynamic Aperture, CERN SPS/89-24 (AMS).
7. A. Bazzani, P. Mazzanti G. Servizi and G. Turchetti ,Normal Forms for Hamiltonian Maps and Nonlinear Effects in a Particle Accelerator, *Il Nuovo Cimento* **102 B, N.1** (1988), pp. 51-80.

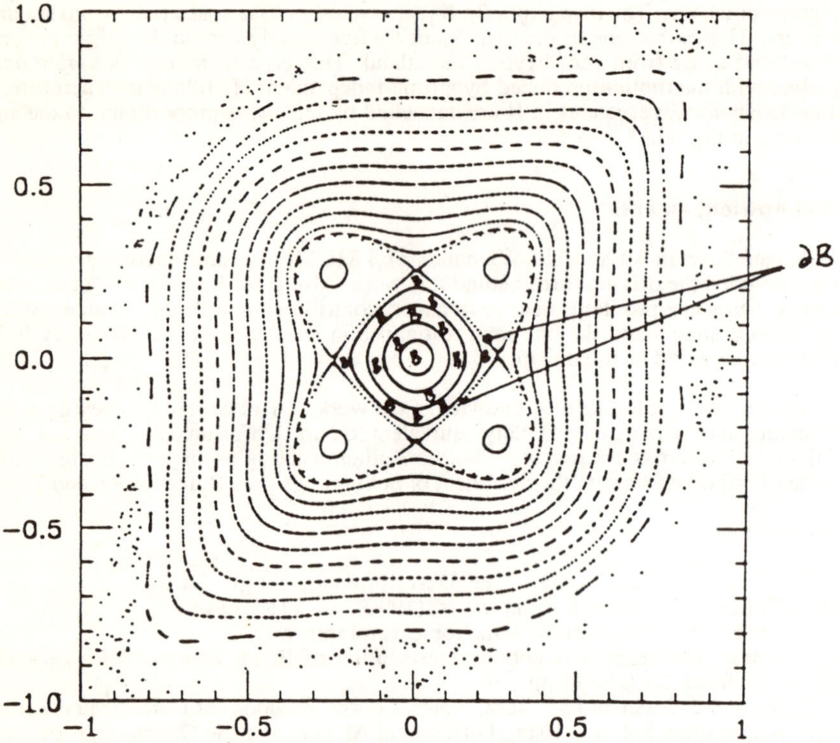

Figure 1. Location of B, ∂B illustrated on a one dimensional map.

EFFECTIVE STABILITY FOR REALISTIC PHYSICAL SYSTEMS

ANTONIO GIORGILLI
Dipartimento di Fisica dell'Università,
Via Celoria 16, 20133 MILANO, Italy

Abstract. The problem of the stability of an elliptic equilibrium of a Hamiltonian system is considered in the framework of Nekhoroshe–like theory. In particular, it is shown how a careful evaluation of some constants by computer allows to obtain realistic results when considering specific physical systems.

1. Introduction

The present talk is concerned with a new insight on the stability problem, in the light of Nekhoroshev–like theory for Hamiltonian systems. Precisely, I will consider a neighbourhood of an elliptic equilibrium, with the aim of giving a quantitative estimate of the time needed in order to reveal a significant deviation of the dynamical evolution of the system from that of an approximated linear system.

The general framework can be briefly described as follows. One considers a canonical system with Hamiltonian

$$H(p,q,\varepsilon) = h(p) + \varepsilon f(p,q) \ , \tag{1.1}$$

where $p \in \mathcal{G} \subset \mathbf{R}^n$ are action variables, with $\mathcal{G}$ an open set, $q \in \mathbf{T}^n$ are angle variables, and ε is a small real parameter. All the functions will be assumed to be analytic. According to Poincaré, this the general problem of dynamics.[1]

The case of an elliptic equilibrium can be considered as a particular case of the system (1.1). Indeed, in such a case the Hamiltonian can be typically given the form

$$H(x,y) = H_0(x,y) + f(x,y) \ , \quad (x,y) \in \mathbf{R}^{2n} \tag{1.2}$$

where

$$H_0(x,y) = \frac{1}{2}\sum_{l} \omega_l (x_l^2 + y_l^2) \tag{1.3}$$

is the Hamiltonian of a system of uncoupled harmonic oscillators with frequencies $\omega \equiv (\omega_1, \ldots, \omega_n) \in \mathbf{R}^n$, and $f(x,y)$ is an analytic function whose power series expansion in a suitable neighbourhood of the origin starts with terms of order at least 3. If one is interested in studying the dynamical evolution of orbits with initial point in a small neighbourhood of the origin, then the size, ϱ say, of that neighbourhood plays the role of the perturbation parameter ε in (1.1).

For $\varepsilon = 0$ the evolution of the system (1.1) is well known: the actions $p_1, \ldots, p_n$ are first integrals, and the flow in phase space is given by

$$p(t) = p^{(0)} \quad , \quad q(t) = \omega(p^{(0)})t + q^{(0)}$$

where $p^{(0)}, q^{(0)}$ are the initial data, and $\omega(p^{(0)}) = \frac{\partial h}{\partial p}(p^{(0)})$ are the unperturbed frequencies of the system. The same holds for the system (1.2), the action variables being $p_l = (x_l^2 + y_l^2)/2$, $(1 \le l \le n)$, and the frequencies being constant. The phase space, in both cases, is foliated into invariant tori $p(t) = p^{(0)}$, parameterized by the initial value of the action variables, and the motion on each torus is either quasi periodic or periodic according to the existence of resonance relations among the frequencies ω, namely relations of the form $k \cdot \omega = 0$ with $0 \ne k \in \mathbf{Z}^n$.

Such a plain picture is no longer valid for the perturbed system, namely the one with $\varepsilon \ne 0$ (or $f(x,y) = 0$). The strongest result for such a system came in 1954 with the celebrated work of Kolmogorov[2], who succeeded in proving that tori with strongly nonresonant frequencies are not destroyed by the perturbation, but are simply distorted. By "strongly nonresonant frequencies" it is meant frequencies satisfying a condition like

$$|k \cdot \omega| \ge \gamma |k|^{-\tau} \ ,$$

with $|k| = |k_1| + \ldots + |k_n|$, for some constants $\gamma > 0$ and $\tau > n-1$. The set of preserved invariant tori is a nowhere dense set of large measure. This can be interpreted as a stability result "in measure", in the sense that the evolution is known for the majority of the initial conditions.

The most unpleasant aspect of the Kolmogorov's theorem is that the set of invariant tori is not an open set, and that, for $n > 2$, the complement of the set of invariant tori is connected. Thus, an orbit starting in the gaps between tori can in principle diffuse through the gaps, so that the values of the action variables can significantly change during the time evolution of the system. Such a phenomenon has been named "Arnold diffusion". Although the existence of such a diffusion has not been rigorously proven till now, nevertheless the existing examples suggest that it can well be a generic phenomenon.

Now, as was realized already by Moser[3] and Littlewood[4], and next fully stated by Nekhoroshev[5], it is interesting to investigate *how much time* is needed in order to observe a significant change in the actions of the system. Indeed, following the Nekhoroshev's formulation, one has

$$|p(t) - p(0)| < \varepsilon^b \text{ for } |t| \le T_* \exp\left(\frac{\varepsilon_*}{\varepsilon}\right)^a \ , \tag{1.4}$$

with positive constants T_*, ε_*, a and b. The exponential growth of the time with $1/\varepsilon$ is the relevant part of the result, since a reasonable decrease of the perturbation can make the time so large to exceed any realistic lifetime of the physical system considered. Moreover, such a result holds for an open set of initial data, so that it may be considered as more suitable, for physical applications, than the Kolmogorov's theorem.

However, despite their mathematical beauty, these theories can hardly be applied to real physical models. The reason is that in all these theorems there appears a threshold, ε_c say, in the sense that for a perturbation larger than ε_c no conclusions can be drawn. Now, the purely analytic estimates of the value of this threshold, as well as of the constants T_*, ε_* and a in the exponential estimate (1.4), give usually ridiculously small values.

The aim of my talk is now twofold. First, I shall illustrate the application of Nekhoroshev like theory to the case of an elliptic equilibrium, namely to a Hamiltonian of the type (1.2); this will be given via a particularly simple method which avoids almost all the technical difficulties. The second goal is to show that, by combining the theory with a numerical computation of the relevant quantities by computer, one can reach realistic estimates of the threshold, which can be hopefully applied to a real system.

As a significant example, I shall illustrate the results in the case of the Lagrangian equilibrium points of the restricted problem of three bodies. Although this example looks quite far from the problem of the dynamics of particles in accelerators, I think that the problems are strictly connected by the fact that one is actually considering the stability of some periodic orbit. Indeed, the essential difference concerns essentially the choice of a natural time unit, since the dynamical evolution of a beam in an accelerator is considerably faster than that of the solar system. But, as I will show, such a difference is not relevant, due to the exponential behaviour of the stability time, according to (1.4).

2. Theoretical scheme

I shall consider here the simplest case of a Hamiltonian

$$H(x,y) = H_0(x,y) + H_1(x,y) \ , \quad (x,y) \in \mathbf{R}^{2n} \ , \tag{2.1}$$

with

$$H_0 = \frac{1}{2} \sum_l \omega_l (x_l^2 + y_l^2) \tag{2.2}$$

and H_1 a homogeneous polynomial of degree 3 in x, y. The general case of an analytic Hamiltonian with an infinite power series expansion does not actually introduce additional difficulties, the extension being just a technical matter. The interested reader will find the complete treatment in ref. [6].

I will also need a restriction on the frequencies. Precisely, I will require that the frequencies be nonresonant, i.e., $k \cdot \omega = 0$ for $k \in \mathbf{Z}^n$ implies $k = 0$. This is the most unpleasant restriction; however, the existence of resonances does not actually affect the development of the theory, since one simply has to use a more complex approach, using the normal form theory. Such a kind of study can be found, for example, in ref. [7].

Starting with the formal aspect of the theory, one tries to build up a set of n independent first integrals of the system (2.1), looking for functions $\Phi_1, \ldots, \Phi_n$ of the form

$$\Phi^{(l)}(x,y) = I_l(x,y) + \Phi_1^{(l)}(x,y) + \ldots \ , \quad 1 \le l \le n \ , \tag{2.3}$$

where

$$I_l(x,y) = \frac{1}{2}(x_l^2 + y_l^2) \tag{2.4}$$

are the actions of the harmonic oscillators, and $\Phi_s^{(l)}$, for $s \ge 1$, are homogeneous polynomials of degree $s+2$. To this end, one substitutes the explicit expansions (2.1) and (2.3) in the well known equation $\{H, \Phi\} = 0$ for a first integral (here, $\{\cdot,\cdot\}$ denotes the Poisson bracket), and gets the infinite system of equations

$$\begin{aligned} L_{H_0} I_l &= 0 \\ L_{H_0}\Phi_1^{(l)} &= -\{H_1, I_l\} \\ L_{H_0}\Phi_s^{(l)} &= -\{H_1, \Phi_{s-1}^{(l)}\} \ , \quad s > 1 \ . \end{aligned} \tag{2.5}$$

Here, the linear operator $L_{H_0}\cdot = \{H_0, \cdot\}$ has been introduced.

The formal aspect of the system (2.5) is well known. By the canonical transformation to complex variables ξ, η

$$x_l = \frac{1}{\sqrt{2}}(\xi_l + i\eta_l) \ , \quad y_l = \frac{i}{\sqrt{2}}(\xi_l - i\eta_l) \ , \quad 1 \le l \le n \ , \tag{2.6}$$

the unperturbed Hamiltonian H_0 is given the form

$$H_0(\xi,\eta) = i\sum_l \omega_l \xi_l \eta_l \ , \tag{2.7}$$

so that the operator L_{H_0} takes a diagonal form, while the order of the homogeneous polynomials is preserved. Denote now by $\Psi_s^{(l)}$ the r.h.s. of eq. (2.5) for $s \ge 1$, and write it in the general form

$$\Psi_s^{(l)} = \sum_{j,k} c_{jk}\xi^j\eta^k \ ,$$

where $j = (j_1, \ldots, j_n)$ and $k = (k_1, \ldots, k_n)$ are integer vectors with non negative components and satisfying $|j+k| = s$, and $\xi^j\eta^k = \xi_1^{j_1} \cdot \ldots \cdot \xi_n^{j_n}\eta_1^{j_1} \cdot \ldots \cdot \eta_n^{j_n}$, while the coefficients c_{jk} are known. Assuming for $\Phi_s^{(l)}$ the same form with unknown coefficients, one has then the formal solution

$$\Phi_s^{(l)} = \sum_{j,k} \frac{c_{jk}}{i(j-k)\cdot\omega}\xi^j\eta^k \ . \tag{2.8}$$

The presence of the so called *small denominators* $(j-k)\cdot\omega$ raises a consistency problem: one has to ensure that the coefficient c_{jk} of the known function $\Psi_s^{(l)}$ vanishes for all j,k for which $(j-k)\cdot\omega = 0$. This is a minor problem in the nonresonant case, since then one can prove that $\Psi_s^{(l)}$ does not contain terms $\xi^j \eta^j$, and for this reason the nonresonance condition has been requested. Thus, one can solve the system (2.5), at least formally, at any desired order.

Let me now come to the quantitative aspect. The main ingredients to be introduced are the *domains* and a suitable *norm* on the homogeneous polynomials.

For what concerns the domains, the most natural choice is to take into account the fact that the actions $I_1, \ldots, I_n$ of the harmonic oscillators are the first approximations of the formal first integrals. This suggests to introduce the domain

$$\mathcal{D}_{\varrho R} = \left\{ (x, y) \in \mathbf{R}^{2n} \; : \; x_l^2 + y_l^2 \leq \varrho^2 R_l^2 \, , \; 1 \leq l \leq n \right\} \, , \tag{2.9}$$

where $R \equiv (R_1, \ldots, R_n)$ is a vector of positive real numbers, and ϱ a nonnegative real parameter. The choice of $(R_1, \ldots, R_n)$ is quite arbitrary, so that they can be adapted to the initial data.

Coming now to the norm, writing a homogeneous polynomial of order s in the general form $f(x,y) = \sum_{j,k} f_{jk} x^j y^k$, one can define

$$\|f\|_R = \sum_{j,k} |f_{j,k}| R^{j+k} \tag{2.10}$$

(with the short notation $R^{j+k} = R_1^{j_1+k_1} \cdot \ldots \cdot R_n^{j_n+k_n}$). With these definitions, it is immediately seen that one has the bound

$$|f(x,y)| \leq \varrho^s \, \|f\|_R \text{ for } (x,y) \in \mathcal{D}_{\varrho R} \; ;$$

thus, the analytic setting so introduced allows to estimate the size of a homogeneous polynomial, and so also, if needed, the convergence properties of a power series.

With the analytic settings above, the formal scheme of construction of the first integrals can be made quantitative. To this end, we use the following estimates.

i. The transformation (2.6) to complex variables changes the norm of a homogeneous polynomial of degree s at most by a factor $2^{s/2}$; the same holds for the inverse of (2.6).

ii. The Poisson bracket between two homogeneous polynomials f_s and g_r of degree s and r respectively is estimated by

$$\|\{f_s, g_r\}\|_R \leq sr\Lambda^2 \|f_s\|_R \|g_r\|_R \, , \tag{2.11}$$

with $\Lambda = (\min_l R_l)^{-1}$.

iii. Given a sequence $\{\alpha_s\}_{s \geq 1}$ of positive numbers such that

$$|k \cdot \omega| \geq \alpha_s \text{ for } 0 < |k| \leq s+2 \, , \quad k \in \mathbf{Z}^n \, , \tag{2.12}$$

the solution (2.8) of the equation (2.5) is estimated by

$$\|\Phi_s^{(l)}\|_R \leq \frac{1}{\alpha_s} \|\Psi_s^{(l)}\|_R \, . \tag{2.13}$$

By ii. and iii. one finds the recursive formula

$$\|\Psi_s^{(l)}\|_R \le \frac{3(s+1)\Lambda^2}{\alpha_{s-1}} \|\Psi_s^{(l)}\|_R \|H_1\|_R \ , \tag{2.14}$$

which holds in complex variables. To complete the quantitative estimates one has to recall that at the very beginning of the procedure the Hamiltonian must be expressed in complex variables ξ, η, and that at the very end the formal integral must be expressed again in real variables (x, y); thus, the property i. must be used in order to estimate the norms. The final estimate is

$$\|\Phi_s^{(l)}\|_R \le 12E\,(12\Lambda^2 E)^{s-1} \frac{(s+1)!}{\prod_{l=1}^{s-1} \alpha_l} \ , \qquad s \ge 1 \ . \tag{2.15}$$

Such an estimate does not allow to prove the convergence of the series for $\Phi^{(l)}$; on the other hand, it is well known that the perturbation series are generally non convergent. However, one can truncate the series expansion to an arbitrary order r, and use these truncated series in order to investigate the dynamics of the system.

3. Use of the truncated first integrals

A first approach consists in using the analytic estimates (2.15) without actually computing the power expansion of the first integrals. Recalling that the harmonic actions $I_1, \ldots, I_n$ are first approximation to such integrals, one looks for a bound on $I_l(t)$ for a time large enough. To this end, denoting by

$$\Phi^{(l,r)} = I_l + \Phi_1^{(l)} \cdots + \Phi_r^{(l)} \ , \quad 1 \le l \le r \ ,$$

the first integrals truncated at an arbitrary order $r \ge 1$, one has

$$|I_l(t) - I_l(0)| \le |I_l(t) - \Phi^{(l,r)}(t)| + |\Phi^{(l,r)}(t) - \Phi^{(l,r)}(0)| + |\Phi^{(l,r)}(0) - I_l(0)| \ . \tag{3.1}$$

This elementary inequality is enough to obtain a qualitative description of the dynamics.

To simplify the discussion, let me put for a moment $r = 1$, and assume moreover that the $\Phi^{(l,1)}$'s are exact first integrals (this is generally false, of course, but helps to interpret the results). Then, it is easily seen that the neighbourhood of the origin is foliated into invariant n–dimensional tori, which are deformations of the tori $I_l = \text{const}$ of the unperturbed system. Thus, if we observe the dynamical evolution of the harmonic actions $I_l(t)$, we find that they are no more constant, but undergo a quasi periodic variation, due to the deformation of the invariant tori. The size of such a variation can be estimated, since it is nothing but $\Phi^{(l,1)} - I_l = \Phi_1^{(l)}$ (recall that $\Phi^{(l,1)}$ is assumed to be an exact first integral). Indeed, it is clear that in a domain $\mathcal{D}_{\varrho R}$ one has the *a priori* estimate $|\Phi_1^{(l)}(x, y)| \le \tilde{C} \varrho^3$, with some constant $\tilde{C}$. On the other hand, it is also clear that the deformation manifests itself in a quite short time, since the time derivative $\dot{I}_l = \{H_1, I_l\}$ turns out to be of order

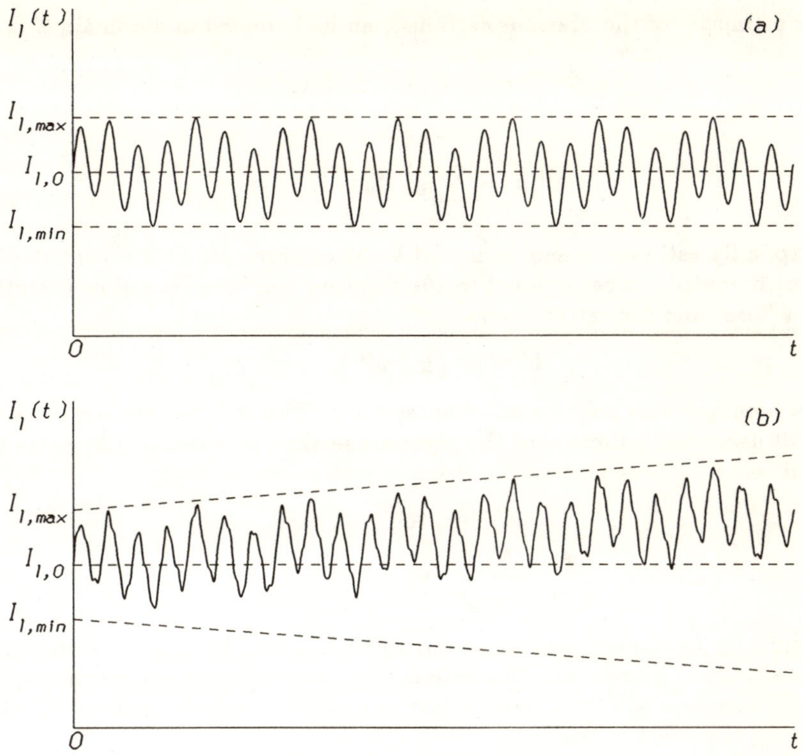

Fig. 1. *Representation of the time evolution of the harmonic action under the effect of the deformation and of the noise. (a): the deformation causes a quasi periodic, bounded oscillation of the action $I_l(t)$. (b): the noise adds further frequencies, and may cause a secular variation (drift), at most linear in time. The figure represents the worst case.*

ϱ^3. Such a situation is illustrated in fig. 1–(a): the harmonic action $I_l(t)$ exhibits a quasi periodic oscillation, but it is confined in the strip $I_{l,\mathrm{min}} \leq I_l(t) \leq I_{l,\mathrm{max}}$, whose width can be estimated.

Let us now take into account the fact that $\Phi^{(l,1)}$ is not an exact first integral, still letting $r = 1$. The time derivative $\dot{\Phi}^{(l,1)}$ is a homogeneous polynomial of order 4, so that in a domain $\mathcal{D}_{\varrho R}$ it can be bounded by $C_1\varrho^4$, with a suitable constant C_1. Thus, for small ϱ, the variation in time of $\Phi^{(l,1)}$ is much slower than that of the harmonic action I_l, but it cannot be bounded *a priori*, so that in the worst case it can be linear in time. This is illustrated in fig. 1–(b).

The same arguments apply to a generic r. The main contribution to the variation of the harmonic actions I_l is due to the *deformation* $\Phi^{(l,r)} - I_l$, which, using the estimates of the previous sections, can be bounded in a domain $\mathcal{D}_{\varrho R}$ by

$$\left|\left(\Phi^{(l,r)} - I_l\right)(x,y)\right| \le \delta_r^{(l)}(\varrho) , \tag{3.2}$$

where

$$\delta_r^{(l)}(\varrho) = \sum_{j=1}^{r} \varrho^{j+2} \|\Phi_j^{(l)}\|_R$$

can be explicitly estimated, and turns out to be of order ϱ^3; such a contribution is fast, but bounded. Superimposed to the deformation, there is a slow evolution of $\Phi^{(l,r)}$, whose time derivative is now

$$\dot{\Phi}^{(l,r)} = \{H_1, \Phi_r^{(l)}\} ,$$

namely, a homogeneous polynomial of order $r+3$. This will be referred to as the *noise*. Still using the estimates of the previous section, in a domain $\mathcal{D}_{\varrho R}$ one has the bound

$$|\dot{\Phi}^{(l,r)}| \le \mathcal{R}^{(l)}(\varrho) , \tag{3.3}$$

where

$$\mathcal{R}^{(l)}(\varrho) = \varrho^{r+3} \|\{H_1, \Phi_r^{(l)}\}\|_R$$

can be explicitly computed, and turns out to be bounded by $C_r \varrho^{r+3}$ with a suitable constant C_r. The relevant fact here is that, according to our estimates, the constant C_r increases with r at least as fast as $r!$, so that some care must be taken in estimating the effect of the noise.

4. Estimate of the stability time

Let me now illustrate how the arguments above can be made quantitative. Let us proceed as follows. Assume that the initial point $(x(0), y(0))$ of an orbit lies in a domain $\mathcal{D}_{\varrho_0 R}$, so that one has $I_l(0) \le \varrho_0^2 R_l^2/2$, and ask $(x(t), y(t)) \in \mathcal{D}_{\varrho R}$ with $\varrho > \varrho_0$ for a finite time interval $|t| \le T$; this is guaranteed provided one has $I_l(t) \le \varrho^2 R_l^2/2$. Using (3.1) and (3.2) one has

$$|I_l(t) - I_l(0)| \le \delta_r(\varrho_0) + \delta_r(\varrho) + |\Phi^{(l,r)}(t) - \Phi^{(l,r)}(0)| ,$$

so that one can define

$$\Delta_r^{(l)}(\varrho_0, \varrho) = \frac{1}{2} R_l^2 (\varrho^2 - \varrho_0^2) - \delta_r(\varrho_0) - \delta_r(\varrho)$$

as the maximum allowed effect for the noise. This gives a first consistency condition on r, ϱ_0 and ϱ, namely

$$\Delta_r^{(l)}(\varrho_0, \varrho) > 0 . \tag{4.1}$$

If this condition is satisfied, then, using (3.3), one has that the orbit is confined to $\Delta_{\varrho R}$ at least for

$$|t| \le \tau_r(\varrho_0, \varrho) = \min_l \frac{\Delta_r^{(l)}(\varrho_0, \varrho)}{\mathcal{R}_r^{(l)}(\varrho)} . \tag{4.2}$$

Such an estimate is still unsatisfactory, since the stability time contains the arbitrary parameter r, while it would be more interesting to give an estimate depending only on the size of the initial domain $\mathcal{D}_{\varrho_0 R}$. To this end, it is natural to look for the best possible estimate by defining the stability time $T(\varrho_0)$ as

$$T(\varrho_0) = \max_r \sup_\varrho \tau_r(\varrho_0, \varrho) , \tag{4.3}$$

where the $\max_r \sup_\varrho$ must be taken over r and ϱ satisfying the consistency condition (1.3).

Let me stress again that all the quantities $\delta_r^{(l)}(\varrho)$, $\Delta_r^{(l)}(\varrho_0, \varrho)$ and $\mathcal{R}_r^{(l)}(\varrho)$ can be explicitly computed provided some estimate on the norms $\|\Phi_1^{(l)}\|_R, \ldots, \|\Phi_r^{(l)}\|_R$ of the expansions of the first integrals is known. Thus, the stability time $T(\varrho_0)$ can be actually estimated by computer. However, it is interesting to check the relevance of the estimate by investigating the analytic dependence of T on ϱ_0, based on the recursive estimates of sect. 2.

Looking in particular at the estimate (2.15), one first needs some analytic form of the sequence $\{\alpha_s\}_{s>0}$ which bounds the small denominators, as requested by (2.12). The usual choice in perturbation theory is

$$\alpha_s = \gamma(s+2)^{-\tau} \tag{4.4}$$

with some constants $\gamma > 0$ and $\tau \ge 0$. Indeed, it is well known that the relative measure of the ω's not satisfying (2.12) is at most proportional to γ, so that it can be made arbitrarily small. With this choice, it is an easy matter to get

$$\begin{aligned} \delta_r^{(l)}(\varrho) &< C_1 \varrho^3 \\ \mathcal{R}_r^{(l)}(\varrho) &< C_2 (r!)^{1+\tau} \left(\frac{\varrho}{\varrho_*}\right)^r \varrho^3 \end{aligned} \tag{4.5}$$

with suitable constants C_1, C_2 and ϱ_* (see ref. [6] for details). To simplify the discussion, let us now put (arbitrarily)

$$\Delta_r^{(l)}(\varrho_0, \varrho) = C_1 \varrho^3 , \tag{4.6}$$

namely, allow the effect of the noise to become as large as that of the deformation. Then, by (4.2) we get

$$\tau_r(\varrho_0, \varrho) \ge \frac{A}{(r!)^{1+\tau}} \left(\frac{\varrho_*}{\varrho}\right)^r , \tag{4.7}$$

with a suitable constant A. Since a relation between ϱ_0 and ϱ is implicitly established by (4.6), we just have to optimize $\tau_r(\varrho_0, \varrho)$ against r. From (4.7) it is easily found that the optimal choice for r is

$$r_{\text{opt}} = \left(\frac{\varrho_*}{\varrho}\right)^{1/(1+\tau)} .$$

Substituting the this value in (4.7) one gets the final estimate of the stability time

$$T(\varrho) = T_* \exp\left(\frac{\varrho_*}{\varrho}\right)^{1/(1+\tau)} , \tag{4.8}$$

with a constant T_*. This is the essential content of the Nekhoroshev theorem when applied to the case of an elliptic equilibrium.

5. An example: the Lagrangian equilibrium points of the restricted problem of three bodies

In order to illustrate the possible physical relevance of the theory above, let me consider the case of the triangular equilibrium solutions of the restricted problem of three bodies.

To briefly recall the model, we consider two mass points (the *primaries*) of mass μ and $1-\mu$ respectively, uniformly rotating on a Keplerian circular orbit around their common center of mass. The problem is to study the motion of a third point of negligible mass (the *planetoid*) under the influence of the gravitational field of the primaries, assuming that the motion of the primaries is not influenced by the presence of the planetoid.

One usually considers a reference frame uniformly rotating with the primaries, so that the position of the primaries is fixed in that frame. It was known to Lagrange that in such a reference frame the Newton equations for the restricted problem of three bodies admit five equilibrium solutions. Three of these solutions lie on the straight line joining the primaries (the *collinear* solutions); the remaining equilibrium points, usually denoted by L_4 and L_5 respectively, are so located that the primaries and the point L_4 (L_5 respectively) are the vertices of an equilateral triangle (the *triangular* solutions).

Concerning the stability, it is well known that the linear approximation of the Newton equations in the neighbourhood of the equilibria always gives instability for the collinear points, while the triangular points are stable for $0 < \mu < \mu_0 = 0.03852089604551\ldots$. However, the problem of the stability of the triangular points with respect to the full (i.e., nonlinear) system is a very hard problem, which has been solved only recently for the planar case (i.e., the planetoid is constrained to move on the orbital plane of the primaries), but is still unsolved in the spatial case.

In the neighbourhood of the point L_4 the Hamiltonian can be expanded in power series as

$$H = \frac{1}{2}(p_x^2 + p_y^2) + yp_x - xp_y + \frac{x^2}{8} - \frac{5}{8}y^2 - axy + \frac{1}{2}(p_z^2 + z^2) + H_1(x,y,z) + H_2(x,y,z) + \dots ,$$

where $a = -3\sqrt{3}(1-2\mu)/4$, and H_1, $H_2, \dots$ are known homogeneous polynomials of degree $3, 4, \dots$ respectively. By a linear change of coordinates the second order Hamiltonian takes the form

$$H_0 = \sum_{l=1}^{3} \frac{\omega_l}{2}(x_l^2 + y_l^2) ,$$

with ω_1, ω_2 and ω_3 roots of the characteristic equation

$$(\omega^4 - \omega^2 + \frac{27}{16} - a^2) \cdot (\omega^2 - 1) = 0 .$$

The sign of the ω's is determined by the linear transformation itself, and it turns out that one has $\omega_1 > 0$, $\omega_2 < 0$ and $\omega_3 = 1$, so that the linear approximation is not sufficient to decide about the stability of the nonlinear system. However, the Hamiltonian has the general form (1.2), so that the theory developed in the previous sections can be applied. (In fact, the Hamiltonian is not truncated at degree three, nor it does satisfy the reversibility condition discussed in sect. 2, but this is just a minor technical problem, which can be easily overcome; so, let me skip these details.)

A particularly interesting case is that of the Sun–Jupiter system. Indeed, it is known that some 20 asteroids are close enough to the point L_4 that their motion could be a libration around that point. A detailed study was performed by Celletti and Giorgilli in ref. [8], and I plan to illustrate here the main result. The value of the parameter μ in this case can be assumed to be $\mu = 0.95387536 \times 10^{-3}$; the frequencies then turn out to be $\omega_1 \simeq 9.9675752552 \times 10^{-1}$, $\omega_2 \simeq -8.0463875837 \times 10^{-2}$ and $\omega_3 = 1$. The expansion of the Hamiltonian and of the formal first integrals has been performed by computer up to $r = 20$, namely up to terms of degree 22. Such an expansion has been used to compute the norms used in estimating the stability time $T(\varrho_0)$, according to the procedure outlined in sect. 3. The final result is illustrated in fig. 2. The four different curves correspond to different choices of the maximal truncation order r, namely $r = 2, 8, 14, 20$ respectively (namely, working with polynomials of degree $4, 10, 16$ and 22 respectively); this gives an idea of how the results improve by increasing the truncation order.

The most striking feature of the graph is the sudden jump of $T(\varrho_0)$ a little below $\varrho_0 = 3 \times 10^{-4}$. This is more evident from the enlargement of fig. 3: it is clear that there is a sharp threshold close to $\varrho_0 = 2.88248 \times 10^{-4}$, where the estimated stability time jumps from 1 to more than 10^4. For smaller values of ϱ_0 the function $T(\varrho_0)$ increases more regularly, roughly as a power of $1/\varrho_0$. Moreover, the value

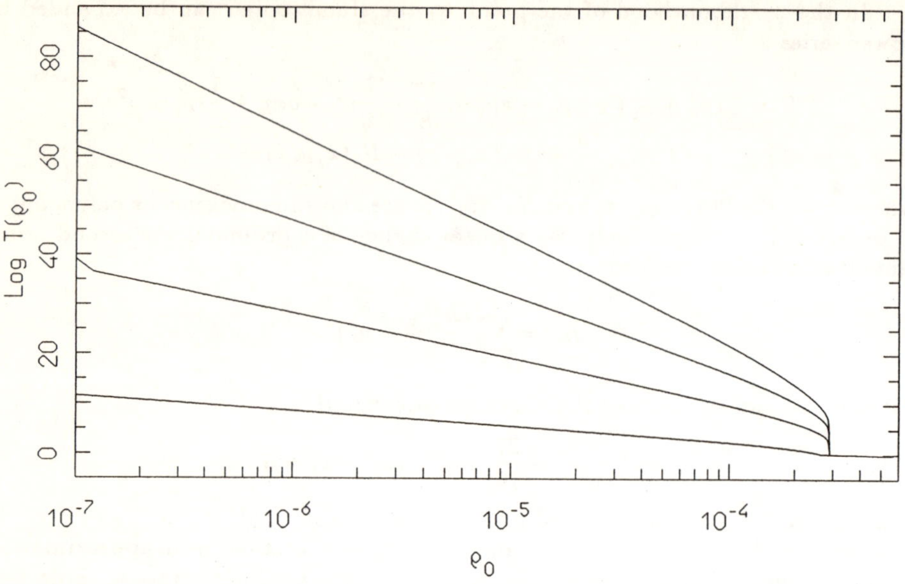

Fig. 2. *Estimated stability time T vs. ϱ_0 for the case of the Sun–Jupiter system. The four curves correspond to truncation of the power expansion of the first integrals at degree* 4 *(bottom curve),* 10, 16 *and* 22 *(upper curve) respectively.*

of this threshold looks well defined even at quite low orders, while the estimate of the time could still improve by further increasing r. Such a fast increase is the most relevant result, since it allows to ensure stability for exceedingly large times.

In order to understand the physical relevance, one can remark that, in our units, the revolution of Jupiter takes 2π time units, so that the estimated age of the universe is of order 10^{10}. If now one asks for stability over a time interval $T = 10^{10}$, one gets that the initial point should lie inside a domain $\mathcal{D}_{\varrho_0 R}$ with $\varrho_0 \simeq 2.58397 \times 10^{-4}$. In physical coordinates, this corresponds roughly to a distance from the point L_4 of about 4×10^{-5} times the distance of Jupiter from the Sun; one has then, roughly speaking, a stability neighbourhood of size 3×10^4 Km. Although not so large (and certainly not optimal), this is a realistic result.

A last remark concerns the possible relevance of such a result for accelerators. As said in the introduction, the naive impression is that this problem is quite different from the problems of stability in Celestial Mechanics, because one has to do with definitely different time scales. However, a more careful consideration shows that this is not the case. Indeed, looking at fig. 2, one sees that increasing the required stability time from 10^{10} to, say, 10^{20} just requires a decrease of the

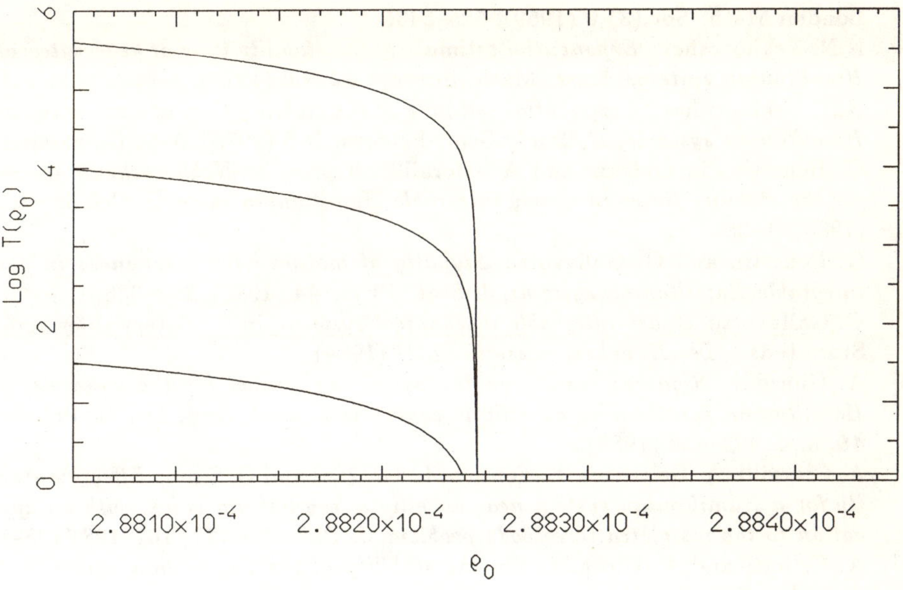

Fig. 2. *Enlargement of fig. 2, in order to put into evidence the existence of a sharp threshold.*

initial size ϱ_0 by a factor less than 2. Thus, a reasonable restriction of the initial domain can well compensate a very large factor in the natural time scale of the physical system.

References

[1] Poincaré, H.: *Les méthodes nouvelles de la mécanique céleste*, Gauthier–Villars, Paris (1892).

[2] Kolmogorov, A. N.: Dokl. Akad. Nauk SSSR, **98**, 527 (1954).
Arnold, V. I.: Usp. Mat. Nauk, **18**, 13 (1963); Russ. Math. Surv., **18**, 9 (1963); Usp. Math. Nauk **18** N.6, 91 (1963); Russ. Math. Surv. **18** N.6, 85 (1963).
Moser, J.: *Proceedings of the International Conference on Functional Analysis and Related Topics*, 60 (Tokio, 1969).

[3] J. Moser: *Stabilitätsverhalten kanonisher differentialgleichungssysteme,* Nachr. Akad. Wiss. Göttingen, Math. Phys. K1 IIa, nr.6 (1955), 87–120.

[4] J. E. Littlewood: *On the equilateral configuration in the restricted problem of three bodies*, Proc. London Math. Soc.(3) **9** (1959), 343–372;
J. E. Littlewood: *The Lagrange configuration in celestial mechanics*, Proc.

London Math. Soc.(3) **9** (1959), 525–543.

[5] N.N. Nekhoroshev: *Exponential estimate of the stability time of near–integrable Hamiltonian systems*, Russ. Math. Surveys **32** N.6 (1977), 1–65.
N.N. Nekhoroshev: *Exponential estimate of the stability time of near–integrable Hamiltonian systems, II*, Trudy Sem. Petrovs. N.5 (1979), 5–50 (in russian).
G. Benettin, L. Galgani and A. Giorgilli: *A proof of Nekhoroshev's theorem for the stability times in nearly integrable Hamiltonian systems*, Cel. Mech **37** (1985), 1–25.
G. Benettin and G. Gallavotti: *Stability of motion near resonances in quasi integrable Hamiltonian systems*, J. Stat. Phys. **44** (1986), 293–338.
G. Gallavotti: *Quasi–integrable mechanical systems*, in K. Osterwalder and R. Stora (eds.), *Les Houches, Session XLIII* (1984).

[6] A. Giorgilli: *Rigorous results on the power expansions for the integrals of a Hamiltonian system near an elliptic equilibrium point*, Ann. Ist. H. Poincaré, **48**, n. 4, 423–439 (1988).

[7] A. Giorgilli, A. Delshams, E. Fontich, L. Galgani and C. Simó: *Effective stability for a Hamiltonian system near an elliptic equilibrium point, with an application to the restricted three body problem*, J. Diff. Eqs. **77**, 167–198 (1989).

[8] A. Celletti and A. Giorgilli: *On the stability of the Lagrangian points in the spatial restricted problem of three bodies*, preprint (1990).

STABILITY OF ORBITS IN NONLINEAR MECHANICS FOR FINITE BUT VERY LONG TIMES*

R. L. WARNOCK and R. D. RUTH
Stanford Linear Accelerator Center,
Stanford University, Stanford, CA 94309, USA

ABSTRACT

In various applications of nonlinear mechanics, especially in accelerator design, it would be useful to set bounds on the motion for finite but very long times. Such bounds can be sought with the help of a canonical transformation to new action-angle variables $(\mathbf{J}, \mathbf{\Psi})$, such that the action $\mathbf{J}$ is nearly constant while the angle $\mathbf{\Psi}$ advances almost linearly with the time. By examining the change in $\mathbf{J}$ during a time T_o from many initial conditions in the open domain Ω of phase space, one can estimate the change in $\mathbf{J}$ during a much larger time T, on any orbit starting in a smaller open domain $\Omega_o \subset \Omega$. A numerical realization of this idea is described. The canonical transformations, equivalent to close approximations to invariant tori, are constructed by an effective new method in which surfaces are fitted to orbit data. In a first application to a model sextupole lattice in a region of strong nonlinearity, we predict stability of betatron motion in two degrees of freedom for a time comparable to the storage time in a proton storage ring (10^8 turns).

1. Introduction

Tracking of single particles, by symplectic numerical integration of Hamilton's equations,[1] provides a direct approach to the study of orbit stability. Unfortunately, computational expense usually limits the time interval of tracking to values much less than the desired storage time of an accelerator beam, and sharply restricts the number of initial conditions that can be tested. Favorable results from tracking are

* Work supported by Department of Energy contract DE–AC03–SF00515.

regarded as a necessary condition for a good storage ring design, but doubt remains about the sufficiency of such results.

We propose a method to derive definite information on long-term stability from short-term tracking of many orbits. The idea derives from the line of argument of the Nekhoroshev Theorem,[2–5] and depends on determination of an action variable $\mathbf{J}$ that is invariant to high accuracy in a certain region of phase space. The residual time variation of $\mathbf{J}$, stronger in some regions than in others, provides a sensitive indicator of unstable behavior.

Nekhoroshev's original argument gives an explicit formula for a lower bound on the time of stability. The bound increases exponentially as the strength of the nonlinear perturbation tends to zero. Although our method does not give an explicit formula, it does yield stability times of practical interest, under realistic conditions. By contrast, the Nekhoroshev estimate gives suitable stability times only for absurdly weak perturbations. It may be possible to improve the Nekhoroshev estimates in simple examples,[6] but sufficient improvement for the complicated Hamiltonians of accelerator physics is not to be expected.

We achieve results for realistic parameters by avoiding Nekhoroshev's perturbative analysis. Our nonperturbative method is necessarily numerical, and for that reason involves some loss of rigor. We shall try to convince the reader that the results can nevertheless be highly reliable for the chosen Hamiltonian. Uncertainties in choosing the Hamiltonian are far greater than any ambiguity due to our technique.

In this brief account we describe only the main ideas, referring the reader to Ref. 7 for details of technique. We begin with the action-angle variables of the underlying linear system, $(\mathbf{I}, \mathbf{\Phi})$. Bold-faced symbols represent $d-$dimensional vectors, where d is the number of mechanical degrees of freedom. We treat an example with $d = 2$, but the method is general. Our discussion is based entirely on the time evolution map $\mathcal{M}$ for N_o turns,

$$\mathcal{M}: \quad (\mathbf{I}, \mathbf{\Phi})_{\theta=0} \mapsto (\mathbf{I}, \mathbf{\Phi})_{\theta=2\pi N_o} \ , \tag{1}$$

where $\theta = s/R$ represents azimuthal location on a closed reference orbit. In the present work we evaluate this map through element-by-element tracking. With sufficient care it should be possible to represent the map by an explicit formula, and thereby enhance the speed of calculations.[8,9]

A canonical transformation to new action-angle variables, $(\mathbf{I}, \mathbf{\Phi}) \mapsto (\mathbf{J}, \mathbf{\Psi})$, is induced by a generating function $G(\mathbf{J}, \mathbf{\Phi}, \theta)$ that is $2\pi-$periodic in $\mathbf{\Phi}$ and θ. The equations relating old and new variables are

$$\mathbf{I} = \mathbf{J} + G_{\mathbf{\Phi}}(\mathbf{J}, \mathbf{\Phi}, \theta) \ , \tag{2}$$

$$\mathbf{\Psi} = \mathbf{\Phi} + G_{\mathbf{J}}(\mathbf{J}, \mathbf{\Phi}, \theta) \ , \tag{3}$$

where subscripts denote partial derivatives. It is sufficient for our purposes to consider the transformation at $\theta = 0$ only. If the transformation is ideal, so that $\mathbf{J}$ is constant, then Eq. (2) gives an explicit representation of an invariant surface (a torus). That is, it gives $\mathbf{I}$ as a $2\pi-$periodic function of $\mathbf{\Phi}$. The invariant action $\mathbf{J}$

plays the role of a parameter to distinguish different tori; it is equal to the average of $\mathbf{I}$ over $\mathbf{\Phi}$. At $\theta = 0$ we adopt the notation

$$\mathbf{u}(\mathbf{J}, \mathbf{\Phi}) = G_{\mathbf{\Phi}}(\mathbf{J}, \mathbf{\Phi}, 0) \quad . \tag{4}$$

For the nonintegrable systems of interest, exact invariant tori exist, if at all, only for values of $\mathbf{J}$ on a strange set of Cantor type. Nevertheless, tori that are approximately invariant exist as smooth functions of $\mathbf{J}$ in open regions of phase space, and they can be computed numerically. A family of approximate invariant tori defines a canonical transformation. If $\mathbf{u}(\mathbf{J}, \mathbf{\Phi})$ is a smooth function of $\mathbf{J}$ in a region Ω, and

$$\mathbf{I}(\theta) \approx \mathbf{J} + \mathbf{u}(\mathbf{J}, \mathbf{\Phi}(\theta)) \ , \quad \theta = 0, 2\pi, 4\pi, \cdots \quad , \tag{5}$$

then the equation $\mathbf{I} = \mathbf{J} + \mathbf{u}(\mathbf{J}, \mathbf{\Phi})$ defines a nearly constant function $\mathbf{J}(\mathbf{I}, \mathbf{\Phi})$. Here we assume that the Jacobian matrix $\mathbf{1} + \mathbf{u_J}(\mathbf{J}, \mathbf{\Phi})$ is nonsingular in the region of interest. It is indeed found to be nonsingular in our calculations. Integrating $\mathbf{u}$ with respect to $\mathbf{\Phi}$, we obtain the generator $G(\mathbf{J}, \mathbf{\Phi}, 0)$ and all the ingredients of a full canonical formalism. The function $\mathbf{u}(\mathbf{J}, \mathbf{\Phi})$, representing a family of approximate invariant tori, is constructed numerically by the method of Sec. 3.

2. A Bound on the Change in J

In a case with $d = 2$, let Ω be the interior of a rectangle in the $\mathbf{J} = (J_1, J_2)$ plane, and let Ω_o be the interior of a smaller, concentric rectangle so that $\Omega_o \subset \Omega$. Denote by ΔJ_i the width of the annulus between Ω_o and Ω as crossed in the ith direction. Suppose that the change in J_i during N_o turns, for any orbit with initial $\mathbf{J}$ in Ω, is less than δJ_i. Then any orbit with initial $\mathbf{J}$ in the smaller region Ω_o cannot reach the outer boundary of Ω in fewer than $N = qN_o$ turns, where

$$q = \min_i \left(\frac{\Delta J_i}{\delta J_i} \right) \quad . \tag{6}$$

This observation is useful if q is sufficiently large. Since the largest tolerable excursion ΔJ_i is sharply restricted by design considerations, a large q is to be achieved by making δJ_i small through a good choice of the canonical transformation.

Note that for practical purposes this is an extremely conservative argument, since it comes from contemplating the worst conceivable case in which the increment of J_i in qN_o turns is just q times the largest possible increment in N_o turns. In reality, the q increments are not likely to add up linearly, so that it will probably take much more than qN_o turns to move from Ω_o to the outer boundary of Ω.

3. Numerical Determination of the Canonical Transformation

To determine the canonical transformation, we expand the function representing a torus in a Fourier series. We write

$$\mathbf{I} = \sum_{\mathbf{m}} \mathbf{u_m} e^{i\mathbf{m}\cdot\mathbf{\Phi}} \quad , \tag{7}$$

and determine the coefficients $\mathbf{u_m}$ so that Eq. (7) is satisfied at points $(\mathbf{I}(\theta), \mathbf{\Phi}(\theta))$, $\theta = 0$ $(mod\ 2\pi)$, all lying on a single nonresonant orbit. The coefficient $\mathbf{u_o}$ of the constant term is identified with the action $\mathbf{J}$, which varies with the choice of initial condition of the orbit. We repeat the process for various initial conditions, thereby obtaining $\mathbf{u_m}(\mathbf{J})$ on a mesh of points $\mathbf{J} = \mathbf{J}_i$, $i = 1, 2, \cdots, s$. We then define $\mathbf{u_m}(\mathbf{J})$ as a smooth function of $\mathbf{J}$ by interpolating the values at mesh points with polynomials. The resulting transformation, $\mathbf{I} = \mathbf{J} + \mathbf{u}(\mathbf{J}, \mathbf{\Phi})$, is mathematically well defined, even though it was obtained numerically.

In determining the $\mathbf{u_m}$ to satisfy Eq. (7), we face the difficulty that the $\mathbf{\Phi}(\theta)$ are scattered unpredictably. For that reason we cannot simply take a discrete Fourier transform, which requires regularly spaced abscissae. Furthermore, a direct solution of the linear equations for the $\mathbf{u_m}$ is impractical, since the matrix is dense and too large for comfort. We avoid these problems by using the values of $\mathbf{I}$ on a regular mesh in $\mathbf{\Phi}$ space as unknowns, rather than the $\mathbf{u_m}$. In one dimension the equations relating values of I at orbit points ϕ_j to values at mesh points $2\pi k/K$ are as follows:

$$I(\phi_j) = \sum_{k=0}^{K-1} D_{jk}\ I(2\pi k/K) \quad , \tag{8}$$

where

$$D_{jk} = \frac{\sin[\pi(x_j - k)]}{K \sin[\pi(x_j - k)/K]} \quad , \tag{9}$$

with

$$\phi_j = 2\pi x_j / K \quad . \tag{10}$$

Here $K = 2M + 1$, where M is the maximum mode number m. To derive these equations, one substitutes in Eq. (7) the approximation to $\mathbf{u_m}$ that is given by the discrete Fourier transform. Reversal of the order of summations, and evaluation of a geometric series, leads to Eq. (8). This system can be solved by iteration provided that we choose to fit only a subset of the data, namely, a set in which there is one and only one ϕ_j in each cell of the mesh. With this selection the matrix D_{jk} is close to the unit matrix, and an iterative method (Gauss-Seidel) converges rapidly. Using the inverse of the discrete Fourier transform, we finally obtain the torus in the form Eq. (7).

This method has three notable properties. First, the resulting torus Eq. (7) passes exactly through the orbit points, provided that the Eq. (8) is solved exactly. This feature may be partly responsible for the high accuracy of the method. Second, the scheme works in any number of dimensions. Third, it provides a useful filter against low-order resonances. In the case of resonant orbits of sufficiently low order, the process of selecting data will fail. Not every cell of the mesh will contain a $\mathbf{\Phi}(\theta)$, no matter how long the orbit, and the iterative solution of Eq. (8) will fail. We fit only nonresonant orbits (invariant tori) or resonant orbits of extremely high order (which are not distinguishable numerically from tori). In Fig. 1 we plot values of $\mathbf{\Phi}$ on a typical resonant orbit. Figure 2 shows values of $\mathbf{\Phi}$ on an apparent invariant

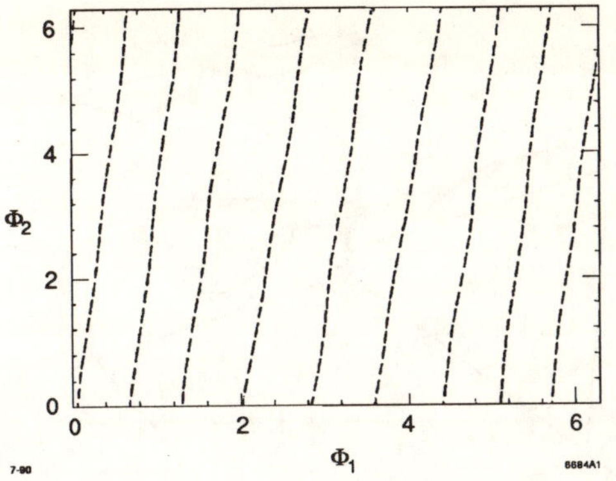

Figure 1: Values of $\mathbf{\Phi} = (\Phi_1, \Phi_2)$ at the surface of section $\theta = 0$ on a resonant orbit.

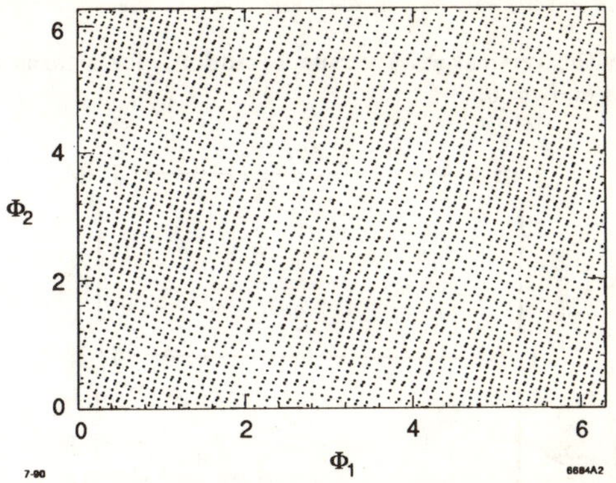

Figure 2: Values of $\mathbf{\Phi} = (\Phi_1, \Phi_2)$ at the surface of section $\theta = 0$ on an apparent invariant torus.

torus. Figure 3 shows points on a nonresonant orbit, at $\theta = 0 \; mod(2\pi)$, plotted in three dimensions. We plot (I_1, Φ_1, Φ_2), with I_1 on the vertical axis. If the motion were linear these points would lie in a plane, $I_1 = constant$.

The polynomial interpolation in $\mathbf{J}$ of the tori forms bridges over resonances. Since there are resonances everywhere, this is an essential feature of any canonical transformation defined as a smooth function of $\mathbf{J}$. It is not merely a feature forced upon us by the use of numerical methods.

4. Computation of δJ_i

To compute a bound it remains to determine δJ_i. Because of practical limits on computation time, there is some uncertainty in this determination, but with some care the uncertainty can be made rather small. It should be emphasized that the

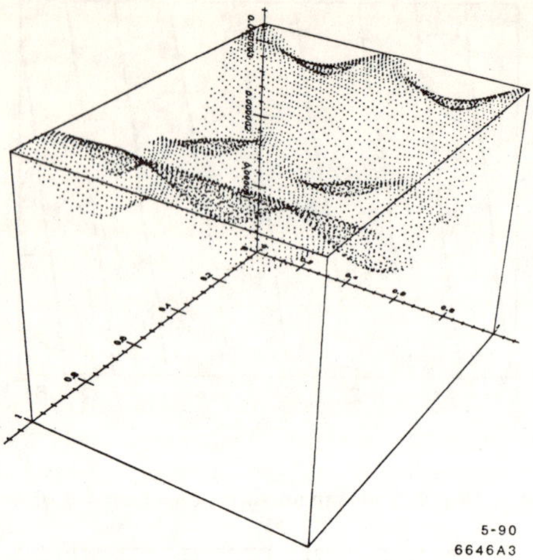

Figure 3: Plot of I_1 (on the vertical axis) versus Φ_1 and Φ_2, for a torus in the region Ω defined in Eq. (13). The origin is at $I_1 = 0$.

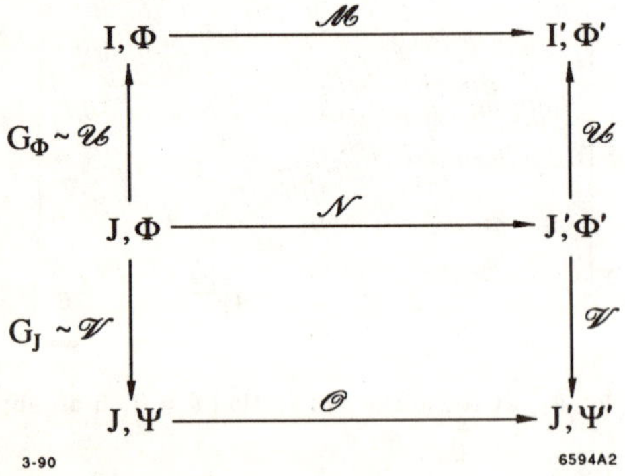

Figure 4: A commutative diagram showing how the maps $\mathcal{N}$ and $\mathcal{O}$ are induced by the original map $\mathcal{M}$ and changes of variable. The transformations $\mathcal{U}$ and $\mathcal{V}$ are given in explicit form through derivatives of the generator; see Eqs. (2) and (3).

only failure of rigor in our argument occurs in the determination of δJ_i.

We denote the increment in $\mathbf{J}$ over N_o turns as

$$\mathbf{J}' - \mathbf{J} = \mathcal{D}(\mathbf{J}, \mathbf{\Phi}, N_o) \ . \tag{11}$$

To compute $\mathcal{D}$ we simply observe the time evolution of $\mathbf{J}$ induced by the map $\mathcal{M}$ through the change of variable $\mathcal{U} : (\mathbf{J}, \mathbf{\Phi}) \mapsto (\mathbf{I}, \mathbf{\Phi})$ and its inverse. Here we need not be concerned with the time evolution of the new angle $\mathbf{\Psi}$. It is enough to deal with the mixed pair $(\mathbf{J}, \mathbf{\Phi})$, which evolves by a map $\mathcal{N}$:

$$\mathcal{N} : (\mathbf{J}, \mathbf{\Phi}) \mapsto (\mathbf{J}', \mathbf{\Phi}') \ . \tag{12}$$

The commutative diagram of Fig. 4 shows how $\mathcal{N}$ is obtained as $\mathcal{N} = \mathcal{U}^{-1} \circ \mathcal{M} \circ \mathcal{U}$. For the computation of $\mathcal{N}$ we have $\mathcal{U}$ in explicit form, since $\mathbf{I} = \mathbf{J} + \mathbf{u}(\mathbf{J}, \mathbf{\Phi})$. To evaluate $\mathcal{U}^{-1}$ we use Newton's method, with $\mathbf{J}$ as the first guess for $\mathbf{J}'$. The first iterate of Newton's method already provides accuracy that is more than adequate for our purposes.

The map $\mathcal{O}$ on the new canonical pair $(\mathbf{J}, \mathbf{\Psi})$, shown in the third level of Fig. 4, is needed only for the work of Sec. 6. For the computation of $\mathcal{O}$ we have $\mathcal{V}$ in explicit form through Eq. (3), but $\mathcal{V}^{-1}$ must be calculated by Newton's method.

In the example studied below, the function $\mathcal{D}$ has many oscillations as a function of $\mathbf{\Phi}$ but relatively few as a function of $\mathbf{J}$ on Ω. It is impractical to follow every oscillation in seeking the upper bound δJ_i of $\mathcal{D}_i$, but we can do random sampling with statistical estimates of sampling error to find a fairly convincing value of δJ_i. The reader may consult Ref. 7 for details on this relatively delicate problem.

5. An Example

To illustrate, we derive a bound for two-dimensional betatron motion in a lattice consisting of one cell of the Berkeley Advanced Light Source (ALS). The lattice parameters are given in Ref. 7. This example involves nonlinear phenomena similar to those in large hadron colliders since the sextupoles are so strong as to drive high-order resonances such as those excited by high-order multipoles in superconducting magnets. We work in a region Ω of substantial nonlinearity, about half-way to the short-term dynamic aperture in the (x_{max}, y_{max}) plane. Orbit points on a typical invariant surface in this region are plotted in Fig. 3. With actions measured in meters the region Ω is given by

$$2.51 \cdot 10^{-6}\ \mathrm{m} < J_1 < 2.82 \cdot 10^{-6}\ \mathrm{m}\ , \qquad 1.34 \cdot 10^{-6}\ \mathrm{m} < J_2 < 1.64 \cdot 10^{-6}\ \mathrm{m}\ . \tag{13}$$

We first determine the leading resonances in this region with the help of our canonical formalism. The tune $\boldsymbol{\nu}$ (winding number) is obtained as a function of $\mathbf{J}$ from Eq. (3). If $\mathbf{J}$ were constant, then the change in $\mathbf{\Psi}$ over one turn would be equal to $2\pi\boldsymbol{\nu}$, and Eq. (3) would give

$$2\pi\boldsymbol{\nu}(\mathbf{J}) = \mathbf{\Phi}(2\pi) - \mathbf{\Phi}(0) + G_{\mathbf{J}}(\mathbf{J}, \mathbf{\Phi}(2\pi), 0) - G_{\mathbf{J}}(\mathbf{J}, \mathbf{\Phi}(0), 0)\ . \tag{14}$$

We take this equation to define $\boldsymbol{\nu}(\mathbf{J})$, even when $\mathbf{J}$ is not precisely constant. It is easy to construct the inverse function $\mathbf{J}(\boldsymbol{\nu})$ by an interpolation technique described in Ref. 7. This function serves to map the resonant tune lines, $m_1\nu_1 + m_2\nu_2 = n$, into the $\mathbf{J}$ plane. Figure 5 shows all resonances in the region Ω of the $\mathbf{J}$ plane with $|m_i| \leq 20$. The stars indicate the mesh points $\mathbf{J}_i$ used to set up the canonical transformation as a smooth function of $\mathbf{J}$. The transformation as represented in Eq. (7) includes up to 20 Fourier modes in each angle Φ_i.

We verify that this method accurately locates resonances in the $\mathbf{J}$-plane, even in the case of narrow resonances of high order. For instance, we follow an orbit by tracking, taking an initial condition on the (17,16) line of Fig. 5. Plotting

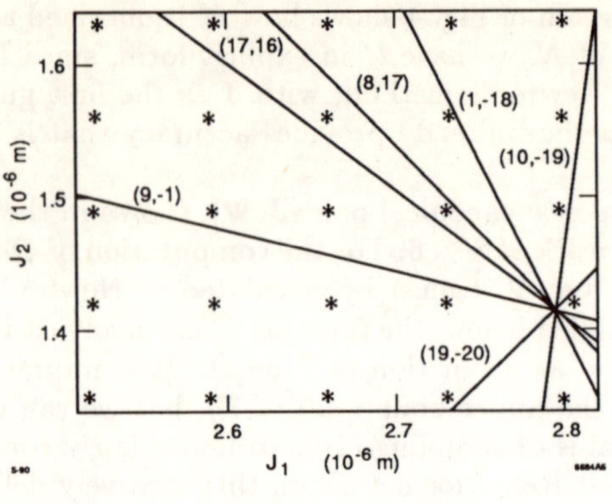

Figure 5: The image in the (J_1, J_2) plane of all resonance lines $m_1\nu_1 + m_2\nu_2 = n$ with $|m_i| \le 20$, for the region Ω defined in Eq. (13). Each line is labeled by (m_1, m_2). The stars indicate the mesh points $\mathbf{J}_i$ used to set up the canonical transform as a smooth function of $\mathbf{J}$.

Table 1: Data on $\mathbf{J}' - \mathbf{J} = \mathcal{D}(\mathbf{J}, \mathbf{\Phi}, N_o)$, the change of $\mathbf{J}$ in N_o turns, for 400 random samples of $(\mathbf{J}, \mathbf{\Phi})$ with $\mathbf{J}$ in subregion Ω_4 as defined in Ref. 7. The brackets $<>$ indicate ensemble averages.

N_0	$\max\lvert\mathcal{D}_1\rvert$	$\max\lvert\mathcal{D}_2\rvert$	$<\lvert\mathcal{D}_1\rvert>$	$<\lvert\mathcal{D}_2\rvert>$	$<\mathcal{D}_1>$	$<\mathcal{D}_2>$
1	$9.9 \cdot 10^{-13}$	$1.9 \cdot 10^{-12}$	$2.4 \cdot 10^{-13}$	$4.7 \cdot 10^{-13}$	$3.8 \cdot 10^{-15}$	$-6.4 \cdot 10^{-15}$
10	$1.5 \cdot 10^{-12}$	$1.7 \cdot 10^{-12}$	$2.7 \cdot 10^{-13}$	$4.7 \cdot 10^{-13}$	$2.1 \cdot 10^{-15}$	$3.8 \cdot 10^{-14}$
100	$1.2 \cdot 10^{-12}$	$2.5 \cdot 10^{-12}$	$3.1 \cdot 10^{-13}$	$5.4 \cdot 10^{-13}$	$-2.7 \cdot 10^{-14}$	$-2.0 \cdot 10^{-14}$
1000	$1.5 \cdot 10^{-12}$	$2.7 \cdot 10^{-12}$	$3.3 \cdot 10^{-13}$	$5.2 \cdot 10^{-14}$	$1.3 \cdot 10^{-14}$	$2.7 \cdot 10^{-14}$
10000	$1.9 \cdot 10^{-12}$	$2.7 \cdot 10^{-12}$	$4.0 \cdot 10^{-13}$	$6.5 \cdot 10^{-13}$	$1.0 \cdot 10^{-14}$	$3.9 \cdot 10^{-14}$

$\mathbf{\Phi}(\theta)$, $\theta = 0, 2\pi, 4\pi, \cdots$, we see apparent curves in the (Φ_1, Φ_2) plane, having 17 intersections with the Φ_1 axis and 16 with the Φ_2 axis; (actually, these are resonant structures of nonzero width, not curves).

Turning now to the computation of δJ_i in the region Ω, we show in Table 1 some typical results for the deviations $\mathcal{D}_i$ of Eq. (11). The results are for a subregion Ω_4 of Ω, roughly one fourth of the full region, and correspond to an ensemble from 400 random samples of $(\mathbf{J}, \mathbf{\Phi})$, $\mathbf{J} \in \Omega_4$, $\Phi_i \in [0, 2\pi]$. The table shows the ensemble maximum of $|\mathcal{D}_i|$, the ensemble average of the same, and the ensemble average of $\mathcal{D}_i$ including sign. It is remarkable that each of these quantities has the same order of magnitude at all N_o up to 10^4. Recalling Eq. (13), we see that $\mathbf{J}$ is constant to about one part in 10^6 for at least 10^4 turns. Moreover, the orbit follows a torus to still greater accuracy in some average sense: the ensemble average of $\mathcal{D}_i$ including sign is considerably smaller than the average absolute value.

The discussion of Ref. 7 yields the following values for the numbers δJ_i that bound $\mathcal{D}_i$, for $N_o = 10^4$:

$$(\delta J_1, \delta J_2) = (2.8, 4.0) \cdot 10^{-12}\ \text{m} \ . \tag{15}$$

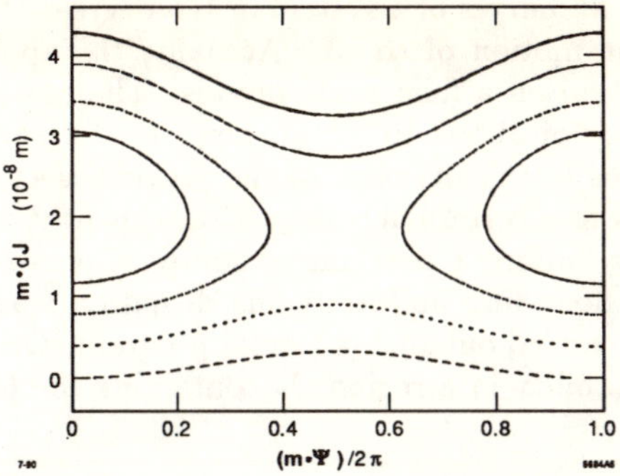

Figure 6: Pseudopendulum motion of the angle-action pair $(\mathbf{m}\cdot\mathbf{J}, \mathbf{m}\cdot\mathbf{\Psi})$ near a $(5,6)$ resonance. Here $\mathbf{m}\cdot\mathbf{\Psi}$ is plotted modulo 2π.

The corresponding values for $N_o = 10^k$, $k = 0,1,2,3,4$, have similar magnitudes. Let us choose ΔJ_i of Sec. 2 so that $q = \Delta J_i/\delta J_i = 10^4$, with $N_o = 10^4$. Then the subset Ω_o of Ω is defined by

$$2.54\cdot 10^{-6}\ \mathrm{m} < J_1 < 2.79\cdot 10^{-6}\ \mathrm{m}\,, \quad 1.38\cdot 10^{-6}\ \mathrm{m} < J_2 < 1.60\cdot 10^{-6}\ \mathrm{m}\,. \quad (16)$$

Any orbit beginning in Ω_o will stay within the slightly larger region Ω for at least $qN_o = 10^8$ turns.

6. Effect of a Strong Resonance

All resonances in the region Ω defined above are weakly excited, and have little effect. The variation of $\mathbf{J}$ on the resonance lines is hardly stronger than elsewhere in the region. In other regions, at comparable amplitudes, we encounter strong resonances that are associated with larger variations of $\mathbf{J}$. This does not necessarily imply a degradation of the time for stability, since oscillations on a well isolated resonance can be stable and not associated with fast transport to nearby resonances, even if the amplitude of oscillation is fairly large. We have studied one such resonance, with $(m_1, m_2) = (5,6)$. In the vicinity of this resonance, the variables $(\mathbf{J}, \mathbf{\Psi})$ follow approximately the pattern expected of action-angle variables in the isolated resonance model.[10] That is to say, $m_2J_1 - m_1J_2$ is nearly constant, while $\mathbf{m}\cdot\mathbf{J}$ and $\mathbf{m}\cdot\mathbf{\Psi}$ behave as action and angle of a physical pendulum. The phase portrait of the latter variables is shown in Fig. 6, with $\mathbf{m}\cdot\mathbf{\Psi}$ plotted modulo 2π.

To establish long-term stability in this situation, we have to limit any movement in the center of oscillation of the pendulum, for any initial condition in the region considered. That can be done by limiting separately the changes parallel to $\mathbf{m}$ and perpendicular to $\mathbf{m}$. For the latter, one can merely put a bound on the change of $m_2J_1 - m_1J_2$, as in our previous bounds on the change of $\mathbf{J}$. For the former, we examine the pendulum motion beyond the separatrix, where full rotations of the pendulum, rather than librations, occur. If the apparent rotation curves of

Fig. 6 behaved like K.A.M. curves of a system in 1-$\frac{1}{2}$ degrees of freedom, they would permanently confine the motion of $\mathbf{m} \cdot \mathbf{J}$. Actually, the apparent curves are not really curves, as inspection on a finer scale reveals. The motion of $\mathbf{m} \cdot \mathbf{J}$ follows a curve within an accuracy of about 10^{-11} m, however, for several thousand turns. This is entirely analogous to the situation of the previous section in which $\mathbf{I}$ follows a two-dimensional torus to a certain accuracy. Consequently, we can proceed in the same way as before, by finding a continuous family of curves that fit the motion to high accuracy, then bounding uniformly the deviation from a curve during N_o turns. We have not yet carried out such a formal program. On the basis of informal estimates, we predict stability in a region Ω_r containing the $(5,6)$ resonance for at least 10^7 turns.

7. Conclusion

We have described a method for theoretical prediction of orbit stability over times comparable to storage times in hadron rings. With further development the method should become practical for machine design. It uses calculational methods that are feasible for elaborate models of large machines, provided that the speed of tracking can be increased substantially. We think that there are good prospects of achieving the necessary speed through construction of full turn maps. Indeed, the preliminary work of Ref. 8 already indicates that very accurate maps can be constructed in a straightforward way.

REFERENCES

1. É. Forest and R. D. Ruth, *Physica D* **43** (1990) 105.
2. N. N. Nekhoroshev, *Russian Math. Surveys* **32**:6 (1977) 1.
3. G. Benettin, L. Galgani, and A. Giorgilli, *Celestial Mech.* **37** (1985) 1.
4. G. Benettin, in *Nonlinear Evolution and Chaotic Phenomena*, eds. G. Gallavotti and P. F. Zweifel, (Plenum, New York, 1988).
5. A. Bazzani, S. Marmi, and G. Turchetti, "Nekhoroshev Estimate for Isochronous Nonresonant Symplectic Maps," University of Bologna preprint, 1990.
6. P. Lochak and A. Porzio, *Ann. Inst. H. Poincaré*, **51** (1989) 199.
7. R. L. Warnock and R. D. Ruth, "Long-Term Bounds on Nonlinear Hamiltonian Motion," SLAC–PUB–5267; submitted to *Physica D.*
8. R. L. Warnock, *Proc. 1989 IEEE Part. Accel. Conf.,* p. 1322. In Table 1 of this paper, the second entry in the second column should have exponent −7.
9. J. Irwin, Superconducting Supercollider report SSC–228.
10. L. Michelotti, *AIP Conference Proceedings* **184**, Vol. 1, *Physics of Particle Accelerators* (Amer. Inst. Phys., New York, 1989).

COMMENT ON ROUND-OFF ERRORS AND ON ONE-TURN TAYLOR MAPS

Y. Yan, T. Sen, A. Chao, and G. Bourianoff
Superconducting Super Collider Laboratory

A. J. Dragt
University of Maryland

E. Forest
Lawrence Berkeley Laboratory

ABSTRACT

Round-off errors in long-term tracking for the Superconducting Super Collider (SSC) with 64-bit precision (single precision in Cray or normal double precision) are found to be insignificant. One-turn Taylor maps are valid for the study of alternating gradient synchrotrons and can potentially be used for fast long-term tracking up to 100 million turns for the SSC.

During the workshop on nonlinear problems in future particle accelerators at Capri, many colleagues expressed concern about round-off errors in long-term tracking and the suitability of Taylor maps for accelerator study, especially for the study of the long-term behavior of particles in an alternating gradient synchrotron such as the SSC. Given the fact that the nonlinear effects in an alternating gradient synchrotron are due to chromaticities, magnet errors and orbit errors and their associated corrections, which must be small enough compared to the leading linear effects in the domain of interest, a one-turn Taylor map with a modest order should be valid. Therefore, the concern about a one-turn Taylor map is actually not whether it works, but whether an efficient numerical method and a powerful computer are available, that allows a practical computation of a one-turn Taylor map at an order high enough to ensure reliability.

There is no doubt that a one-turn Taylor map such as for the SSC and its associated analysis via such methods as Dragt-Finn factorization[1] or normal forms[2,3] would be valid for the study of short-term behavior of particles. The map can be used to calculate the smear, tune shifts, resonances and other quantities of interest. However, it has always been a controversial issue as to whether a one-turn Taylor map is valid for the study of the long-term behaviour of particles in an alternating gradient synchrotron such as the study of the SSC dynamic aperture.[4]

Recently, an "IMSL"-style differential-algebra library "ZLIB"[5] has been developed for mapping studies that could clarify this issue. "ZLIB" uses dynamic memory and is both vectorized and parallelized (multi-tasked) besides being scalarly optimized to allow fast performance for map extraction, tracking, and analysis. Associated with "ZLIB" is a program "Zmap"[6] for the extraction of a differential-algebra map in a beam line. In particular, "Zmap" can extract a one-turn map from a post-Teapot[7] tracking program "Ztrack"[4]. Extraction of a six-dimensional one-turn map for the SSC up to the 16th order is practically possible. The cpu time needed in each processor to extract a one-turn map for the SSC injection lattice on a eight-processor Cray-2 (C machine in NERSC) is about 30 hours for a 16th-order map and about 1.5 hours for an 11th-order map.

Shown in Figure 1 are two survival plots, in the same frame, comparing tracking data for approximately a million turns from "Ztrack" element-by-element tracking (shown in "⌑" symbol) for an SSC injection lattice and from its associated 11th-order one-turn-Taylor-map tracking (shown in "+" symbol) using routines in "ZLIB". Although the map tracking result and the element-by-element tracking result are not identical, they agree in the global behaviour, as well as in many details up to one million turns. Including the time for extracting the one-turn Taylor map, to obtain a survival plot with resonable detail up to 10 million turns (the required proton-coasting time of the SSC injection lattice), an 11th-order Taylor-map tracking is about 40 times faster than the corresponding "Ztrack" element-by-element tracking for the SSC injection lattice. It would be reduced to only about 8 times faster if a 16th-order map is used.

For particles with small (relative to the dynamic aperture) initial transverse and longitudinal amplitudes, there is little doubt that an 11th-order Taylor-map tracking for the SSC is as accurate as the element-by-element tracking for all practical purposes. However, interest is in those particles with reasonable longitudinal amplitudes and with transverse amplitudes approximately equal to the radius of the dynamic aperture, if the ultimate goal is to determine the dynamic aperture. To be precise, Figure 1 shows only that the 11th-order Taylor-map tracking is valid up to a million turns in this particular case. However, considering the general property of the alternating gradient synchrotron and that increasing an order higher for the Taylor map will generally increase its one-turn-tracking accuracy by 5 to 10 times (nearly one more digit accuracy) for particles under interest (see Figure 2), it seems reasonable to expect that a one-turn Taylor map with an order higher than the 11th but less than the 16th will be valid for the SSC long-term tracking for more than 100 million turns.

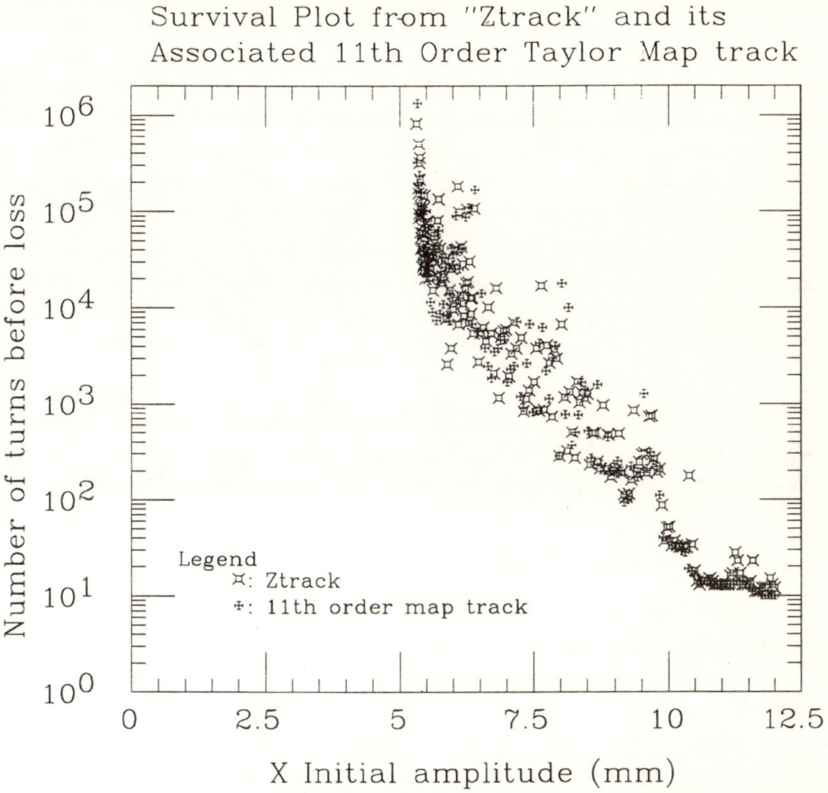

Figure 1. Million-turn survival plots for a 2-TeV injection lattice of the SSC, comparing the data from an 11th-order Taylor-map (extracted with "Zmap") tracking with the data from its associated "Ztrack" element-by-element tracking. The two sets of data match quite well, showing that: (a) round-off errors (64-bit precision) for long-term tracking in the SSC is of no concern since "Ztrack" element-by-element tracking is accomplished at an accuracy of 11 digits in one turn, while the 11th-order Taylor-map tracking is less accurate by 3 to 4 digits (7 to 8 digits of accuracy) in one turn for the selected region of interest, due to truncation of higher orders, and still generate the same result; and (b) Increasing the order of the Taylor map by one or two would allow reliable fast tracking up to 10 million turns for the SSC (the required proton-coasting time of the SSC injection lattice) since increasing one order higher in the Taylor map will enhance nearly 10 times as much accuracy (about one more digit accuracy) in one turn tracking.

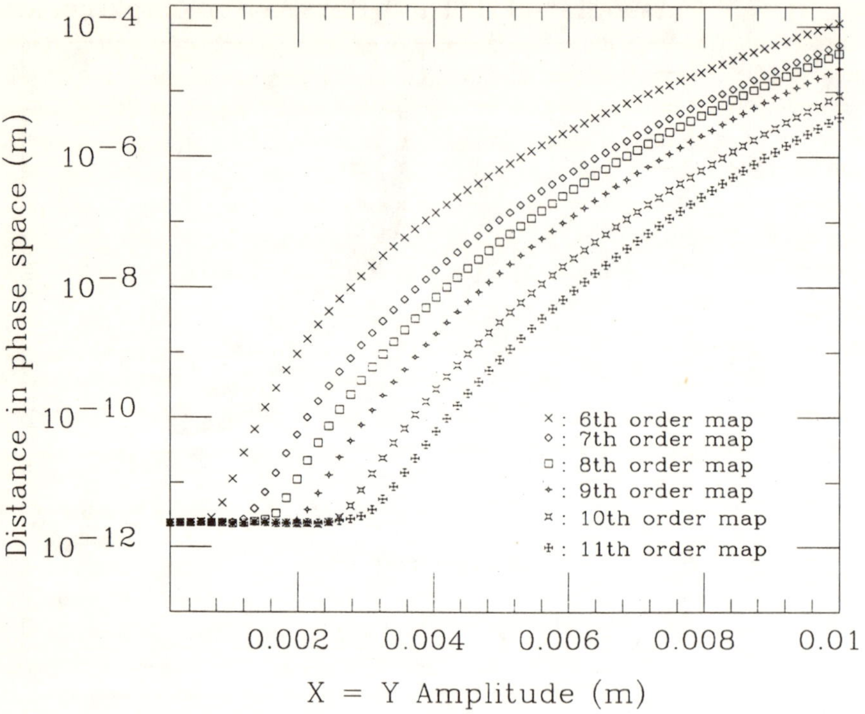

Figure 2. Error analysis of a one-turn Taylor-map tracking for a 2-TeV injection lattice of the SSC (the same lattice as used for the survival plots in Figure 1). The horizontal axis represents the transverse amplitude of the particles while the vertical axis represents the weighted distance D between the positions in the four-dimensional transverse phase space, obtained with an-element-by element tracking (Ztrack) and with a Taylor-map tracking after one turn for each particle. $D = [(x^Z - x^M)^2 + \beta_x^2(p_x^Z - p_x^M)^2 + \frac{\beta_x}{\beta_y}(y^Z - y^M)^2 + \beta_x\beta_y(p_y^Z - p_y^M)^2]^{1/2}$, where $\beta_x' = 0.0 = \beta_y'$ at the measuring point. The deviation in energy for each of the particles is $\delta = 4.2\text{x}10^{-6}$. The distance at small amplitudes is dominated by the round-off errors. The region of interest for long-term tracking is around 0.005m in transverse amplitude where the above figure shows the accuracy of the 11th order map to be between 7 to 8 digits after one turn.

The one-turn-tracking accuracy for particles under interest is about 7 to 8 digits for an 11th-order Taylor map, due to truncation of higher orders, while it is about 11 digits for an element-by-element tracking using "Ztrack" due to round-off errors with 64-bit precision. Therefore, Figure 1 also shows that the effects due to round-off errors for the SSC long-term tracking are negligible with 64-bit precision. Furthermore, round-off errors will eventually dominate the errors from the truncation of higher orders for a Taylor-map tracking, if one increases the order of the map by a few orders over the 11th. Such a Taylor map should be regarded as symplectic from a numerical view point if the extraction of the map is from a symplectic tracking program, which happens to be true for the "Zmap". However, a truncated Taylor map is not symplectic in general.

Recently Irwin[8] and Dragt[9] have proposed methods of converting a Dragt-Finn factorization map[1,10] into kicks for particle tracking. Since a one-turn Taylor map is valid for long-term tracking, a factorization kick map should also be valid, provided: (a) the same order of Taylor map is used to extract the factorization kicks; and (b) high-order (larger than the order of the Taylor map) spurious terms are reasonably suppressed. Research in this direction is in progress. The method proposed by Irwin[8] has been programmed to arbitrary order for some preliminary studies[11] while the method proposed by Dragt[9] with less kicks is currently under implementation in the tracking program "Marylie".[12]

A modest-order one-turn differential-algebra Taylor map being valid for long-term tracking also implies that the interpolation Taylor map proposed by Warnock[13] should also be valid provided that such an interpolation map is constructed up to a suitable order.

One (YY) of the authors thanks J. Irwin, and R. Warnock for many valuable conversations during the workshop at Capri.

References

1. A. J. Dragt and J. M. Finn, "Lie Series and Invariant Functions for Analytic Syplectic Maps" J. Math. Phys. 20, 2649 (1979)

2. E. Forest, M. Berz, and J. Irwin, "Normal Form Methods for Complicated Periodic Systems," Particle Accel. 24, 91 (1989)

3. A. Bazzani, M. Giovannozzi, G. Servizi, G. Turchetti, and E. Todesco, "Normal Forms for Area Preserving Maps—Resonances and Dynamic Aperture," CERN SPS/89-24 (1989).

4. Y. Yan, "Supercomputing for the Superconducting Super Collider," SSCL-239, September 1989; also to be published in Energy Sciences Supercomputing by DOE National Energy Research Supercomputer Center (NERSC).
5. Y. Yan, "ZLIB—A Numerical Library for Differential Algebra," Superconducting Super Collider Laboratory Report No. SSCL-300, 1990.
6. Y. Yan, "Zmap: A Differential-Algebra Map Extraction Program using ZLIB," SSCL-299, 1990.
7. L. Schachinger and R. Talman, "Teapot: A Thin-Element Accelerator Program for Optics and Tracking," Particle Accelerators, Vol. 22, 1987.
8. J. Irwin, "A Multi-Kick Factorization Algorithm for Nonlinear Maps," SSC-228, 1989.
9. A. Dragt, "Method for Symplectic Tracking," presented in the workshop on nonlinear problems in future particle accelerators, Capri, Italy (1990).
10. A. Dragt, et al., "Lie Algebraic Treatment of Linear and Nonlinear Beam Dynamics," Ann. Rev. Nucl. Part. Sci., Vol 38, pp 455-496 (1988)
11. E. Forest and Y. Yan, private communication.
12. A. Dragt, et al., "Release of Marylie 3.0," in Proceedings of the 1989 IEEE Particle Accelerator Conference, Vol. 2, p. 1283 (1989).
13. R. L. Warnock, "Construction of Symplectic Full-Turn Maps by Application of an Arbitrary Tracking Code," in Proceedings of the 1989 IEEE Particle Accelerator Conference, Vol. 2, p. 1322 (1989).

DISCRETE DYNAMICS FOR STRONGLY CHAOTIC SYSTEMS

FRANCO VIVALDI

School of Mathematical Sciences
Queen Mary and Westfield College
University of London
Mile End Road
London E1 4NS, ENGLAND

ABSTRACT

The phase space of strongly chaotic hamiltonian systems can be naturally discretized. The resulting discrete dynamical systems can be studied theoretically using methods of algebraic number theory, and numerically using integer arithmetic, which is faster, error-free and limited only by the size of one's computer.

1. Introduction

Since Newton, dynamics has been based on the continuum and on differentiable structures, being concerned with smooth differential and difference equations over smooth manifolds. Such problems have been treated largely (if not overwhelmingly so) with continuum mathematics.

Recent studies in chaotic dynamics are exposing two structural weaknesses of this traditional view. Firstly, the developements of contemporary dynamics is undoubtely the direct result of that of digital computers (e.g. the Henon-Heiles problem, the Lorentz attractor, the Feigenbaum universality, fractal geometry, etc.). Computers perform only discrete operations, yet they are routinely forced to mimic the continuum. Secondly, chaos, in the final analysis, originates from the randomness of the continuum real number system, which poses unsormontable difficulties and causes esoteric logical paradoxes[1].

The divorce of dynamics from discrete mathematics has profound roots. While the mathematics of the continuum has often found motivations in the solution of applied problems (e.g. Gauss invented differential geometry in order to survey the land of the Duke of Brunswick), much of discrete mathematics (number theory, ring theory, field theory, theory of discrete groups, etc.) was developed quite independently from the physical sciences. As a result, many dynamicists make little use of a vast arsenal of mathematical tools, some of which essential to computer-oriented theories.

Attempts to overcome this impasse have been few and scattered, and, on the whole, lacking a recognizable direction. One of the earliest studies is that of Rannou[2], who showed that chaos is not just a numerical artifact. A similar theme has reappeared more recently[3]. Discretization has been performed in order to mimic quantum effects[4], or to control computer roundoff in a delicate numerical experiment[5]. Discrete phase spaces occur naturally in certain billiard systems[6].

I am interested in the discrete phase spaces that arise from the discrete arithmetic of the computers. These phase spaces are lattices, whose points are spaced logarithmically in floating point arithmetic, and linearly in integer arithmetic. The latter are considerably more natural and interesting.

The computer discretization is problematic in many ways. Firstly, the discrete and continumm representations of a dynamical systems differ considerably. Plainly, most continuum orbits cannot be reproduced on a discrete set. For instance, there cannot be quasiperiodic or homoclinic motions in a discrete system.

Secondly, the dynamical properties of a discrete system depend sensitively on the discretization. This means that as the discretization becomes finer and finer, the asymptotic properties of the system will exhibit large fluctuations (this behaviour is rather common in the asymptotics of arithmetical functions[7]).

Furthermore, algebraically speaking, these lattices are very simple objects. Their elements can be summed, subtracted, and multiplied by an integer ($\mathbf{Z}$-modules), but not multiplied together. As a result, they are unhappily unsuitable for both continuum and discrete mathematics.

However, lattices can be provided with richer algebraic structures. For instance, a lattice on $\mathbf{R}^n$ or on the n-dimensional torus can be made into a *ring*, by defining a rule for multiplying its elements, much in the same way as one defines multiplication in lattices of complex numbers. This procedure opens the doors to some methods of discrete mathematics (arithmetic, in particular).

2. Strongly Chaotic Systems

The theory of fully chaotic hamiltonian systems is the one branch of dynamics where discrete mathematics holds a legitimate place. The Anosov diffeomorphism of the n-torus (i.e. the most chaotic among the smooth hamiltonian maps of such manifolds) can be studied as discrete dynamical systems, using methods of algebraic number theory. (For an introduction to arithmetic, see refs. 8 and 9.)

The link between dynamics and arithmetic derives from the possibility of representing these systems with integral matrices (up to a conjugacy[10]), and by exploiting a theory relating integral matrices and algebraic numbers[11]. The periodic orbits of an integral matrix M acting on a torus coincide with points having rational coordinates, which makes discretization natural. The set of rational points sharing the same denominator g is a dynamical invariant, which can be transformed into an integral lattices with coordinates taken modulo g.

It turns out that these lattices possess a natural ring structure, which is induced by the eigenvalues of the matrix representing the system. We illustrate this point in the two dimensional case. Let M be a two by two integral matrix acting on the 2-torus, and let $f(x) = x^2 + 2bx + c$ be its characteristic polynomial. Then b and c are integers (for simplicity, the middle coefficient has been assumed even). One eigenvalue of M is given by

$$\lambda = -b + n\sqrt{D},$$

where $n^2 D = b^2 - c$ and D has no square divisors (the choice of the eigenvalue is immaterial here). Numbers of the form $x + y\sqrt{D}$ form a ring, because the sum, subtraction and multiplication of two numbers of this type yield a number of the same type, and these operations obey the usual associative, commutative and distributive properties. This ring, denoted by $\mathbf{Z}[\sqrt{D}]$, can be put in one-to-one correspondence with the set of integer pairs (x, y), which are points on a two dimensional lattice[11]. For instance, for $D = -1$ we obtain a two dimensional lattice of complex numbers.

It can be shown that the action of M on $\mathbf{Z}^2$ can be transformed into multiplication by λ in $\mathbf{Z}[\sqrt{D}]$ (this step is somewhat delicate[12]). The dynamical problem is thus reduced to a problem of multiplicative arithmetic in $\mathbf{Z}[\sqrt{D}]$, where methods of algebraic number theory provide a wealth of results and computational techniques[13,14,15]. Equally important, these results provide deep insight into the arithmetical nature of chaotic motions, which is also relevant to quantum mechanics[16].

Results with a similar flavour have been recently obtained in the study of dynamical systems over *finite fields,* which are a class of multiply periodic rings that can be interpreted as discrete phase spaces on tori[17,18]. On these phase spaces, all dynamical systems can be represented by polynomials[19].

Using integer dynamics, the numerical experiments on chaotic systems assume a different connotation, being faster, error-free and reproducible, and can be used confidently to provide support to the theoretical development (and to become part of it). It must be noted however, that exponential instability of motion is an intrinsic property of dynamical chaos, and the computational problems it entails are not eliminated by discretization. Chaos in discrete dynamics manifests itself with the appearance of integers that are exponentially large in the period of the orbits. Thus the limits to computation are very sharply determined by the size of one's computer.

In closing it is worth mentioning that, quite apart from discrete systems, number theoretical methods are commonly employed in chaotic dynamics, for instance in the study of periodic motions of geodesic flows on surfaces of negative curvature[20,21].

Acknowledgements

This research was supported in part by S.E.R.C.

References

1. J. Ford, How random is a coin toss?, *Physics Today* **36** (1983) 40–47.
2. F. Rannou, Numerical studies of discrete plane area-preserving mappings, *Astron. Astrophys.* **31** (1974) 289–301.
3. K. Kaneko, Symplectic cellular automata, *Phys. Lett. A* **129** (1988) 9–16.
4. B. V. Chirikov, F. M. Izrailev, and D. L. Shepelyansky, Dynamical stochasticity in classical and quantum mechanics, *Soviet Scientific Reviews C, Vol.2*, Gordon and Bridge (1981) 209–267.
5. C. F. F. Karney, *Physica* **8**D (1983) 360.
6. F. Vivaldi and A. V. Shaidenko, Global stability of a class of discontinuous dual billiards, *Commun. Maths. Phys.,* **110** (1987) 625–640.
7. T. A. Apostol, *Introduction to analytic number theory,* Springer-Verlag, New York (1984).
8. H. Davenport, *The higher arithmetic,* Hutchinson & Co. Ltd., London (1952). (Reprinted by Dover, New York (1983).)
9. H. Cohn, *A second course in number theory*, John Wiley & Sons, New York (1962). (Reprinted as *Advanced number theory*, Dover, New York (1980).)
10. A. Manning, Classification of Anosov diffeomorphisms on tori, *Lecture Notes in Mathematics*, No. 468, Springer-Verlag, New York (1975).
11. O. Taussky, Introduction into connections between algebraic numbers and integral matrices, appendix to H. Cohn, *A classical invitation to algebraic numbers and class fields*, Springer-Verlag, New York (1978), p 305.
12. F. Vivaldi, The arithmetic of chaos, *Chaos, Noise and Fractals*, Adam Hilger, Bristol (1987) 187–199.
13. I. C. Percival and F. Vivaldi, Arithmetical properties of strongly chaotic motions, *Physica* **25** *D* (1987) 105–130.
14. F. Vivaldi, Arithmetical theory of Anosov diffeomorphisms, *Proc. R. Soc. Lond. A* **413** (1987) 97–107.
15. M. Bartuccelli and F. Vivaldi, Ideal orbits of toral automorphisms, *Physica D* **39** (1989) 194–204.
16. J. P. Keating, Semiclassical properties of the cat maps, in *Number Theory and Physics,* J.-M. Luck, P. Moussa, and M. Waldschmidt editors, Springer Proceeding in Physics No. 47, Springer-Verlag, Berlin (1990) 302–310.
17. E. Jen, Linear cellular automata and recurring sequences in finite fields, *Commun. Math. Phys.* **119** (1988) 13–28.
18. F. Vivaldi, The geometry of linear maps over finite fields, Preprint QMW-DYN 90-3, Queen Mary and Westfield College (1990). (Submitted to *Nonlinearity.*)
19. R. Lidl and H. Niederreiter, *Finite fields*, Encyclopedia of Math. and its Appl., vol. 20, Addison-Wesley, Reading, Mass. (1983).
20. C. Series, The modular surface and continued fractions, *J. London Math. Soc.* (2)**31** (1985) 69–80.
21. R. Aurich and F. Steiner, On the periodic orbits of a strongly chaotic system, *Physica D* **32**, (1988) 451–460.

PERFORMANCES OF THE LHC LATTICE IN THE PRESENCE OF MAGNETIC AND ALIGNMENT ERRORS

F. GALLUCCIO and W. SCANDALE
CERN — SL Division
CH-1211 Geneva 23

ABSTRACT

The systematic and random multipolar imperfections in the magnetic field of the superconducting dipoles and quadrupoles may strongly affect the performances of large proton synchrotrons, especially at injection energy. In the case of the Large Hadron Collider (LHC) two main methods, bore-tube windings and lumped multipolar elements, have been studied and compared to compensate the effect of systematic multipolar field errors; particle-tracking simulations have shown that both methods improve the dynamical properties of the machine to an acceptable level. Ordering the magnets with respect to their random sextupolar contents while taking into account their two-in-one design, has proved to help increasing both the linear and the dynamic apertures of the two LHC rings. Methodical tracking studies have been performed for on- and off-momentum particles, with both random and systematic field errors, with and without compensations, including also the residue after correction of the closed orbit distortion due to alignment errors. In this paper the results of this study are discussed.

1 Introduction

The future hadron colliders will operate in the range of energy of some TeV per beam, therefore they will have to be built with superconducting magnets. Obvious cost considerations suggest to have magnet apertures as small as possible, but then the imperfections of the magnetic field are likely to be sizably larger than in a conventional warm accelerator. This is the case of the LHC, the Large Hadron Collider to be installed in the LEP tunnel, above the LEP ring itself, in the late nineties.

The perturbations of the beam motion due to the non-linearity of the guiding magnetic field are big enough to affect the performances of the accelerator. Criteria are needed to estimate these perturbations, and compensation techniques to minimize them.

This paper summarizes the studies done over the last two years to evaluate the dynamical properties of the LHC, to determine adequate correction schemes, and ultimately to provide prescriptions on the magnetic field quality necessary to achieve the requested machine performances.

2 The LHC Lattice Model

The LHC lattice used here is that of Ref. 1, retuned at $Q_x = 70.28$, $Q_y = 70.31$. It is made of eight arcs and eight insertions, each of them including two dispersion suppressors and one long straight section. An arc contains 49 regular half-cells with four dipoles each, and a dispersion suppressor contains four pseudo-half-cells with three dipoles each. Therefore there are 1760 dipoles in total, and all of them are 9.54 m long.

There are 384 quadrupoles in the regular cells, split in two families known as QF and QD respectively, all of them 3.08 m long. The insertion quadrupoles, 160 in total, are of special design and lengths. Sextupoles, powered in two families, are included to correct chromaticity and are placed next to the main quadrupoles.

The injection optics considered consists of four insertions with β^* values fixed at $\beta_x^* = \beta_y^* = 6.5$ m, placed in the even straight sections, one dump-dedicated insertion in straight section No. 3, and three identical insertions with $\beta_x^* = \beta_y^* = 4$ m at the interaction point in the other three odd straight sections.

The collision optics, instead, is non-symmetric, with two different interaction points: one with $\beta^* = 0.25$ m in insertion No. 1, and the other with $\beta^* = 0.5$ m in insertion No. 5.

Lumped multipolar correctors up to decapole are implemented in the regular cells to compensate the effects of the systematic multipolar imperfections of the main dipoles; they are located close to the main quadrupoles and halfway between them. The systematic dodecapole imperfections in the quadrupoles are assumed to be corrected by bore-tube windings.

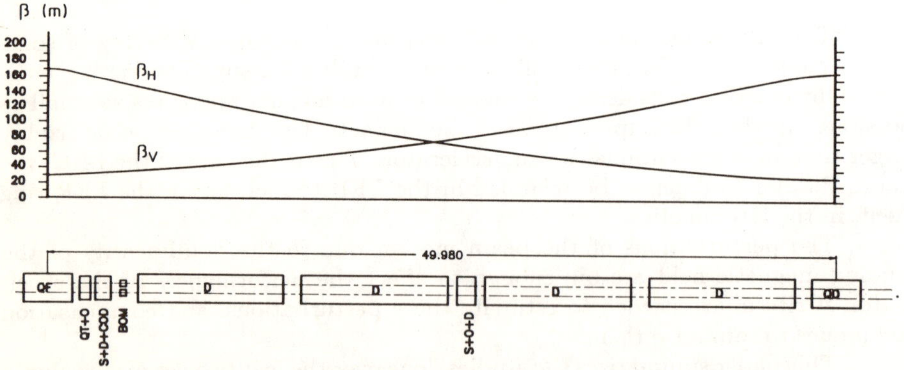

D:dipole magnets, Q:quadrupoles, QT+O:combined tuning quadrupole and octupole corrector BOM:beam observation monitor, S+D+COD:combined sextupole, decapole and dipole corrector, S+O+D:combined sextupole, octupole and decapole corrector.

Figure 1: Layout of the LHC Standard Half-Cell

The realistic layout of a standard half-cell including also the lumped correctors is shown in Figure 1, together with its lattice functions. The three central correctors are combined in one single block of length $\ell = 1.18$ m, which contains, from the outside going inwards, sextupole, octupole, and decapole windings. The correctors close to the main quadrupoles are instead split into two blocks: the sextupole and the decapole are in the same block of length $\ell = 1$ m, whilst the octupole is enclosed in the tuning quadrupole which is 0.72 m long.

Crossing angles varying from 96 to 300 μrad are considered for both injection and collision optics; for this reason the particle trajectory in the inner triplets of quadrupoles is displaced off-axis.

The idea of using a cell lattice model of the machine, especially in view of the long-term tracking studies, has also been analyzed, but it has been abandoned, since the results of the particle tracking simulations, shortly described in Appendix A, are too far from those of the more realistic lattice with insertions discussed in the next sections.

3 Lattice Imperfections

The random and systematic multipolar magnetic errors which are present in the LHC superconducting magnets and their sources are extensively discussed elsewhere[2], therefore here we will limit ourselves to recall their values and to discuss the assumptions made during the simulations.

The magnetic field can be expressed by

$$B_y + iB_x = B_o \sum_{i=1}^{\infty} (b_n + ia_n) \left(\frac{x + iy}{R_r} \right)^{n-1} , \tag{1}$$

where
B_o is the dipole field in the vertical direction,
R_r is the reference radius,
b_n and a_n are the normal and skew multipolar coefficients,
$2 \times n$ is the number of poles.

At injection two sets of values of systematic multipolar coefficients have been used for the dipoles, and are shown in Table 1: the first set (1987 errors) is

Table 1: Multipolar Coefficients for Systematic Errors in the Dipoles (in units of 10^{-4} at $R_r = 1$ cm) at Injection ($B_o = 0.56$ T)

	b_3	b_4	b_5	b_7	b_9
1987 errors	−4.05	0.05	0.56	−0.03	0.010
1990 errors	−4.05	±0.05	0.56	0.13	0.024

Table 2: Multipolar Coefficients for Systematic Errors in the Dipoles (in units of 10^{-4} at $R_r = 1$ cm) at Collision ($B_o = 10.0$ T)

	b_3	b_4	b_5	b_7	b_9
1987 errors	−1.95	−0.45	−0.16	−0.037	0.010

reported also in Table 5.4 of Ref. 3; the second set (1990 errors) is rather more pessimistic about the higher order components.

Only the normal components of systematic errors are taken into account, and the quadrupolar coefficient b_2, mostly due to the saturation of the yoke, is neglected because it can be substantially reduced by proper shaping of the iron, and finally corrected by the tuning quadrupoles.

The octupolar coefficient, b_4, changes sign every octant, owing to the two-in-one design of the LHC dipoles; this was not included in the 1987 errors.

At collision, only one set of errors has been used in the dipoles and it is shown in Table 2.

Table 3: Multipolar Coefficients for Systematic Errors in the Quadrupoles (in units of 10^{-4} at $R_r = 1$ cm) at Injection

	b_2	b_6	b_{10}	b_{14}
1989 errors	0.36	−0.75	0.03	5×10^{-5}
1990 errors	0.32	−0.43	0.01	1×10^{-5}

Systematic errors in the quadrupoles have been considered only at injection; two sets of them are given in Table 3: the first one (1989 errors) is that of Ref. 4 and holds for quadrupoles with coil inner diameter of 50 mm, whilst the second one (1990 errors) takes into account the increase of the coil inner diameter from 50 to 56 mm.

The r.m.s. values of the random magnetic imperfections assumed in each dipole and quadrupole are given in Table 4. We disregarded the normal and skew quadrupole components, and we assumed the IR-triplets to be largely compensated by bore-tube windings up to the dodecapole term.

For the misalignment and the field errors we assumed the values listed in Table 5, based on the experience of LEP.

The average closed-orbit distortion and the related maximum excursion of the particle trajectory, after correction in both planes are:

$$X_{rms} = 0.26 \text{ mm}, \quad X_{peak} = 1.2 \text{ mm},$$

$$Y_{rms} = 0.25 \text{ mm}, \quad Y_{peak} = 1.1 \text{ mm}.$$

These values were obtained by averaging over 10 different distributions of the errors of Table 5.

Table 4: R.M.S. Coefficients of Random Multipolar Errors (in units of 10^{-4} at $R_r = 1$ cm)

Order	Dipole		Quadrupole		IR–Triplet	
	b	a	b	a	b	a
1	0.0	0.0	0.0	0.0	0.0	0.0
2	0.0	0.0	0.0	0.0	0.0	0.0
3	1.5	0.5	2.28	2.10	0.1	0.1
4	0.15	0.2	0.34	0.72	0.1	0.1
5	0.2	0.07	0.16	0.17	0.1	0.1
6	0.0	0.0	0.12	0.06	0.1	0.1
7	0.02	0.04	0.02	0.02	0.022	0.022
8	0.0	0.0	0.0	0.0	0.016	0.016
9	0.005	0.002	0.007	0.006	0.013	0.013
10	0.0	0.0	0.005	0.004	0.009	0.009

Table 5: R.M.S. Values of Misalignment and Field Errors

	Dipoles	Quadrupoles
Horizontal misalignment	0.14 mm	0.14 mm
Vertical misalignment	none	0.14 mm
Tilt (DPSI)	0.24 mrad	0.24 mrad
Relative field error	5×10^{-4}	5×10^{-4}

4 Compensation of Systematic Imperfections

The systematic multipoles of Table 1 mainly produce a large amplitude- and momentum-dependent tune shift which results in a sensible reduction of both linear and dynamic apertures. Octupole errors mostly induce an amplitude-dependent tune shift[5], whilst decapole errors give a stronger contribution to the momentum-dependent tune shift[6,7]. The higher order multipole errors perturb the beam predominantly at the larger amplitudes, mainly affecting the dynamic aperture, as will be seen in the comparison of the results obtained with the two different sets of systematic errors.

The main purpose of any correction scheme must be to reduce the tune shift to a sufficiently low value over a sufficiently large range of amplitude and momentum.

Our aim here is to find a satisfactory compensation for sextupolar, octupolar, and decapolar components, and to evaluate the effects of the higher order ones (14-pole and 18-pole), possibly indicating a maximum tolerable value.

Our quality criterion, inspired by the operational experience with the $p\bar{p}$

Collider at the SPS is that:

$$|\Delta Q_{x,y}| \leq 0.005 \tag{2}$$

$$\begin{array}{llll} \text{for} & a \leq r_o = b/2 & \text{if} & \Delta p/p = 0 \\ \text{or for} & a \leq r_o' = r_o/\sqrt{2} & \text{if} & \Delta p/p = \pm\delta \end{array}$$

where:
$b = 20$ mm is the radius of the vacuum chamber, and at injection $b/2$ corresponds to the amplitude of a particle at 10σ of a 15π mm mrad emittance beam at a maximum-β location in the regular cell,
δ is the bucket half-height; at injection $\delta = 1.25 \times 10^{-3}$.

4.1 Correction by Bore-Tube Windings

Additional multipolar windings placed in the bore of each dipole are the most natural way that one can think of to cancel systematic errors, but not always the easiest from the hardware point of view. In particular, in the case of the LHC superconducting magnets, in the dipole bore tube there is not enough space for all the multipolar coils (6-, 8- and 10-polar) one would like to install to achieve a really "local" error compensation; instead, at most one of them may be foreseen per half-magnet, therefore allowing only a "quasi-local" correction.

Amongst all the possible arrangements achievable with only one kind of corrector per half-dipole, we have selected the one in which sextupolar windings are implemented in every dipole, whilst octupolar and decapolar windings are implemented every other dipole, alternatively, and the strength of each multipolar corrector integrated over a cell equals that of the corresponding systematic error.

The compensation for the sextupolar errors is obviously "more local" than for octupoles and decapoles, and indeed, at larger amplitudes the tune shift starts to increase more rapidly; nevertheless, this simple setting, without any further optimization, allows the dynamical behaviour of the machine to be improved enough to largely satisfy the criterion of Eq. (2), and thus to consider this correction method well acceptable.

4.2 Correction by Lumped Multipoles

The significant problems presented by the practical implementation of the bore-tube windings in the LHC two-in-one dipoles pushed us to explore a different correction technique, based on lumped correction elements by which, instead, a "global" compensation can be achieved.

Different dispositions of correction elements can be thought of, and, of course, the larger the number of correctors, the better the compensation.

The most efficient arrangement with a small number of elements seems to be that proposed by Neuffer[8], and we have chosen to implement it in the regular cells for the compensation of the sextupolar, octupolar, and decapolar systematic field components of the LHC dipoles: to correct each kind of multipolar error to first order three elements per half-cell are used, two of them are placed at the ends of the half-cell and one in the middle. Actually end correctors of contiguous half-cells can be combined, giving a total of four correctors per cell: two, powered together (except for the sextupoles), next to the cell quadrupoles, and two, also powered together, halfway between them.

Three strategies can be employed to set the strengths of the lumped multipoles at their optimum values, two minimizing the amplitude-dependent tune shift, and the third minimizing the momentum-dependent tune shift; in spite of the very different starting points, the three approaches give fairly similar results, producing three sets of correctors whose values are rather close to each other; they all allow the condition of Eq. (2) to be satisfied, and the compensation scheme to be considered adequate.

Figure 2 gives a very impressive representation of the power of the lumped correction method: particles of nominal momentum have been tracked for 100 turns at injection with starting coordinates on a regular mesh in the transverse physical plane X–Y at a focusing quadrupole ($\beta_x = 169$ m, $\beta_y = 30$ m), and the horizontal and vertical tune shifts are plotted against transverse coordinates for (a) the ideal bare LHC (chromaticity corrected), (b) the machine with systematic errors and only chromaticity correction, (c) the machine with systematic errors and lumped multipole correctors. The effect of the systematic errors is evident in the strong distortion of the tune surfaces in Figure 2b; yet, these curves are almost perfectly smoothed out and brought back to the original bare machine shape (Figure 2a) by the action of the lumped correctors (Figure 2c).

The three different optimization methods are shortly described below. We should point out that since the sextupoles next to the main quadrupoles are also chromaticity sextupoles, the first-order compensation of systematic sextupoles is always done at the same time as the correction of chromaticity.

4.2.1 Amplitude-dependent Tune Shift Minimization

The first set of correctors, called *Simpson Correctors*, was obtained[6] by minimizing the amplitude-dependent tune shift by means of a "trial and error" tracking procedure. Particles over a wide range of amplitude were tracked, starting along the diagonal, the horizontal, and the vertical direction in the transverse space X–Y to simulate a round, a flat horizontal, and a flat vertical beam respectively; the optimization was done implementing the systematic error components one by one and tuning their correcting elements: sextupolar errors were corrected with sextupolar and octupolar elements, octupolar errors were corrected with octupolar elements, and decapolar errors were corrected with decapolar correctors; the quadratic sum of the horizontal and vertical tune shifts at half of the vacuum

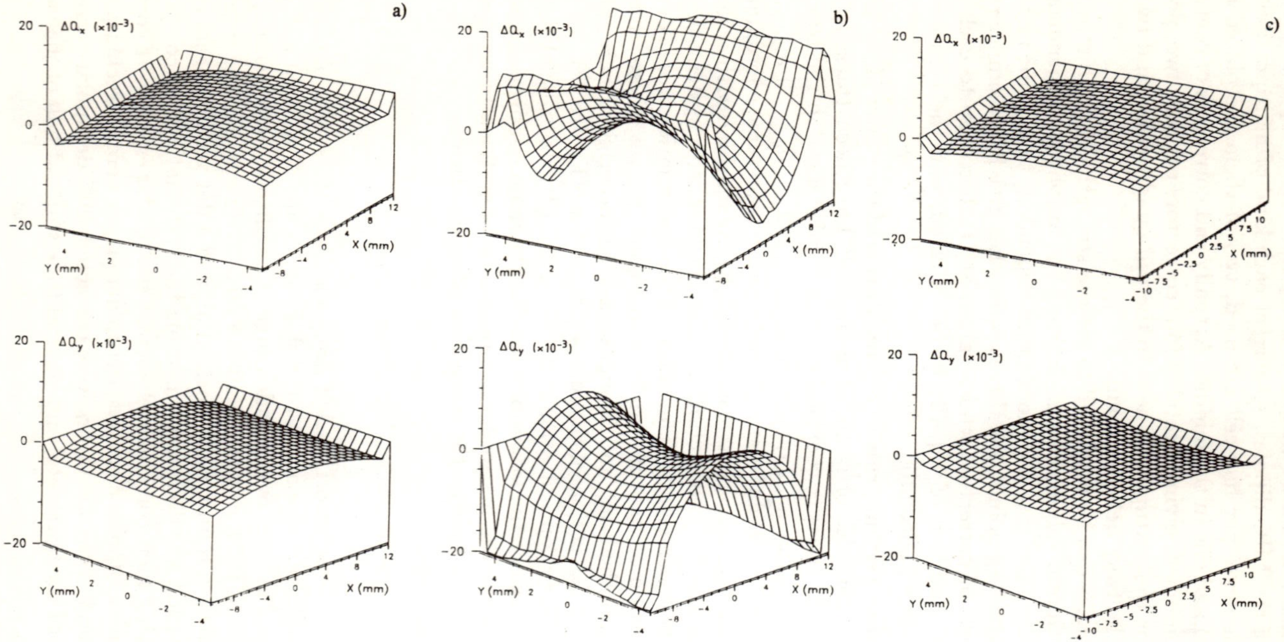

Figure 2: Tune Shift against X and Y in the LHC: a) Bare, b) with Systematic Errors, and c) with Systematic Errors and Lumped Multipole Correctors

chamber radius along the three directions was minimized, taking care that the behaviour of the tune shifts was monotonic along any direction and that up to 3/4 of the vacuum chamber their values were still below 0.01. On the strengths of the correctors the constraint was applied that the integrated gradients of the horizontal and vertical octupoles (decapoles) were identical, and, according to the Simpson's Rule, they were just half of the central octupole (decapole) gradient. As a first guess the corrector strengths were fixed in such a way that their integrated gradient over one cell was equal and opposite to the integrated gradient of the relative systematic multipolar field. The dependence of tune shift on momentum was not considered, but only checked a posteriori: as will be shown in the next section, this scheme of correction is efficient also for off-momentum particles.

An alternative set of correction multipole strengths, called here *Normal Forms Correctors*, was calculated using a method which employs normal forms techniques to compute the quasi-invariants of the non-linear motion and to minimize the tune shifts. By this method, which is extensively reviewed in Ref. 9, the constraint on the ratio of the integrated gradients of the central and lateral octupole (decapole) is removed, whilst the constraint on the equality of the lateral elements, which in principle could be released as well, has been retained owing to cost considerations. Although the minimization is only done on the lower order coefficients of the tune shifts, this method has proved to give reasonably good compensation for the tune shifts over a large range of amplitudes and momenta.

4.2.2 Momentum-dependent Tune Shift Minimization

A completely different approach was used to compute the strengths of the *Non-Linear Chromaticity Correctors*[7]; the contribution to chromaticity from the systematic sextupole errors is mainly linear with a small quadratic part, the contribution from octupole errors is mainly quadratic, and that from decapole errors is mainly cubic, as it can be seen in Figure 3.

Then, sextupole correctors, including the central one, have been used to set the linear part of chromaticity to zero, whilst octupoles have been used to cancel its second-order derivative and decapoles to cancel its third-order derivative. Since only two free parameters are necessary for each kind of corrector, the condition that the lateral octupoles (decapoles) had the same strength was retained. This procedure was applied successfully around $\Delta p/p = 0$, but the correctors were found to minimize rather well the linear and non-linear chromaticities also at very large momenta. Furthermore, even if the amplitude dependence of the tunes was not taken into account in the optimization, this scheme was recognized as being very efficient also at large amplitudes.

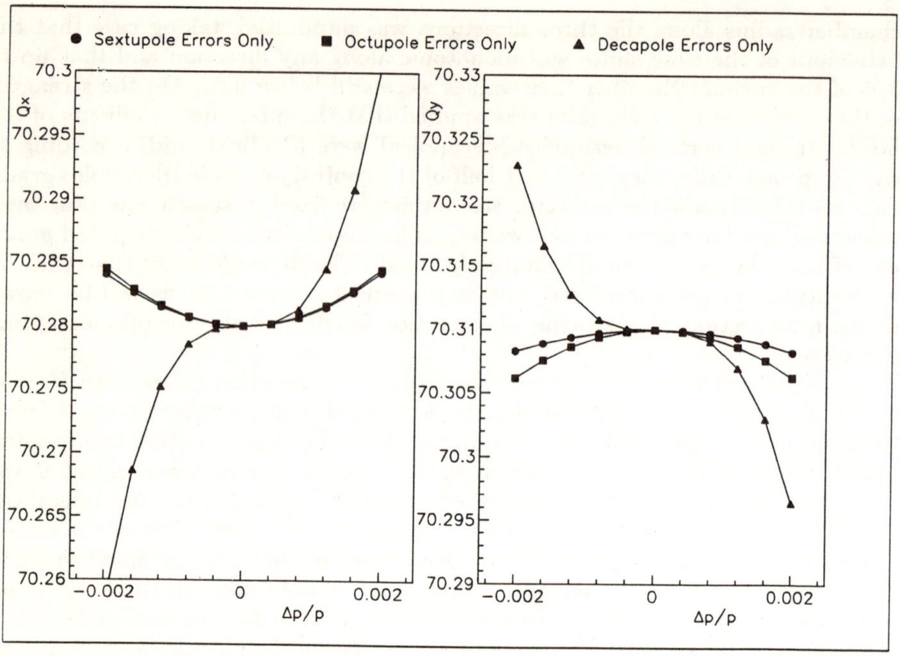

Figure 3: Chromatic Curves for Different Uncorrected Multipole Errors

4.3 Results

On- and off-momentum particle tracking has been performed with the program MAD[10] and used to compare the performances of the local and the global correction methods, the latter with its three sets of lumped correctors. In Table 6 the normalized linear apertures based on tune shift and on smear,

$$Sm_I = \frac{1}{\sqrt{N-1}} \frac{\sqrt{\sum_{i=1}^{N} (I_i - < I >)^2}}{< I >}, \tag{3}$$

where:

I $= I_x, I_y$ (Courant–Snyder invariant),
N = No. of turns,

and the normalized short-term dynamic aperture are displayed for these four configurations of the LHC, and compared with an ideal bare machine and with the machine with systematic errors and no compensation other than for chromaticity, for the two sets of errors available.

Table 6: Linear and Dynamic Apertures in the Presence of Systematic Errors in the Dipoles

Systematic Multipolar Errors in the Dipoles APERTURE with Different Criteria — Injection — Round Beam Normalized Amplitudes in mm ±0.02 mm						
	LINEAR APERTURE				DYNAMIC AP.	
CRITERION	$\|\Delta Q_{x,y}\| \leq 0.005$		$Sm_{x,y} \leq 0.10$		400 turn over-flow	
$\|\Delta p/p\|$ ($\times 10^{-3}$)	0	1.25	0	1.25	0	1.25
Ideal Bare Machine	1.49	1.43	1.01	0.96	3.71	3.40
<u>1987 Systematic Errors</u>						
Errors + Chromaticity Correct.	0.59	>	0.53	>	1.35	0.75
+ Coil Correction	1.39	0.65	0.93	0.74	1.50	1.25
+ Normal Forms Correction	1.54 †	0.57	1.20	0.70	1.54	1.13
+ Simpson Correction	1.46 †	0.93	1.05	0.74	1.46	1.19
+ Non-Linear Chromaticity Co.	1.10	1.05	1.33	1.04	1.54	1.23
<u>1990 Systematic Errors</u>						
Errors + Chromaticity Correct.	0.62	>	0.98	0.15	1.19	0.77
+ Coil Correction	1.09	0.52	1.03	0.54	1.46	0.77
+ Normal Forms Correction	0.81	0.55	1.13	0.56	1.23	0.69
+ Simpson Correction	1.08	0.57	1.04	0.53	1.19	0.85
+ Non-Linear Chromaticity Co.	0.97	0.59	1.20	0.63	1.31	0.77

†linear aperture threshold is not reached, particles are lost before; dynamic apert. is shown.
>tune-shift or smear values are above threshold for all amplitudes.

The change of the octupole error sign from arc to arc in the 1990 data of Table 1 is beneficial, as the average effect of b_4 on the machine becomes smaller.

We see that all the correction schemes allow recovery of most of the linear aperture of the ideal bare machine and fully satisfy the criterion of Eq. (2), but they cannot do as much for the dynamic aperture.

The detrimental effect of the new b_7 and b_9 is also evident: they strongly lower the power of all the correcting methods, substantially equalizing their performances; both linear and dynamic aperture are deteriorated and the effect is particularly severe for the off-momentum particles.

Table 7: Recommended Gradients for Lumped Correctors

	Sextupole		Octupole		Decapole	
	ℓ (m)	$B^{(2)}$ (Tm^{-2})	ℓ (m)	$B^{(3)}$ (Tm^{-3})	ℓ (m)	$B^{(4)}$ (Tm^{-4})
Lateral	1.00	4500	0.72	100000	1.00	32×10^6
Central	1.18	650	1.18	75000	1.18	60×10^6

The correction has been tested also at collision energy[6]: the tracking results show that the method is still applicable and that, owing to the smaller dimensions of the beam, the problems are less critical. The strengths of the correctors at high energy are those determining the specification displayed in Table 7.

The results relative to the systematic quadrupole imperfections are summarized in Table 8.

The two available sets of errors have been considered. In both cases tune shift with amplitude as well as with momentum is generated, whose value is larger than that due to the imperfections in the dipoles.

Bore-tube windings are foreseen to take care of the systematic 12-pole in the quadrupoles and they are assumed to bring its strength down of a factor 10 at least; the results obtained with two compensation schemes are presented here: the first in which correction windings are implemented only in the Interaction Region Triplets, and the second in which correction is applied in all quadrupoles.

Quadrupoles with a larger bore (1990 errors) must be used, with dodecapolar correction windings, in order to satisfy the criterion of Eq. (2).

Table 8: Linear and Dynamic Apertures in the presence of Systematic Errors in the Quadrupoles

Systematic Multipolar Errors in the Quadrupoles APERTURE with Different Criteria — Injection — Round Beam Normalized Amplitudes in mm ±0.02 mm						
	LINEAR APERTURE				DYNAMIC AP.	
CRITERION	$\|\Delta Q_{x,y}\| \leq 0.005$		$Sm_{x,y} \leq 0.10$		400 t. over-flow	
$\|\Delta p/p\|$ ($\times 10^{-3}$)	0	1.25	0	1.25	0	1.25
<u>1989 Systematic Errors</u>						
Errors + Chromaticity Correction	0.34	0.27	0.50	0.42	0.70	0.68
+ 12-pole Corr. in IRT Quadrup.	0.56	0.53	0.58	0.60†	0.60	0.60
+ 12-pole Corr. in All Quadrup.	0.54	0.54	0.54	0.54	0.68	0.62
<u>1990 Systematic Errors</u>						
Errors + Chromaticity Correction	0.57	0.53	0.50	0.47	0.73	0.78
+ 12-pole Corr. in IRT Quadrup.	0.70†	0.72	0.70†	0.67	0.70	0.72
+ 12-pole Corr. in All Quadrup.	0.70†	0.66†	0.70†	0.66†	0.70	0.66

†linear ap. threshold is not reached, particles are lost before; dynamic ap. is shown.

5 Compensation of Random Imperfections

On the assumption that the random multipole components of the magnetic imperfections do not depend on the manufacturing sequence, we aim to order the LHC magnets in such a way that the magnetic errors are approximately self-compensated in the two rings.

The ordering rules are based on the following very demanding hypotheses:

1. all the main dipoles as well as the main quadrupoles are of a modular design and are thus interchangeable;

2. at least the sextupolar component of the magnetic field in the main dipoles and quadrupoles is known with high precision;

3. all the magnets of a full octant are available at a time in a large storage area, and can be installed in the LEP tunnel in any possible sequence.

Each octant will be separately ordered, thus 220 dipoles, 24 QF and 24 QD cell quadrupoles will be concerned at a time. The magnets will be ordered according to the value of their sextupolar random error. The compensation is required to be as local as possible. The basic rules are:

1. two consecutive dipoles have approximately equal and opposite sextupolar content;

2. two dipoles 2π apart in phase have approximately equal and opposite sextupolar content;

3. dipoles without a partner 2π apart in phase have approximately zero sextupolar content;

4. QF (QD) quadrupoles 2π apart in phase have approximately equal and opposite sextupolar content.

To fulfil these rules the dipoles and the quadrupoles are classified in families depending on the value of the sextupole error in the two magnetic channels (two-dimensional binning). There are 16 families of dipoles, four of QF, and four of QD. The particular sequence in which the magnets are extracted has been optimized in Refs. 11 and 12 by a "trial and error" method. Considerations that support our choice of the ordering rules are presented in Ref. 13.

The insertion quadrupoles are assumed to be non-interchangeable, and are thus in random sequence. The separating dipoles near each crossing point are assumed to be perfect.

5.1 Computational Tools

The particle-tracking simulations were carried out along the test lattice with imperfections using the code FASTRAC[14] to identify the dynamic aperture of a round beam over 400 revolutions. The linear aperture based on smear was also determined; in each plane the threshold was chosen, as before, at 10% r.m.s. relative variation.

No physical aperture limit was used around the beam axis. Appropriate routines to generate samples of random errors and to assign them to the dipoles and quadrupoles in a given ordered sequence, were coded and included in FASTRAC. The random components considered were set with random Gaussian distributions centered around zero and truncated at $\pm 3\sigma$. Random multipole components of different orders, or located in different rings, were supposed to be statistically uncorrelated. To improve the statistical meaning of our results we considered 20 different samples of the random multipoles components, generated with 20 different seeds in the random number generator routine. Tracking for off-momentum particles was done with fixed $\Delta p/p$, excluding synchrotron oscillations. Also in this case we chose $\Delta p/p = \delta = 1.25 \times 10^{-3}$ at injection. The magnetic imperfections in each dipole and quadrupole were represented by a set of thin-lens multipoles, all located in the middle of the magnetic length.

The dynamical quantities of interest were evaluated by the average value and the spread over the 20 samples.

The values of the "ensured linear aperture", namely the lower edge of the spread of the linear aperture over the 20 samples, is considered to be the best indicator of the detrimental effect of the non-linearities. Alternatively, the value of the "ensured dynamic aperture" can be used. These two quantities, indeed, are representative of the behaviour of the machine in the most pessimistic conditions that can be expected.

5.2 Results of the Simulations

The beneficial effects of ordering the LHC magnets are substantially impaired by the two-in-one design of the magnets, by the presence of higher-order random multipole imperfections, and to a smaller extent by the imprecise knowledge of the random sextupole imperfections. This has been shown in Refs. 11 and 12, by tracking on-momentum particles with perfect closed orbit, and by neglecting the systematic errors and the multipolar correctors. An additional reduction of the dynamic and linear aperture appears for off-momentum particles and to a smaller extent for finite closed orbit. The feed-down effects of the systematic imperfections and of the multipolar correctors in combination with the finite closed orbit are small.

5.2.1 On-Momentum, On-Axis Particle Tracking

In a single bore ring, with only sextupolar magnetic errors in the injection optics, the ensured dynamic aperture of on-momentum particles with perfect closed orbit is increased by a factor of two, from 0.9 to 1.8 mm, by applying the ordering rules. With higher-order multipolar imperfections and two-in-one magnets, the improvement is reduced to about a factor 1.25, from 0.8 to 1.1 mm, but it holds in both rings. Changing the tune working point can be beneficial; for instance with $Q_x = 70.15$, $Q_y = 70.18$, the ensured dynamic aperture increases by about a factor of 1.2, both for random and ordered installation of the magnets.

When the collision optics is considered, the dynamic aperture is determined by the imperfections of the inner triplets, where the value of the β-functions is maximum. With perfect quadrupoles the ensured aperture is 0.7 mm and becomes 0.9 mm by applying the ordering rules. With realistic quadrupoles the ensured aperture drops dramatically down to 0.1 mm, and the ordering rules can only increase it to 0.2 mm.

Changing the crossing angle α can also have a strong effect, both with injection and collision optics. The previous results, obtained with $\alpha = 96$ μrad, are practically the same as with head-on crossing; larger values of α, instead, enhance the effect of the errors in the inner triplets and reduce the dynamic aperture: at 200 and 300 μrad the reduction factors are 0.9 and 0.8 respectively for both optics. To reduce the long-range beam–beam tune shift a crossing angle of 300 μrad is deemed necessary[15]; therefore, this value has been used for the simulations of the next section.

5.2.2 *Methodical Particle Tracking*

Methodical tracking simulations have been performed with injection optics, two-in-one magnets with 50 mm coil inner diameter and crossing angles of $\alpha = 300$ μrad, the results of which are summarized in Table 9 (particles circulate clockwise and anticlockwise in ring No. 1 and ring No. 2 respectively). An ideal bare machine, chromaticity corrected, has a very large value (3.7 mm) of the dynamic aperture, well outside the vacuum chamber, whose radius of 20 mm corresponds to a normalized amplitude of 1.54 mm at $\beta = \beta_{max} = 169$ m, and a linear aperture of 1 mm. By including the systematic and the random multipole components in all the dipoles and quadrupoles, the detrimental effect on the particle motion is dramatic, especially off-momentum (Figure 4). When the systematic errors are neglected, the situation is slightly improved and the effect of the ordering rules is significant (compare Figure 5 and Figure 6). Similar results are obtained when the systematic errors are compensated by the lumped Simpson correctors (Figure 7). A finite closed orbit provokes a further reduction of the ensured linear and dynamic aperture (Figure 8). The size of the reduction seems to be significantly related to the size of the peak orbit distortion. For instance the ensured linear aperture changes by less than 0.10 mm and less than 0.16 mm for on- and off-momentum particles respectively, for a peak orbit distortion of $1.3/\sqrt{\beta_{max}} = 0.1$ mm and a peak dispersive distortion $D_{max}\delta/\sqrt{\beta_{max}} = 0.18$ mm.

The superposition of the quadrupole systematic errors to the machine with

Table 9: Linear and Dynamic Apertures in Different Configurations of the LHC.

MACHINE DEFINITION	LINEAR APERTURE based on SMEAR Normalized Amplitudes in mm ±0.02 mm									Short-Term DYNAMIC APERTURE Normalized Amplitudes in mm ±0.02 mm								
	$\Delta p/p = -1.25 \times 10^{-3}$			$\Delta p/p = 0$			$\Delta p/p = +1.25 \times 10^{-3}$			$\Delta p/p = -1.25 \times 10^{-3}$			$\Delta p/p = 0$			$\Delta p/p = +1.25 \times 10^{-3}$		
	A_{min}	A_{max}	A_{ave}	A_{min}	A_{max}	A_{ave}	A_{min}	A_{max}	A_{ave}	A_{min}	A_{max}	A_{ave}	A_{min}	A_{max}	A_{ave}	A_{min}	A_{max}	A_{ave}
All Multipole Errors																		
Rand. <u>not</u> Sorted, Syst. <u>not</u> Comp.																		
Ring 1	<0.17	0.30	**0.24**	0.32	0.61	**0.47**	<0.15	0.21	**0.17**	0.34	0.59	**0.48**	0.63	0.76	**0.70**	0.38	0.71	**0.50**
Ring 2	<0.20	0.29	**0.25**	0.34	0.68	**0.52**	0.15	0.22	**0.17**	0.40	0.59	**0.51**	0.61	0.78	**0.69**	0.36	0.78	**0.50**
Random Multipole Errors in Dipoles and Quadrupoles																		
Not Sorted — Ring 1	0.32	0.63	**0.47**	0.33	0.61	**0.48**	0.34	0.62	**0.47**	0.48	0.76	**0.67**	0.63	0.72	**0.68**	0.49	0.79	**0.65**
Not Sorted — Ring 2	0.38	0.62	**0.52**	0.41	0.64	**0.52**	0.44	0.64	**0.51**	0.59	0.82	**0.68**	0.55	0.74	**0.66**	0.51	0.74	**0.63**
Sorted — Ring 1	0.52	0.69	**0.62**	0.53	0.71	**0.62**	0.53	0.71	**0.60**	0.63	0.79	**0.70**	0.63	0.76	**0.70**	0.59	0.76	**0.68**
Sorted — Ring 2	0.53	0.69	**0.60**	0.49	0.71	**0.62**	0.48	0.67	**0.59**	0.55	0.82	**0.69**	0.63	0.78	**0.69**	0.59	0.76	**0.68**
+ Systematic Multip. Errors in Dipoles - (Random Sorted)																		
Not Compensated — Ring 1	<0.24	0.36	**0.28**	0.58	0.70	**0.63**	<0.15	0.35	**0.23**	0.49	0.68	**0.58**	0.63	0.85	**0.72**	0.46	0.75	**0.62**
Not Compensated — Ring 2	<0.25	0.35	**0.28**	0.56	0.75	**0.64**	<0.15	0.35	**0.22**	0.50	0.68	**0.58**	0.63	0.82	**0.72**	0.44	0.74	**0.60**
Compensated — Ring 1	0.44	0.64	**0.57**	0.49	0.70	**0.61**	0.48	0.65	**0.57**	0.55	0.68	**0.63**	0.63	0.82	**0.72**	0.59	0.79	**0.68**
Compensated — Ring 2	0.52	0.63	**0.57**	0.43	0.72	**0.61**	0.51	0.66	**0.59**	0.59	0.68	**0.64**	0.59	0.82	**0.69**	0.59	0.74	**0.67**
+ Closed Orbit Residue (Random Sorted, Syst. Comp.)																		
Ring 1	0.37	0.59	**0.50**	0.39	0.64	**0.55**	0.38	0.61	**0.51**	0.46	0.66	**0.58**	0.49	0.74	**0.63**	0.49	0.74	**0.61**
Ring 2	0.36	0.60	**0.50**	0.37	0.67	**0.56**	0.38	0.62	**0.51**	0.44	0.74	**0.58**	0.47	0.74	**0.65**	0.48	0.69	**0.61**
Systematic Multipole Errors in Quadrupoles (Random Sorted, Syst. Comp.)																		
Ring 1	0.48	0.63	**0.55**	0.50	0.61	**0.57**	0.36	0.59	**0.51**	0.58	0.66	**0.62**	0.55	0.70	**0.62**	0.51	0.67	**0.58**
Ring 2	0.47	0.63	**0.56**	0.51	0.61	**0.58**	0.41	0.58	**0.52**	0.57	0.68	**0.62**	0.55	0.70	**0.62**	0.48	0.66	**0.58**
+ Closed Orbit Residue																		
Ring 1	0.34	0.59	**0.50**	0.34	0.63	**0.50**	0.32	0.56	**0.47**	0.44	0.74	**0.58**	0.44	0.66	**0.58**	0.42	0.61	**0.55**
Ring 2	0.31	0.59	**0.50**	0.32	0.63	**0.51**	0.31	0.54	**0.48**	0.40	0.74	**0.59**	0.42	0.69	**0.58**	0.40	0.65	**0.56**

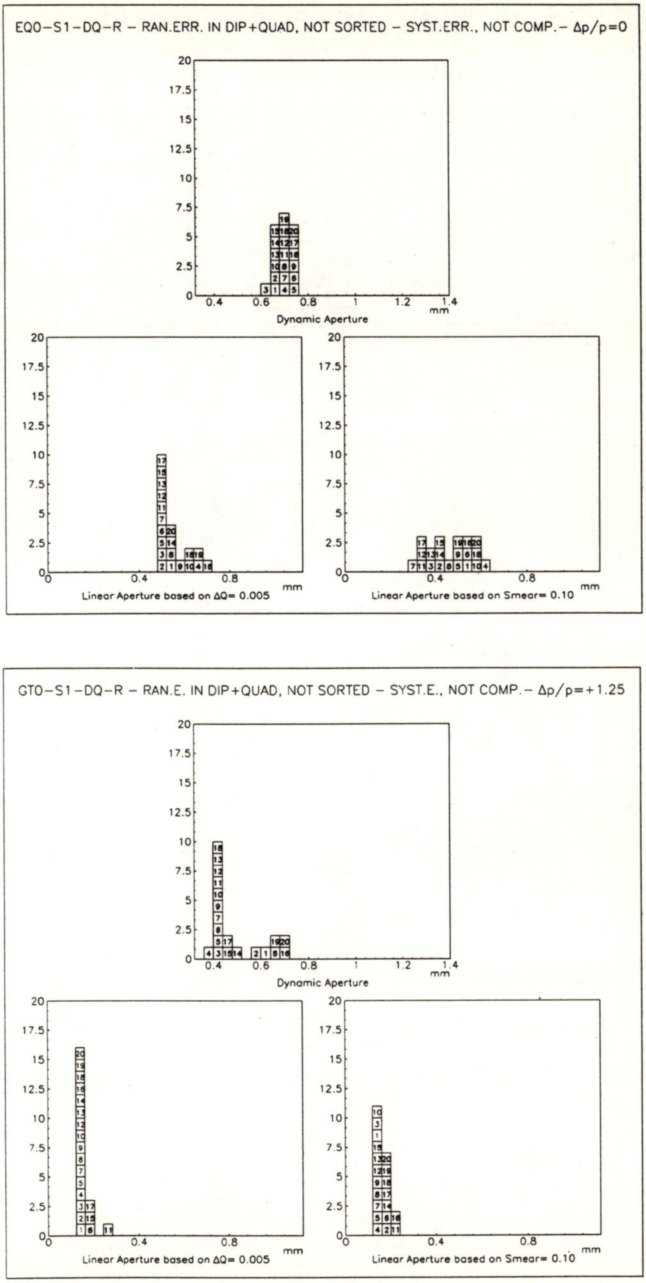

Figure 4: Dynamic and Linear Apertures for 20 Machines with Random and Systematic Errors not Compensated (nr = threshold not reached)

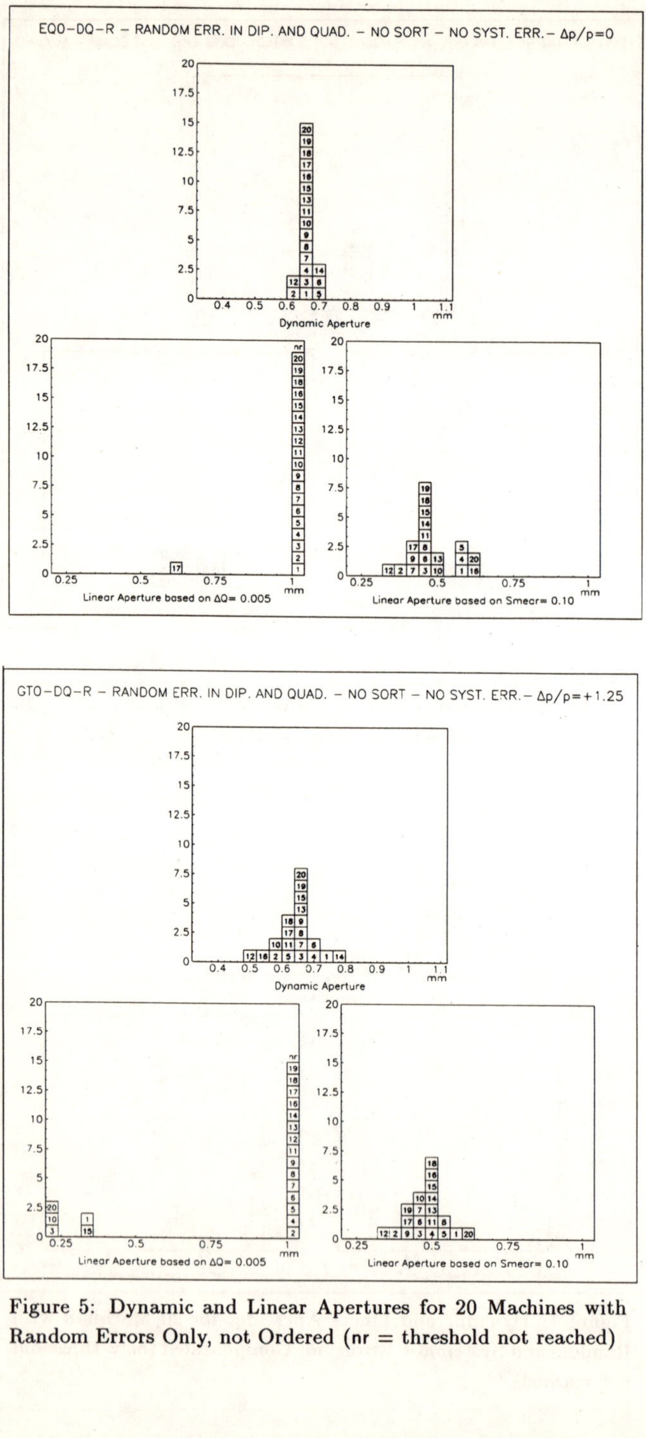

Figure 5: Dynamic and Linear Apertures for 20 Machines with Random Errors Only, not Ordered (nr = threshold not reached)

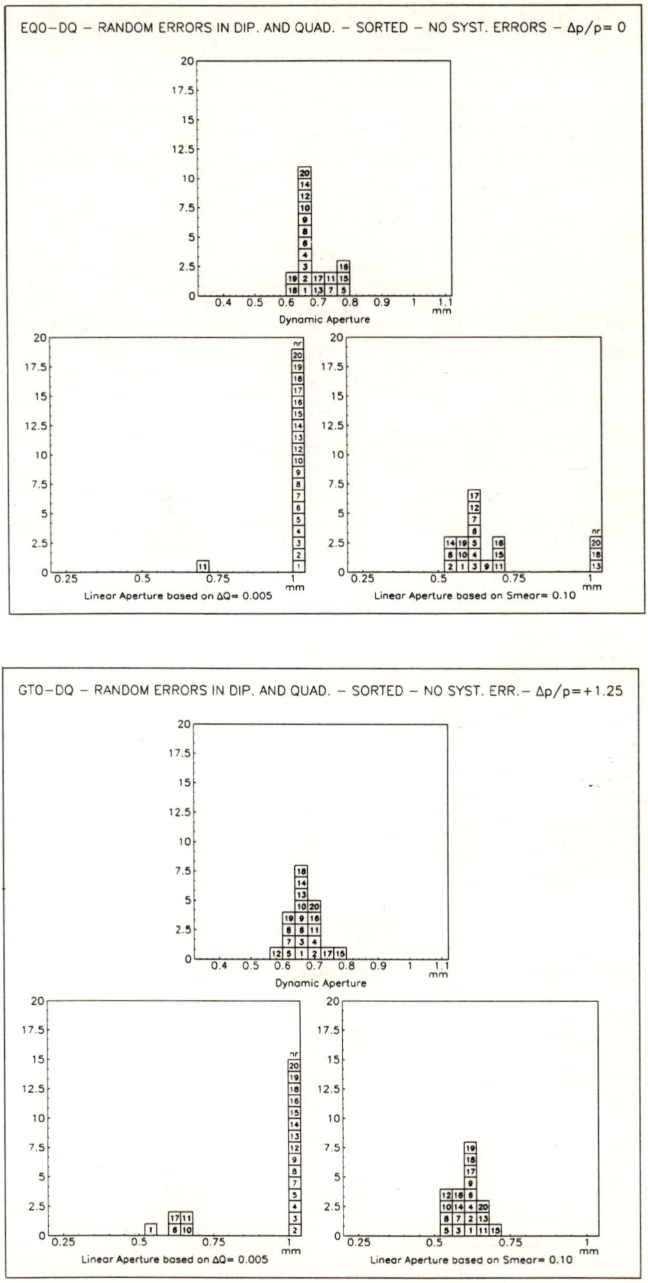

Figure 6: Dynamic and Linear Apertures for 20 Machines with Random Errors Only, Ordered (nr = threshold not reached)

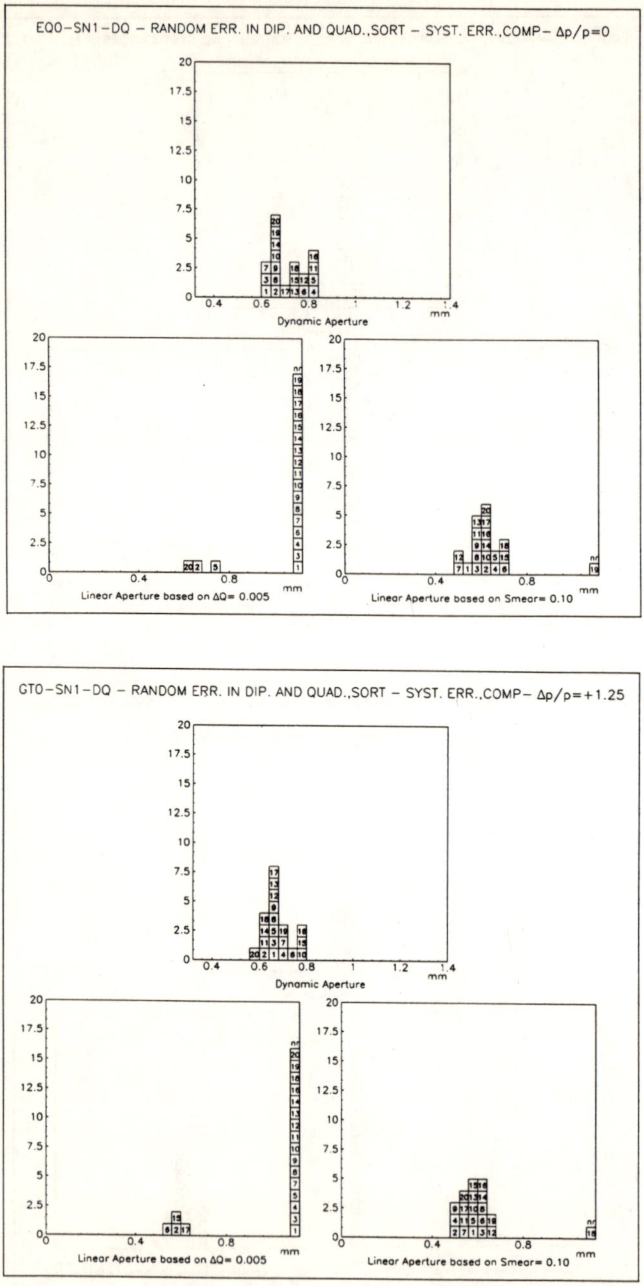

Figure 7: Dynamic and Linear Apertures for 20 Machines with Ordered Random Errors and Compensated Systematic Errors (nr = threshold not reached)

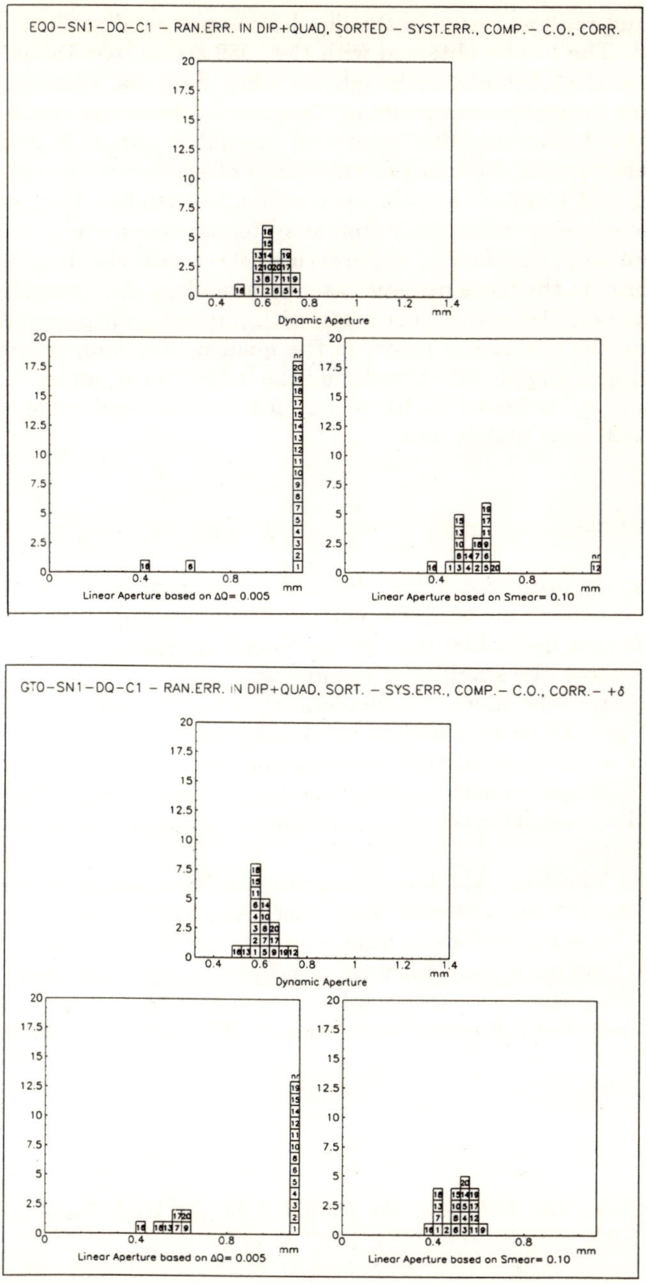

Figure 8: Dynamic and Linear Apertures for 20 Machines with Ordered Random Errors and Compensated Systematic Errors, plus Closed Orbit Residue (nr = threshold not reached)

systematic and random errors in the dipoles compensated at their best has also been studied. The results obtained with the 1989 errors (see Table 3) are shown in the third part of Table 9: although the effect does not look so severe like in Table 8 where the systematic errors in the quadrupoles where studied alone, still a substantial reduction in both linear and dynamic aperture is noticeable. The same simulations with the more recent values of systematic errors (1990 errors of Table 3) might yield more optimistic results, nevertheless local compensation by bore-tube windings of the dodecapolar systematic component has to be done.

In summary, the effect of the ordering rules is still visible in all the considered situations. In the more realistic case, with random and systematic imperfections in all the dipoles and quadrupoles, with lumped multipolar compensation, and assuming the systematic errors in the quadrupoles compensated up to the point of having negligible effect, ordering the magnets may improve the ensured linear aperture by at least a factor of 1.16 for both on- and off-momentum particles, from 0.31 mm to 0.38 mm.

6 Conclusions

The study presented follows from the short-term analysis of the particle motion in the LHC. This is by far not an exhaustive approach, although some powerful concepts used, like that of the linear aperture, give very interesting informations about the stability of the motion.

At injection the short term dynamic aperture needed in the LHC is of the order of 7σ, i.e. 3σ to accomodate the beam and 4σ to take into account the imperfections of the injection process (position, angle, phase and energy errors). This has to be compared with the computed short term aperture of on-momentum particles without closed orbit errors of Table 9, which is about 6.7σ, and then almost adequate.

On the long term, the linear aperture available taking into account orbit and momentum deviation is of at least 4.3σ. Although no safety margin is left, this seems to be barely sufficient to accomodate the proton beam.

It still remains to be confirmed, also by means of long-term tracking simulations, whether the aperture available is sufficient to ensure a safe operation of the LHC over the full operational cycle, and, if not, how the errors can be diminished.

References

1. W. Scandale, *The Lattice of the CERN–LHC for High Luminosity pp Collisions*, CERN–SPS 88–6 (AMS), LHC Note 68.

2. J. Gareyte, *The LHC: General Parameters and Magnetic Errors*, these Proceedings, and also CERN SL/90–90 (AP), LHC Note 130.

3. G. Brianti and K. Hübner eds., *The Large Hadron Collider in the LEP Tunnel*, CERN 87–05, 1987.

4. T. Tortschanoff and R. Wolf, *Effects of the Persistent Currents Computed for the LHC Lattice Quadrupoles*, CERN Group Report SPS/EMA 89–11, LHC Note 99.

5. W. Scandale, *Numerical Results on the Effect of the Parasitic Multipoles in the Superconducting Dipoles of the Large Hadron Collider*, CERN–SPS 87–28 (AMS), LHC Note 57.

6. F. Galluccio and W. Scandale, *Compensation of the LHC Lattice in the Presence of Systematic Multipolar Errors in the Dipoles*, CERN–SPS 89–51 (AMS), LHC Note 121.

7. F. Galluccio, *Compensation of Systematic Multipolar Errors in the LHC from the Operational Point of View*, SL/AP/Note/90–12, LHC Note 133.

8. D. Neuffer, *Multipole Correction in Large Synchrotrons*, in Proceedings of Second Advanced ICFA Beam Dynamics Workshop, CERN 88–04, 1988, p. 179.

9. W. Scandale, F. Schmidt and E. Todesco, *Multipolar Correctors for the LHC: an Analytical Approach with Normal Forms*, in Proceedings of the 2nd European Particle Accelerator Conference, Nice, June 12–16, 1990.

10. F.C. Iselin and J. Niederer, *The MAD Program Version 7.2, User's Reference Manual*, CERN/LEP–TH/88–38.

11. W. Scandale and L. Wang, *Towards a Procedure for Ordering LHC Dipoles in Small Groups*, CERN SPS/89–22 (AMS), LHC Note 95.

12. W. Scandale and L. Wang, *Installing Two-in-one LHC Magnets in Ordered Sequences*, CERN SPS/89–41 (AMS), LHC Note 107.

13. S. Ohnuma, *Frequency-Domain Considerations in Magnet Sorting*, in Proceedings of Second Advanced ICFA Beam Dynamics Workshop, CERN 88–04, 1988, p. 218.

14. E. Forest and B.T. Leemann, *Brief Description of the Tracking Codes FASTRAC and THINTRAC*, SSC–133, 1987.

15. W. Herr, *Tune Shifts and Spread Due to Short and Long Range Beam-Beam Interaction in the LHC*, CERN/SL 90–06, LHC Note 119.

A Realistic Lattice and Cell Lattice: a Comparison

Tracking simulations in such a large and complicated machine like the LHC can be very computer time consuming.

The possibility of using a simplified lattice, with a high level of symmetry, namely a so-called "Cell Lattice", is therefore very attractive to speed up the tracking process.

A cell lattice for the LHC has been built, starting from the realistic lattice described in Section 2, taking away the insertion lines with also the dispersion suppressors, and slightly readjusting the dipole bending angle to make the machine close, and the quadrupole gradients to match the fractional part of the tunes.

With this choice it is not possible to match the integer part of the original tune and consequently its position with respect to structure resonances: to do this, at least another 80 cells with 640 more dipoles should be added, but this would lead to an excess of error sources in the ring.

The resulting lattice is made of 196 identical cells and its tunes are $Q_x = 50.28$ and $Q_y = 50.31$.

It has been tested by tracking and compared with the so-called "realistic" lattice of Section 2 with several error configurations.

In Table A.1 the results of this comparison are shown for the cases of bare machine, of systematic errors in the dipoles, of random errors in the dipoles and in the quadrupoles, and for the case of both systematic and random errors in the dipoles and random errors in the quadrupoles.

There is no clear relation between the apertures found with the two lattice, but in all cases those obtained with the cell lattice are largely overestimated, and especially the dynamic aperture, which in the bare machine case is even 3.5 times bigger than with the realistic lattice.

The results obtainable by using a cell lattice are therefore unreliable in the case of the LHC, and such a model has been abandoned.

Its use might indeed lead to too optimistic, and therefore dangerous, specifications of the tolerances of the LHC magnet fields.

Table A.1: Comparison Between Realistic and Cell Lattice: Linear and Dynamic Apertures

	LINEAR APERTURE based on SMEAR Normalized Amplitudes in mm ±0.02 mm									Short-Term DYNAMIC APERTURE Normalized Amplitudes in mm ±0.02 mm								
MACHINE DEFINITION	$\Delta p/p = -1.25 \times 10^{-3}$			$\Delta p/p = 0$			$\Delta p/p = +1.25 \times 10^{-3}$			$\Delta p/p = -1.25 \times 10^{-3}$			$\Delta p/p = 0$			$\Delta p/p = +1.25 \times 10^{-3}$		
	A_{min}	A_{max}	A_{ave}	A_{min}	A_{max}	A_{ave}	A_{min}	A_{max}	A_{ave}	A_{min}	A_{max}	A_{ave}	A_{min}	A_{max}	A_{ave}	A_{min}	A_{max}	A_{ave}
Bare Machine																		
Realistic Lattice			**1.00**			**1.00**			**1.01**			**3.62**			**3.90**			**3.97**
Cell Lattice			**1.64**			**1.64**			**1.64**			**14.0**			**14.0**			**14.0**
System. Errors in Dipoles																		
Realistic Lattice			**<0.35**			**0.38**			**<0.35**			**1.02**			**1.37**			**1.02**
Cell Lattice			**0.22**			**0.61**			**<0.20**			**1.61**			**1.96**			**1.52**
Random Multipole Errors in Dipoles and Quadrupoles																		
Realistic Lattice — Ring 1	0.32	0.63	**0.47**	0.33	0.61	**0.48**	0.34	0.62	**0.47**	0.48	0.76	**0.67**	0.63	0.72	**0.68**	0.49	0.79	**0.65**
Realistic Lattice — Ring 2	0.38	0.62	**0.52**	0.41	0.64	**0.52**	0.44	0.64	**0.51**	0.59	0.82	**0.68**	0.55	0.74	**0.66**	0.51	0.74	**0.63**
Cell Lattice — Ring 1	0.42	0.99	**0.68**	0.41	1.05	**0.69**	0.40	1.04	**0.69**	0.89	1.38	**1.17**	0.95	1.47	**1.21**	0.95	1.37	**1.20**
Cell Lattice — Ring 2	0.43	0.99	**0.68**	0.42	1.05	**0.70**	0.41	1.05	**0.69**	0.89	1.38	**1.19**	0.94	1.51	**1.22**	0.92	1.41	**1.21**
Rand. Err. in Dip.+Quad. System. Errors in Quadrup.																		
Realistic Lattice — Ring 1	<0.17	0.30	**0.24**	0.32	0.61	**0.47**	<0.15	0.21	**0.17**	0.34	0.59	**0.48**	0.63	0.76	**0.70**	0.38	0.71	**0.50**
Realistic Lattice — Ring 2	<0.20	0.29	**0.25**	0.34	0.68	**0.52**	0.15	0.22	**0.17**	0.40	0.59	**0.51**	0.61	0.78	**0.69**	0.36	0.78	**0.50**
Cell Lattice — Ring 1	0.22	0.35	**0.27**	0.37	0.64	**0.51**	0.14	0.35	**0.21**	0.44	0.98	**0.64**	0.70	1.28	**0.90**	0.44	0.81	**0.66**
Cell Lattice — Ring 2	0.22	0.35	**0.28**	0.38	0.64	**0.51**	0.14	0.35	**0.20**	0.47	0.92	**0.66**	0.70	1.31	**0.89**	0.43	0.81	**0.66**

STRONG AND WEAK INSTABILITIES IN A 4–D MAPPING MODEL OF FODO CELL DYNAMICS

Tassos Bountis
Mathematics Department, University of Patras
Patras 26110, Greece

and

Stathis Tompaidis
Physics Department, University of Texas at Austin
Austin, Texas 78712

ABSTRACT

Periodic solutions of a 4-dimensional (4–D) mapping model of a FODO cell are obtained and their stability is studied. It is found that near such *unstable* periodic orbits of low period, there exist *chaotic* regions of strong instability in the 4-dimensional space x_n, x_{n+1}, y_n, y_{n+1}, through which orbits escape very quickly to infinity. On the other hand, near 2–dimensional orbits in the x_n, x_{n+1} plane (i.e., in the vicinity of "flat" beams with $y_n \equiv 0$) stability conditions are obtained for which the y_n oscillations do not grow appreciably even if x_1, x_0 are chosen within the chaotic layer of an associated unstable 2–D orbit. In the latter case, evidence of weak instabilities, or *Arnol'd diffusion* is found and diffusion coefficients are calculated ($\sim 10^{-11}$) and compared with the ones obtained ($\sim 10^{-19}$), when x_1, x_0 are chosen within a region of oscillatory (quasiperiodic) motion.

I. Introduction

The stability of particle beams in high energy accelerators is a problem of great practical concern to which the methods and techniques of Nonlinear Dynamics are expected to make a significant contribution[1−6].

Two are the major types of nonlinear phenomena that can cause serious blow–up effects and significantly decrease the beams' luminosity, over short (strong instabilities) or long times (weak instabilities): The beam–beam interaction and sextupole (or higher multipole) nonlinearities due to residual currents in the super–conducting magnets.

The effects of the beam–beam interaction have been discussed by J. Tennyson and many other authors, in this and other volumes [1−4]. Particle transport through resonances, flip–flop, and other related phenomena have been analyzed, yielding appropriate tune and tune–shift values at which the beam–beam interaction should not pose a major threat to the safe operation of intersecting storage rings[1−4,6].

Particularly in the case of colliding $p - \bar{p}$ beams, (where radiation damping and quantum fluctuation effects may be considered negligible), even *weak instabilities* produced by Arnol'd diffusion[7], are not expected to significantly affect beam lifetimes[8]. *Strong instabilities* on the other hand (i.e., rapid escape of orbit, to infinity via large chaotic regions), simply do not occur in the beam–beam problem, since the beam–beam force drops to zero (as r^{-2}) away from the origin and particle orbits remain bounded apparently for all time[6,8].

In this paper, we shall treat the effect of sextupole nonlinearities on hadron beams passing through a FODO cell, composed of a dipole and two quadrupoles, focusing the particles' motion in the horizontal $(-x)$ and vertical $(-y)$ direction[5]. Unlike the beam–beam interaction, the (quadratric) nonlinear forces here increase "monotonically" in magnitude away from the origin and may cause strong (as well as weak) instabilities limiting significantly the beams' dynamical aperture.

Treating the sextupole nonlinearity as concentrated at *one point* in the cell, the dynamics of $p(\bar{p})$ particles passing through the cell, can be described by the Hamiltonian

$$H = \frac{1}{2}(p_x^2 + p_y^2 + q_x^2 x^2 + q_y^2 y^2) + \epsilon(\frac{x^3}{3} - xy^2)\delta_{2\pi}(t), \tag{1}$$

where q_x, q_y are the betatron frequencies (or "tunes") in the x and y directions, ϵ is the strength of the nonlinearity and $\delta_{2\pi}(t)$ the 2π–periodic δ-function

$$\delta_{2\pi}(t) = \frac{1}{2\pi} \sum_{k=-\infty}^{\infty} \cos kt \; . \tag{2}$$

In this work, we shall actually work with the 4–dimensional (4–D) mapping to which Hamilton's equations derived from Eq. 1 *rigorously* reduce[9]:

$$x_{n+1} = 2x_n \cos\omega_x - x_{n-1} - (\epsilon \sin\omega_x/q_x)(x_n^2 - y_n^2) \tag{3a}$$

$$y_{n+1} = 2y_n \cos\omega_y - y_{n-1} + (2\epsilon \sin\omega_y/q_y)x_n y_n \tag{3b}$$

describing the x_n, y_n displacements of the particle after its nth passage through the cell ($\omega_x = 2\pi q_x$, $\omega_y = 2\pi q_y$).

In section 2, we construct some fundamental m-periodic orbits of Eqs. 3 ($m = 3, 4, 5, \ldots$) in 4 dimensions, and find that when they are *unstable* (with respect to small perturbations) they have *large* chaotic regions about them through which particles can *escape to infinity* after a very small number of iterations.

On the other hand, in section 3, we show that weak instabilities can also occur near unstable 2-D orbits of Eq. 3a ($\hat{x}_n = \hat{x}_{n+m}, \hat{y}_n \equiv 0$). In particular, we find at some specific tune values that placing our initial x_1, x_0 within a chaotic layer of one such orbit, (small) y_n oscillations exhibit a slow amplitude growth that may be characterized as Arnol'd diffusion [7]. Diffusion coefficients for such a growth were calculated and found to be 8-9 orders of magnitude *larger* than the corresponding ones, obtained when x_1, x_0 are located near a stable 2-D orbit.

Finally, in section 4, we offer some concluding remarks and point out that the instabilities discussed in this paper need to be further studied in the presence of an additional factor, which is expected to *enhance* them and thus increase their damaging effect on the beams' life–time: This factor is the so–called *synchrotron oscillations*[1–4] occuring in the *longitudinal direction* (i.e., along the beam) due to the particles being accelerated in that direction by the rf cavities.

These oscillations are actually seen to produce a slow *modulation* in the horizontal and vertical oscillation frequencies of Eq. 1, which can be modelled by

$$q_{x,y} = q_{1,2}(1 + \lambda \cos \Omega t), \ |\lambda| << 1, \ \Omega << 1 \ . \tag{4}$$

Thus, their overall effect may be studied by the theoretical methods of *modulation diffusion*, developed in recent years by Chirikov and co–workers[10]. This approach is currently under investigation and results are expected to appear in future publications[11].

2. 4–Dimensional Periodic Orbits and Stability

The transfer map for a particle's horizontal $(-x)$ and vertical $(-y)$ position and momentum variables, as it passes through a single FODO cell, may be written in the form[5]

$$\begin{pmatrix} x' \\ p'_x \\ y' \\ p'_y \end{pmatrix} = \mathbf{M} \begin{pmatrix} x \\ p_x + \frac{1}{2} k l_d (x^2 - y^2) \\ y \\ p_y - k l_d x y \end{pmatrix} \tag{5}$$

where k is the strength and l_d the length of the cell's dipole

$$\mathbf{M} = \mathbf{U}\,\mathbf{R}\,\mathbf{U}^{-1} \ , \ \mathbf{U} = \begin{pmatrix} \mathbf{U}_x & \mathbf{O} \\ \mathbf{O} & \mathbf{U}_y \end{pmatrix} \ , \ \mathbf{R} = \begin{pmatrix} \mathbf{R}(\omega_x) & \mathbf{O} \\ \mathbf{O} & \mathbf{R}(\omega_y) \end{pmatrix}$$

$$\mathbf{U}_{x,y} = \begin{pmatrix} \beta_{x,y}^{\frac{1}{2}} & 0 \\ \gamma_{x,y} & \beta_{x,y}^{-\frac{1}{2}} \end{pmatrix} , \ \mathbf{R}(\alpha) = \begin{pmatrix} \cos\alpha & -\sin\alpha \\ \sin\alpha & \cos\alpha \end{pmatrix} \tag{5a}$$

$\beta_{x,y}(s)$ are the betatron functions (s is the coordinate *along* the particle's ideal circular) path around the ring, $\gamma_{x,y} = \beta'_{x,y}(s)/2\beta_{x,y}^{-\frac{1}{2}}$, and

$$\omega_x = 2\pi q_x \ , \ \omega_y = 2\pi q_y \ , \tag{6}$$

q_x, q_y being the betatron frequencies or *"tunes"* of the x- and y-oscillations respectively, caused by the (linear) quadruple fields. Clearly, we have assumed in Eq. 5 that the sextupole nonlinearity is concentrated at the midpoint of the cell and serves to alter the momentum (but not the position) of the particle, by an instantaneous "kick".

Substituting Eq. 5a in Eq. 5 and eliminating the momentum variables p_x and p_y, we arrive after a little algebra, at the *second order* difference equations

$$\begin{aligned} x_{n+1} &= 2x_n \cos\omega_x - x_{n-1} - \frac{1}{2} k\, l_d \beta_x \sin\omega_x (x_n^2 - y_n^2), \\ y_{n+1} &= 2y_n \cos\omega_y - y_{n-1} + k\, l_d \beta_y \sin\omega_y x_n y_n \ , \end{aligned} \tag{7}$$

$n = 0, 1, 2, \ldots$, yielding the *new* x_{n+1} , y_{n+1} displacements after the nth passage of the particle through the cell (and its nth rotation around the ring).

Even though the betatron functions β_x, β_y are periodic functions of the distance s along the ring, they do not significantly vary about their mean values, so we may take

$$\beta_{x,y} = L/(2\pi q_{x,y}) = R_{eff} q_{x,y}^{-1} \tag{8}$$

where L is the length of the circumference and R_{eff} the effective radius of the ring. Defining now

$$c_{x,y} \equiv \cos\omega_{x,y} \quad , \quad s_{x,y} \equiv \sin\omega_{x,y} \ , \tag{9}$$

and

$$A = 2(k\, l_d \beta_y\, s_y)^{-1} \quad , \quad B \equiv 2(k^2 l_d^2 \beta_x \beta_y s_x s_y)^{-\frac{1}{2}} \ , \tag{10}$$

we can scale our x_n, y_n variables to

$$x_n = AX_n \quad , \quad y_n = BY_n \tag{11}$$

and rewrite our mapping equations in the simplified form

$$X_{n+1} = 2c_x X_n - X_{n-1} - \rho X_n^2 + Y_n^2 \ , \tag{12a}$$
$$Y_{n+1} = 2c_y Y_n - Y_{n-1} + 2X_n Y_n \ , \tag{12b}$$

with

$$\rho \equiv \beta_x s_x / \beta_y s_y \ . \tag{12c}$$

Note that Eqs. 12 constitute a 4–dimensional mapping

$$\mathbf{X}_{n+1} = T(\mathbf{X}_n) \quad , \quad \mathbf{X}_n \equiv (X_n, X_{n-1}, Y_n, Y_{n-1}) \; , \tag{13}$$

with only two parameters q_x, q_y.

The variational equations of this mapping about an orbit $\hat{\mathbf{X}}_n$ are found by substituting $\mathbf{X}_n = \hat{\mathbf{X}}_n + \mathbf{\Delta X}_n$ in Eq. 13 and linearizing

$$\mathbf{\Delta X}_{n+1} = \mathbf{J}(\hat{\mathbf{X}}_n)\mathbf{\Delta X}_n \; , \tag{14}$$

where $\mathbf{J}(\hat{\mathbf{X}}_n)$ is the Jacobian matrix of T. Clearly, since $|c_{x,y}| \leq 1$, cf. Eq. 9, the origin $\hat{\mathbf{X}}_n = \mathbf{O}$ is (linearly) stable for all q_x, q_y.

We now seek periodic orbits of Eq. 12 in the form of Fourier polynomials[12],

$$X_n = \sum_k A_k e^{i\omega kn} \quad , \quad Y_n = \sum_k B_k e^{i\omega kn} \; , \tag{15}$$

with frequency

$$\omega/2\pi = m_1/m_2 \quad , \quad -\pi < \omega \leq \pi \; , \tag{16}$$

where k in Eq. 11 takes integer values such that

$$-\pi < k\omega \leq \pi \quad , \text{ or } \quad -\frac{m_2}{2} < km_1 \leq \frac{m_2}{2} \; .$$

Substituting Eq. 15 in Eq. 12 we obtain algebraic equations for the coefficients A_k, B_k:

$$\begin{aligned}(\cos k\omega - \cos\omega)A'_k &= A_k(c_x - \cos\omega) - \frac{\rho}{2}\sum_l A_l A_{k-l} + \frac{1}{2}\sum_l B_l B_{k-l}, \\ (\cos k\omega - \cos\omega)B'_k &= B_k(c_y - \cos\omega) + \sum_l A_l B_{k-l} \; ,\end{aligned} \tag{17}$$

which may be solved recursively from the left had side for A'_k, B'_k for all $k \neq 1$. Of course, Eqs. 17 can also be solved in their original form ($A'_k = A_k$, $B'_k = B_k$) by an appropriate Newton algorithm.

Since we shall be interested in low period solutions, $m_1 = 1$, $m_2 = m$ in Eq. 16, and

$$\omega = 2\pi/m \quad , \quad m = 3, 4, 5 \ldots \tag{18}$$

we will solve Eq. 17 by iterative schemes[12,13], which are generally rapidly convergent. For example, for period $m = 3$ orbits we shall have to solve 4 equations for A_0, $|A_1|$, B_0 and $|B_1|$ of the form:

$$3A_0' = A_0(2c_x+1) - \rho(A_0^2 + 2|A_1|^2) + B_0^2 + 2|B_1|^2,$$
$$3B_0' = B_0(2c_y+1) + 2[A_0B_0 + 2|A_1|\,|B_1|\cos(\theta-\phi)],$$
$$0 = |A_1|(2c_x+1) - 2\rho A_0|A_1| + 2\sigma_1^{-1}B_0|B_1| - \rho\sigma_2|A_1|^2 + \sigma_3|B_1|^2,$$
$$0 = |B_1|(2c_y+1) + A_0|B_1| + \sigma_1|A_1|B_0 + \sigma_3|A_1|\,|B_1|, \tag{19}$$

where

$$A_1 = |A_1|e^{i\theta} \quad , \quad B_1 = |B_1|e^{i\phi} \tag{20a}$$

and

$$\sigma_1 = e^{i(\theta-\phi)} \quad , \quad \sigma_2 = e^{-3i\theta} \quad , \quad \sigma_3 = e^{-i2\phi-i\theta} \tag{20b}$$

Now, there are several choices of ϕ, θ that correspond to *real* σ_1, σ_2 and σ_3 in Eq. 19 and some of them yield different period–3 orbits of the mapping. For example, for

$$q_x = 0.425 \quad , \quad q_y = 0.344 \tag{21}$$

(i) $\theta = 0, \phi = \pi/2$, yields the *stable* periodic orbits at the centers of the "islands" in Figure 1.

(ii) $\theta = \phi = \pi/3$, gives the *unstable* period–3 orbits between the "islands" in Fig. 1

The stability of these orbits is determined by the eigenvalues of the return Jacobian matrix

$$\mathbf{J}_m = \prod_{n=1}^{m} \mathbf{J}(\hat{X}_n) \tag{22}$$

$m = 3$. As is well known, these eigenvalues must all lie on the unit circle for the m–periodic orbit to be stable[14].

There is a *strong instability* associated with these simple periodic orbits (or, low order *resonances*) of Eq. 12. When particles enter into their chaotic regions they are seen to escape very quickly to infinity. Thus the locations of these low period unstable orbits provide useful estimates of distances from the origin (in the X_n, X_{n+1}, and Y_n, Y_{n+1} planes) where this strong instability occurs.

We have similarly constructed period 4 ($\omega = \pi/2$) solutions of the mapping given by Eqs. 12. One of them, for example, was found to be stable for

$$q_x = 0.235 \quad , \quad q_y = 0.230 \tag{23}$$

at significantly large distances from the origin in both the X_n, X_{n+1} and Y_n, Y_{n+1} planes (see Figure 2a). This orbit was also obtained with the choice $\theta = 0$, $\phi = \pi/2$

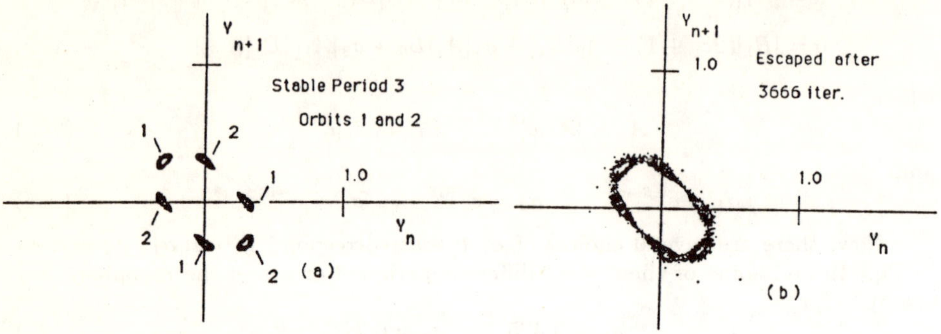

Figure 1(a). Stable period 3 orbit of the 4–D map at the centers of the "islands" marked by 1 (and its symmetric one, $Y_n \to -Y_n$, marked by 2), for initial conditions $X_0 = -.065$, $X_1 = .057$, $Y_0 = .02$, $Y_1 = -.31$; (b) Changing to $Y_0 = .04$, $X_1 = .06$ leads to rapid escape through the chaotic regions of the unstable period 3 orbits lying between the "islands". In both cases, $q_x = .425$, $q_y = .344$.

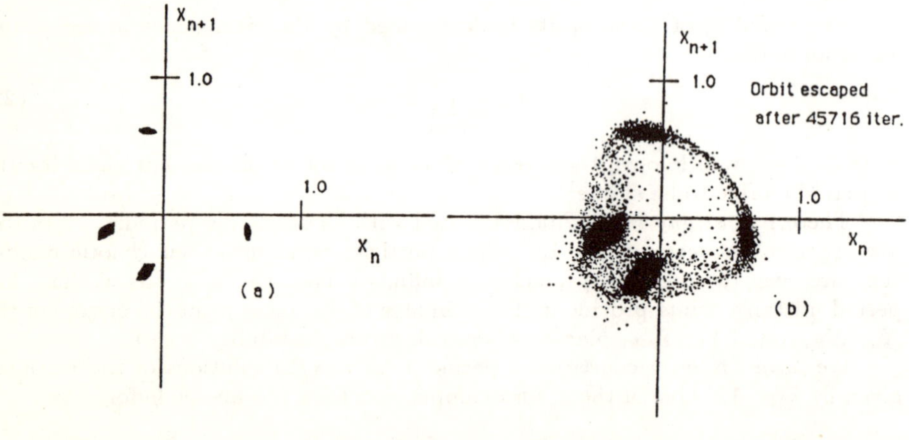

Figure 2. Same as Fig. 1 for motion near 4–D orbits of period 4 at $q_x = .235$, $q_y = .23$. (a) $X_0 = .6$, $X_1 = -.1$, $Y_0 = .01$, $Y_1 = -.5$ and no escape up to 10^6 iterations is observed; (b) Same as (a) but with $Y_0 = .05$: Escape occured a few hundred iterations after the orbit entered the chaotic region.

in its main Fourier coefficients of Eq. 20a, and has $B_0 = B_2 = 0$. As in the case of the period 3 orbit, when the initial conditions were such that particles eventually entered the chaotic region between the "tori", orbits were soon thereafter seen to escape to infinity, as shown, e.g. in Fig. 2b.

It is interesting to note what the intersections (projections) of 4–D tori about these periodic orbits look like in the X_n, X_{n+1} and/or Y_n, Y_{n+1} planes: In Figures 3 and 4 we show some of these "tori" associated with the period 3 and 4 orbits discussed above. As initial conditions are chosen further and further away from the periodic orbits these 4–D tori projections (which look remarkably like 2–dimensional tori!) grow in size and become more "jagged" and complicated in structure. It will take, however, further study before reliable statements can be made about how (and whether!) these tori actually "break-up" and "join" allowing orbits to rapidly run away to infinity.

3. Arnol'd Diffusion Near an Unstable 2–D Periodic Orbit

Long term orbital stability is not observed only near the origin and stable 4–D periodic orbits of our Eqs. 12. It can also be found near *stable 2–D m*-periodic solutions of Eq. 12a, $\{\hat{X}_n\}$, with $\hat{X}_n = \hat{X}_{n+m}$ and $\hat{Y}_n \equiv 0$. At the q_x values, however, where such solutions exist, q_y must be so chosen that small Y_n perturbations about these solutions *do not grow* exponentially via the parametric driving of $\hat{X}_n$ in Eq. 12b.

Suppose $\{\hat{X}_n\}$ is an m-periodic orbit of Eq. 12a, with $Y_n \equiv 0$. It is known that it will be (linearly) stable in 2–D as long as the following condition is satisfied [15]

$$|2 + \det \mathbf{H}_x| < 2 \tag{24}$$

where $\mathbf{H}_x$ is the $m \times m$ matrix with:

$$\begin{aligned}(\mathbf{H}_x)_{ii} &= 2c_x - 2\rho\hat{X}_i \quad , \quad i = 1, 2, \ldots, m\,, \\ (\mathbf{H}_x)_{i,i+1} &= (\mathbf{H}_x)_{i+1,i} = (\mathbf{H}_x)_{1,m} = (\mathbf{H}_x)_{m,1} = -1\,,\end{aligned} \tag{24a}$$

and all other elements zero.

This criterion was derived[15] from a Floquet type analysis of the *linearized* Eq. 12a (with $Y_n = 0$) about the m-periodic orbit $\{\hat{X}_n\}$. But the second equation of our 4–D mapping, Eq. 12b, is also *linear* in the Y_n and may be viewed, for small Y_n, as parametrically driven by the 2–D m-periodic solution $\{\hat{X}_n\}$ of Eq. 12a. We can therefore apply the same criterion as above to ensure that small Y_n oscillations about this 2-D periodic orbit remain bounded for very long times,

$$|2 + \det \mathbf{H}_y| < 2\,, \tag{25}$$

where $\mathbf{H}_y$ is the same matrix as as $\mathbf{H}_x$, except for its diagonal elements:

$$(\mathbf{H}_y)_{ii} = 2c_y + 2\hat{X}_i \quad , \quad i = 1, 2, \ldots, m. \tag{25a}$$

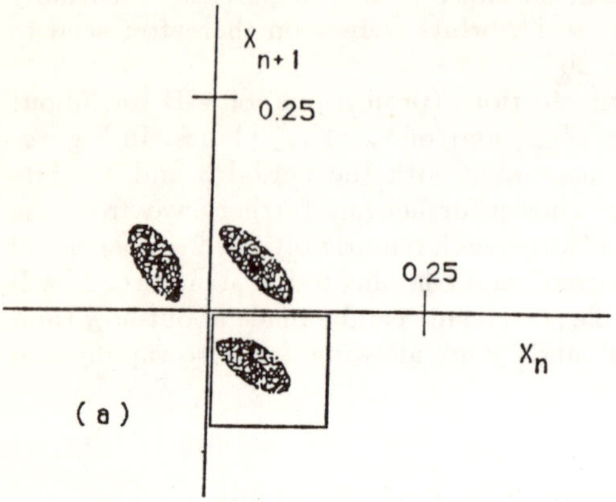

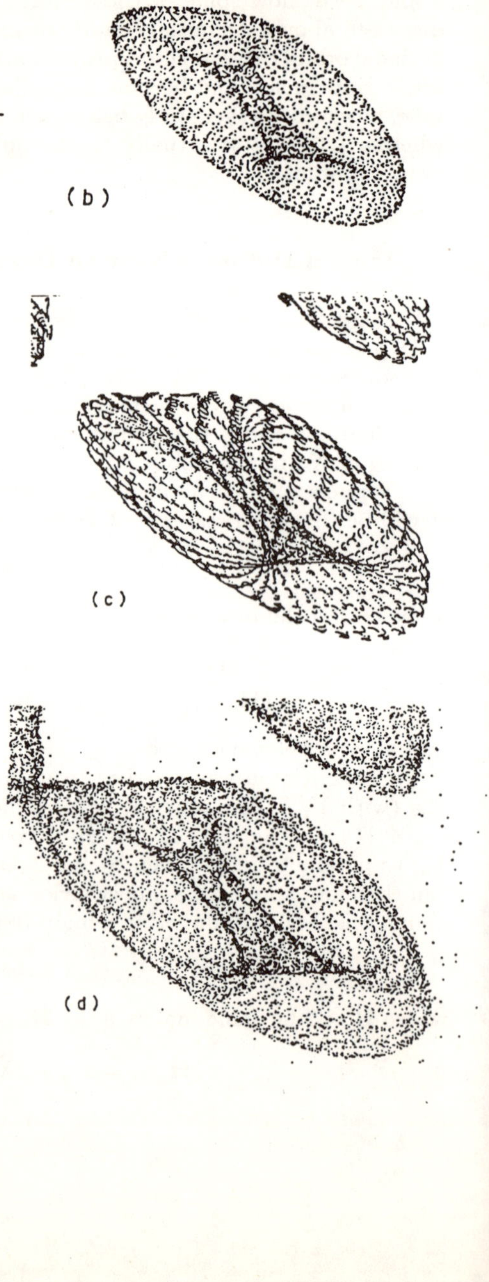

Figure 3.

(a) The stable period 3 orbit marked by 1 in Fig. 1(a), at $q_x = .425$, $q_y = .344$, in the X_n, X_{n+1} plane.

(b) A magnification of the 4–D torus in the box of (a), at the same initial conditions as in Fig. 1(a).

(c) Increasing only the Y_0 initial condition to $Y_0 = .03$ we observe that the tori grow in size and become more "irregular" in appearance.

(d) At $Y_0 = .0345$, the orbit stays for a long time on the tori, filling them out in an even more irregular and non-uniform way than (c) and eventually escapes after 28527 iterations of the map.

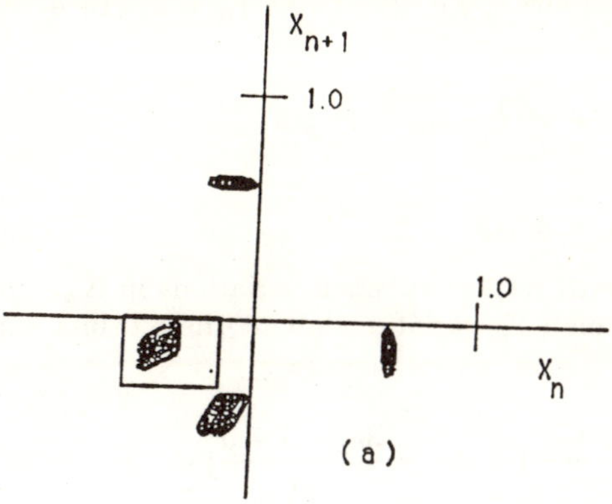

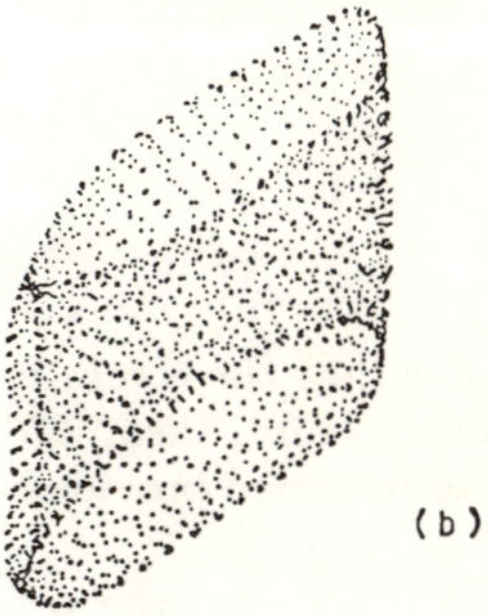

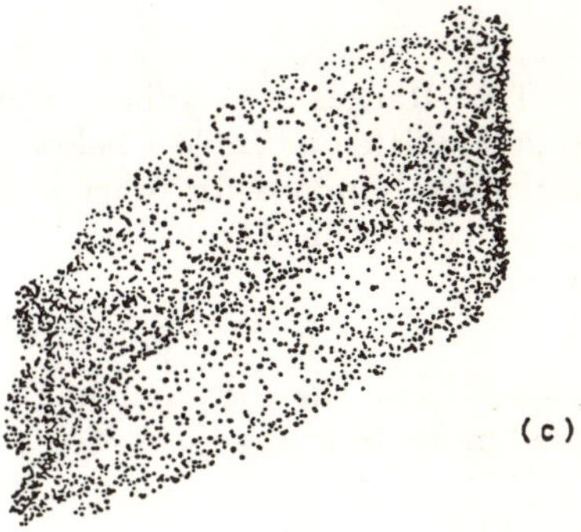

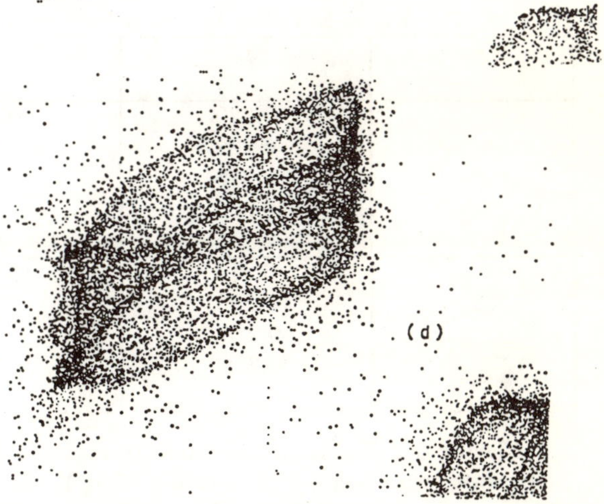

Figure 4. Same as Fig. 3 for a 4–D stable periodic orbit of period 4, at $q_x = .235$, $q_y = .23$.

(a) Starting with initial conditions $X_0 = .61$, $X_1 = -.13$, $Y_0 = .05$, $Y_1 = -.48$ we have no escape for 10^6 iter.

(b) Magnification of the torus in the box shown in (a).

(c) The same torus after changing only one in.cond. to $Y_1 = -.52$. Note the more "jagged" appearance of the torus and the "spotty" distribution of points over it.

(d) $Y_1 = -.5215$ and escape occurs at iteration 58924, after the orbit has wandered for a few hundred iterations in the chaotic region between the tori.

Thus, all we need to do to ensure long term stability in this case is look for q_x, q_y values such that the above two Ineqs. 24 and 25 are simultaneously satisfied.

Consider, for example, period 3 solutions of Eq. 12a (with $Y_n = 0$). Ineq. 24 gives in this case

$$1 < (4c_x^2 - 8c_x - 3)(1 + \sqrt{c_x^2 - 2c_x - 1}) < 3$$

or

$$0.318 \leq q_x \leq 0.333 \quad , \tag{26}$$

in order that these solutions be stable with respect to small variations in X_n. On the other hand, Ineq. 25 implies that small Y_n variations about these orbits will remain bounded provided

$$|4\left(c_y + \frac{c_x - a}{\rho}\right)^2 \left(c_y + \frac{1 + c_x + a}{\rho}\right) - 3c_y - \frac{3c_x + 1 - a}{\rho}| < 1 \, , \tag{27}$$

$a \equiv (c_x^2 - 2c_x - 1)^{\frac{1}{2}}$. Ranges of q_y values obtained from Ineq. 27 for different q_x satisfying Ineq. 26 are listed in Table 1 below.

Similar results are obtained for other m-periodic 2–D solutions of Eq. 12a: For $m = 5$, we have determined from Ineq. 24 that stability with respect to X_n-variations requires

$$0.21 \leq q_x \leq 0.229 \tag{28}$$

while for Y_n-boundedness, the corresponding intervals of q_y values obtained from Ineq. 25 are listed in Table 1 also.

Table 1

q_x, q_y Stability Intervals Near m–Periodic 2–D Orbits

m	q_x	q_y–Interval
3	0.320	(0, 0.12) and (0.387, 0.473)
	0.322	(0.390, 0.468)
	0.326	(0.396, 0.461)
	0.330	(0.400, 0.456)
5	0.210	(0.12, 0.16) and (0.2, 0.363)
	0.220	(0.195, 0.330)
	0.225	(0.185, 0.312)
	0.229	(0.180, 0.295)

Placing our X_1, X_0 initial conditions for the above q_x, q_y values near the stable 2–D orbits, we observed that orbits with Y_1, Y_0 small enough (typically $|Y_{0,1}| \sim 10^{-2}$) remained *bounded*, i.e., $X_n^2 + Y_n^2 < 1$, for 10^6 iterations of Eqs. 12. We also observed that, for each q_x, the $Y_{\max} = \max\{|Y_0| + |Y_1|\}$ for stability varies over the corresponding q_y interval, attaining its largest values ($Y_{\max} \sim 0.1$) near the center of that interval.

And now we come to our evidence of *weak instabilities*, or Arnol'd diffusion in this model: Choosing initial conditions for X_0, X_1 at the point A, within the chaotic layer of the unstable 2–D period 5 orbit of Fig. 5a, we found that the Y_n oscillations—no matter how small their Y_1, Y_0—kept *growing* in amplitude (on the average) producing in the Y_n, Y_{n+1} plane a set of point *scattered* over a figure of intersecting "rings", as shown in Figs. 5b,c.

On the other hand, an orbit starting at a point B in Fig. 5a, within the period 5 islands, exhibited considerably *weaker* outward diffusion properties and produced a set of intersections lying on a much more clearly defined "ring"–pattern in the Y_n , Y_{n+1} plane (see Fig. 5d).

In an attempt to quantify these observations we followed a method due to Chirikov et al.[7,16] and divided our total number of iterations $N = 10^6$ into

1) $N_1 = 100$ subintervals of length $\Delta N_1 = 10,000$, and
2) $N_2 = 10$ subintervals of length $\Delta N_2 = 100,000$.

Then for each of these cases we computed the *diffusion coefficient*

$$D_k = \frac{2}{N_k(N_k-1)} \sum_{m>l} \frac{[\bar{Y}(m) - \bar{Y}(l)]^2}{(\Delta N_k)(m-l)} , \qquad k = 1,2 \tag{29}$$

$\bar{Y}(m)$ being the average of Y_n over the mth subinterval, m, $l = 1,2,\ldots N_k$. (In Eq. 29 we have used $\bar{Y}(m)$ instead of the more common average of the Hamiltonian $\bar{H}(m)$[7,16] because, in our system, diffusion phenomena are more pronounced in the Y_n direction).

For a true diffusion process it should not matter if the motion is averaged over different numbers of subintervals, and hence one should expect:

$$D_1 \approx D_2 \qquad \text{(for Arnol'd diffusion)} \tag{30}$$

On the other hand, if initial conditions are chosen within "islands" of stable oscillatory motion: $[\bar{Y}(m) - \bar{Y}(n)] \propto (\Delta N_1)^{-1}$, whence

$$\frac{D_2}{D_1} \propto \frac{(\Delta N_1)^3}{(\Delta N_2)^3} = 10^{-3} \qquad \text{(for oscillations)} \tag{31}$$

Placing our initial X_0, X_1 at the point A, inside the separatrix of Fig. 5a, and starting with Y_0, Y_1 selected among the values -.01, -.005, 0.0, 0.005, .01 we computed after $N = 10^6$ iterations, on the average,

$$D_1 = .72 \times 10^{-11} \quad , \quad D_2 = 0.4 \times 10^{-11}$$

or

$$D_2/D_1 = 0.6 \qquad \text{(for diffusion)} \tag{32}$$

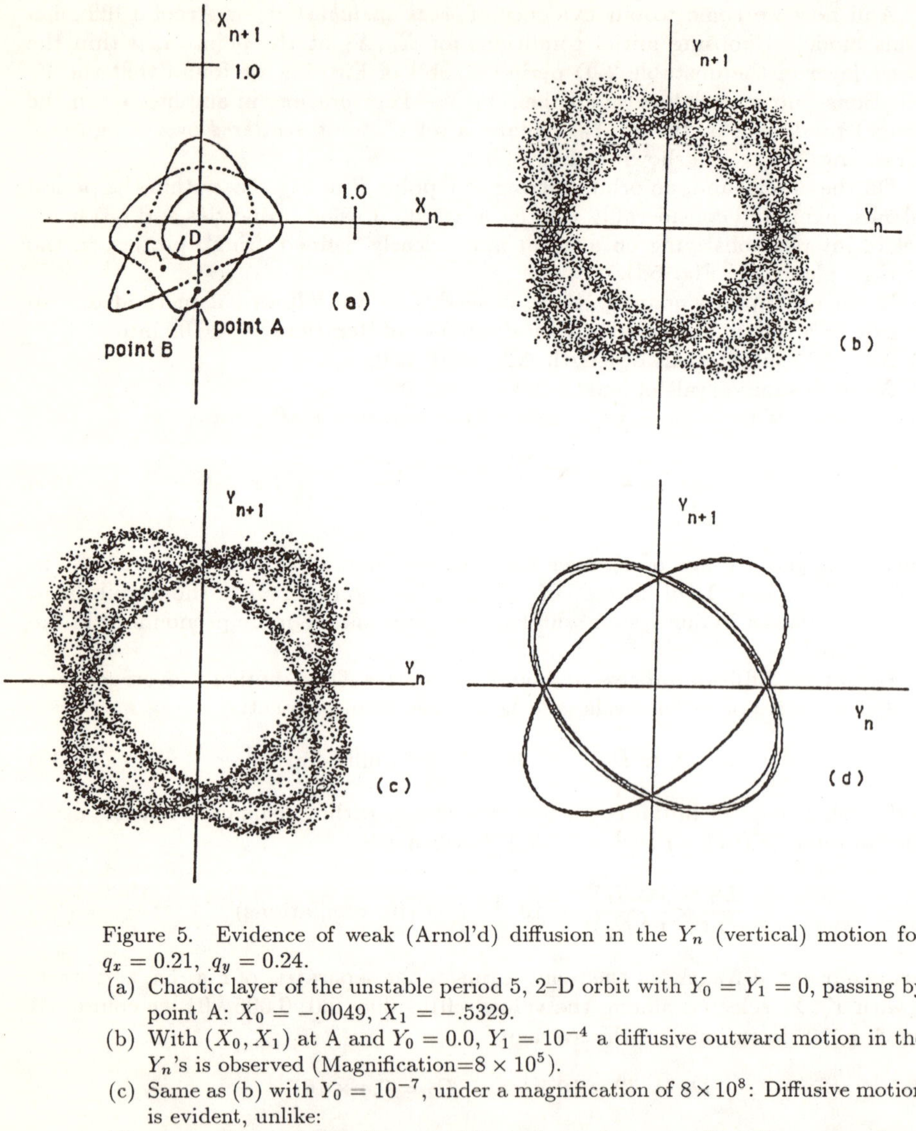

Figure 5. Evidence of weak (Arnol'd) diffusion in the Y_n (vertical) motion for $q_x = 0.21$, $q_y = 0.24$.

(a) Chaotic layer of the unstable period 5, 2–D orbit with $Y_0 = Y_1 = 0$, passing by point A: $X_0 = -.0049$, $X_1 = -.5329$.

(b) With (X_0, X_1) at A and $Y_0 = 0.0$, $Y_1 = 10^{-4}$ a diffusive outward motion in the Y_n's is observed (Magnification=8×10^5).

(c) Same as (b) with $Y_0 = 10^{-7}$, under a magnification of 8×10^8: Diffusive motion is evident, unlike:

(d) where $X_0 = -.0019$, $X_1 = -.4569$ are located at point B at the center of one of the islands of (a).

Starting, finally, with X_0, X_1 at a point B inside one of the islands of Fig. 5a and averaging over a similar set of Y_0, Y_1 initial conditions as above gives

$$D_1 = .32 \times 10^{-18} \quad , \quad D_2 = .17 \times 10^{-19} \ ,$$

and

$$D_2/D_1 = 5.3 \times 10^{-2} \qquad \text{(for oscillations)} \tag{33}$$

Thus, not only are the diffusion rates 7–8 orders of magnitude *smaller* in the oscillatory regimes, the ratio D_2/D_1 is also a lot lower than in the diffusive case (32), as expected. Eq. 33 is, of course, quite larger than predicted in (31) (and one order of magnitude higher than found by Chirikov et al. in Ref. 16). We have checked, however, our program in other oscillatory regimes, starting with X_0, X_1 at points C, D in Fig. 5a and have found results

$$\text{for point } C: \quad D_1 = 10^{-16.7} \ , \ D_2 = 10^{-18.7} \ \rightarrow \ D_2/D_1 = 10^{-2}$$
$$\text{for point } D: \quad D_1 = 10^{-16.95} \ , \ D_2 = 10^{-19.37} \ \rightarrow \ D_2/D_1 = 3.8 \times 10^{-3}$$

which are a lot closer to the expected value of Eq. 31 and clearly distinguish Eq. 32 as demonstrating the presence of a weak (Arnol'd) diffusion process.

4. Concluding Remarks

We have studied a 4–D mapping model of the dynamics of hadron beams passing repeatedly through a FODO cell containing sextupole nonlinearities (concentrated at one point), a dipole field and 2 focusing quadrupoles, in the storage ring of a high energy accelerator.

We found that low period 4–D periodic orbits of the map, when *stable*, have 3–D "tori" around them on which the orbits execute bounded oscillations for 10^6 mapping iterations and beyond. However, particles entering the large chaotic regions about such *unstable* low period orbits experience *strong instabilities* that quickly (over a few hundred iterations) lead them to *infinite* distances away from the origin of the map.

We have also discovered "tune" values (betatron frequencies) for which nearly "flat" beams ($|y_n| << |x_n|$) execute bounded oscillations (at least up to 10^7 passages through the cell) near *stable* 2–D periodic orbits of the x_n–mapping ($y_n \equiv 0$). However, when the initial x_0, x_1 are placed inside the chaotic layer of one such unstable 2–D orbit, a slow outward diffusion in the y_n motion is observed, with diffusion coefficient $D \leq 10^{-11}$ for $N = 10^6$ mapping iterations.

Comparing with the results of other researchers we have verified that this *weak instability* is indeed evidence of *Arnol'd diffusion*, which typically occurs at much faster rates than one finds when x_1, x_0 lie within oscillatory regimes ($D \leq 10^{-19}$).

These computations must, of course, be carried out for longer times ($N = 10^7$ and beyond...) and compared with analytical diffusion estimates existing in the literature[7,16]. Moreover, synchrotron oscillations (causing a slow modulation in the q_x, q_y betatron frequencies) also need to be included in our model and their effect on beam lifetimes analyzed numerically as well as analytically[10]. Such studies are currently under way, and results will be reported in a forthcoming paper[11].

5. Acknowledgments

One of the authors (T.B.) is indebted to the organizers of this conference, and in particular to G. Turchetti, for many informative discussions on the model studied in this paper. He also gratefully acknowledges the hospitality of the Center for Statistical Mechanics and Complex Systems and the Institute of Fusion Studies of The University of Texas at Austin, where this research was carried out, while he was a visiting professor in the spring semester of 1989-90.

6. References

1. *Nonlinear Dynamics and the Beam–Beam Interaction*, M. Month and J.C. Herrera eds., A.I.P. Conf. Proc. **57** (A.I.P., New York, 1979).
2. *Physics of High Energy Accelerators*, R.A. Carrigan, F.R. Huson and M. Month eds., A.I.P. Conf. Proc. **87** (A.I.P., New York, 1982); see esp. the articles by A. Dragt and J.L. Tennyson.
3. *Proceeding of the Beam–Beam Interaction Seminar*, SLAC Publ. 2624, Conf. 8005102 (SLAC, Stanford, 1980).
4. See several articles in *Workshop on Orbital Dynamics and Applications to Accelerators*, Particle Accelerators **19** (1986).
5. See e.g. A. Bazzani, P. Mazzanti, G. Servizi and G. Turchetti, Il Nuovo Cim. **102B**(1) (1988) p. 51.
6. T. Bountis, C.R. Eminhizer and N. Budinsky, Nucl. Instr. Meth. Phys. Res. **227** (1984) p. 205.
7. B.V. Chirikov, Phys. Rep. **52** (1979) p. 265.
8. A. Ruggiero, in Ref. 4, p. 157.
9. R.H.G. Helleman in *Long Time Prediction in Dynamics*, C.W. Horton, L.E. Reichl and V. Szebehely eds. (J. Wiley, New York, 1982) p. 95.
10. F. Vivaldi, Rev. Mod. Phys. **56** (4) (1984) p. 737.
11. T. Bountis, G. Mahmoud, G. Servizi and G. Turchetti, in preparation.
12. R.H.G. Helleman in *Statistical Mechanics and Statistical Methods*, U. Landman ed. (Plenum, New York, 1978).
13. R.H.G. Helleman and T. Bountis, Lect. Notes in Phys. **93** (Springer, New York, 1978).
14. See e.g. J.E. Howard and R.S. MacKay, J. Math. Phys. **28** (1986) p. 1036; H–T. Kook and J.D. Meiss, Physica **35D** (1989) p. 65.

15. T. Bountis and R.H.G. Helleman, J. Math. Phys. **22** (9) (1981) p. 1867.
16. B.V. Chirikov, J. Ford and F. Vivaldi, in Ref. 1.

THE BEAM-BEAM LIMIT IN SYMMETRIC AND ASYMMETRIC LEPTON COLLIDERS*

J.L. Tennyson
Stanford Linear Accelerator Center, Stanford University, Stanford 94305

ABSTRACT

A partial theory of the beam-beam limit in symmetric and asymmetric electron-positron colliders is reviewed. It is shown that there are three separate beam-beam limits: one due to beam blow-up, one due to the flip-flop instability, and one due to particle loss. A detailed treatment of the flip-flop limit shows that this phenomena corresponds to the appearance of a critical point, or "cusp catastrophe". Several examples from the simulation of a round beam collider are used to illustrate the various limits and their properties.

1. Introduction

The luminosities of most electron-positron colliding beam facilities are limited by the beam-beam interaction. When a particle of one beam passes through a bunch of the opposing beam, it feels a sharp transverse impulse due to the relativistically enhanced electro-magnetic field of the opposing bunch. If the current in the opposing beam is high enough, consecutive beam-beam impulses can disrupt the dynamical stability of the particle, causing its betatron amplitude to diffuse or increase resonantly. This, in turn, can result in increased transverse beam size or loss of particles to the aperture. The former generally causes a drop in specific luminosity while the latter may increase background levels in the detectors or shorten beam lifetime. The beam-beam limit is often represented as a graph of luminosity vs. beam current, as shown in figure 1. At low currents, the beam dimensions are independent of current. For a symmetric flat beam collider, the specific luminosity is given by the formula

$$L = \frac{N^2}{4\pi\sigma_x\sigma_y} \tag{1}$$

The low current luminosity thus increases as N^2 . When the current exceeds a certain value, known as the "blowup threshold", the vertical beam size σ_y begins to increase approximately linearly with N (actually, the dependence is usually a little

⋆ Work supported by Department of Energy contract DE-AC03-76SF00515.

less than linear). The high current luminosity is thus approximately proportional to N. Finally, when the current reaches a second threshold called the *beam-beam limit*, beam lifetime becomes unacceptably short, causing the integrated luminosity to fall below the level necessary for practical operation. The end of the curve in figure 1 corresponds to the beam-beam limit. Maximum luminosity is usually found at the beam-beam limit. This paper reviews the three ways that the beam-beam interaction can limit the amount of current in the beam. A more detailed description of these phenomena is presented in reference [1] . A specific treatment of the flip-flop instability is given in reference [2].

2. Parameter Space

There are a large number of parameters that affect the maximum attainable luminosity. The luminosity as expressed in (1) depends on the beam currents and transverse beam dimensions. In order to include asymmetric beams in the discussion, but still keep expressions relatively simple, we drop the assumption that the two beams have equal transverse dimensions but add the simplifying constraint that both beams are round at all times. The luminosity in this case is given by

$$L = \frac{N_1 N_2}{2\pi\left(r_1^2 + r_2^2\right)} \tag{2.1}$$

The beam-beam limit places upper bounds ξ_{lim1}, ξ_{lim2} on the values of the tuneshifts

$$\xi_1 = \frac{r_e \beta_1 N_2}{4\pi\gamma_1 r_2^2} < \xi_{\ell im1} \tag{2.2}$$

$$\xi_2 = \frac{r_e \beta_2 N_1}{4\pi\gamma_2 r_1^2} < \xi_{\ell im2} \tag{2.3}$$

Maximum luminosity is always found when both beams are running at their limiting tune shifts. Equations (2.2) and (2.3) can be used to elminate r_1 and r_2 in (2.1). The maximum luminosity with respect to variation of the beam emittances is

$$L = \frac{2}{r_e\left(\frac{\beta_1}{\gamma_1 \xi_{\text{lim1}} N_1} + \frac{\beta_2}{\gamma_2 \xi_{\text{lim2}} N_2}\right)} \tag{2.4}$$

The maximum luminosity thus depends on

a) The nominal energies γ_1, γ_2

b) The minimum attainable beta functions at the IP β_1, β_2

c) The maximum numbers of particles per bunch N_1, N_2

d) The limiting tune shifts ξ_{lim1}, ξ_{lim2}

The limiting tune shifts usually lie somewhere between $\xi_{lim} = .03$ and $\xi_{lim} = .08$, but their actual values depend on several auxiliary parameters

e) The damping decrements δ_1, δ_2

f) The transverse machine tunes ν_1, ν_2

g) The longitudinal machine tunes ν_{s1}, ν_{s2}

h) The bunch lengths σ_{s1}, σ_{s2}

i) The chromaticities C_1, C_2

j) The dispersions at the IP D_1, D_2

k) The orbit displacements at the IP Δ_1, Δ_2

The beam-beam lifetime is assumed to depend on all of the above parameters plus

l) The size of the hard aperture

and possibly others, including the passing separation of bunches in the wings of the IR, so-called "parasitic crossings", and lattice nonlinearities (which determine the dynamic aperture). One of the problems in colliding beam design is to find the dependence of the limiting tune shift ξ_{lim} on the auxiliary parameters *e)-l)*. Part of this determination must be performed either by experiment or computer simulation since it's not possible to calculate the limiting tuneshift by analytic methods alone. However, some very useful characteristics of this dependence can be inferred from theoretical consideration. These characteristics are reviewed in the following section.

3. Three Sources of the Beam-Beam Limit

The three limitations on luminosity come from beam blow-up, particle loss to the aperture, and the flip-flop instability.

Blow-up Limit

The blow-up limit is imposed by a transverse enlargement of the beam resulting from an accumulation of beam-beam kicks at high tuneshift. In figure 2(a), a representative family of curves shows the dependence of the radius of beam #1 on the radius of beam #2 for a number of different "natural radii" of beam #1 (assuming the current in beam #2 remains constant). There's a limit to how large the luminosity can be, as is demonstrated by the quarter-circle contours of L. A similar limit exists for beam #1, as shown in figure 2(b). A superposition of the two blow-up "envelopes" gives the acual luminosity limition, as illustrated in figure 2(c).

Particle Loss Limit

A second luminosity limit is imposed by particle loss to the aperture. As with blow-up, there is a separate particle loss limit for each beam. Examples of lifetime limits can obtained from simulation as shown in figure 3. The parameter values for these simulation runs are:

Beam radii	$r = \sigma_x = \sigma_y$ (varied)
Tunes	$\nu_x = \nu_y = .08$
Chromaticities	$C_x = C_y = 10$ units
Dispersions	$\eta_x = \eta_y = 0$
Particles/Bunch	$N = 1.7\text{x}10^{11}$
Energy spread	$\Delta p/p = .001$
Synchrotron Period	$P_s = 25$ turns
Energy	$E_1 = 9$ GeV
Beta functions	$\beta_x = \beta_y = 6$ cm
Apertures	$A = .05$ cm at IP
No lattice nonlinearities or long range forces	
One IP and one bunch per beam	

The lifetime limit shown in figure 3(a) corresponds to a beam with a damping time of $T = 1200$ turns. The black triangles mark points in the (r_1, r_2) plane where lifetime is longer than one minute (assuming a ring radius of 2200 meters). Hollow squares mark points where the lifetime is less than one minute. A dashed line is drawn separating the two families of points. A similar limit is shown in figure 3(b) for a damping time of $T = 20000$ turns. The maximum attainable luminosity for a collider with asymmetric beams, T_1 = 2000 turns and T_2 = 20000 turns, is indicated by the intersection of these two lines, as shown in figure 3(c). This assumes that the currents are limited but that the emittances of the two beams are not. Note that because the lifetime limits of the two beams are uncoupled, figure 3(c) can also be used to find the luminosity of the symmetric cases $T_1 = T_2$ = 20000 and $T_1 = T_2 = 1200$.

The lifetime limit curves are fairly straight but sloped. The area of highest luminosity therefore sits at a "vertex" characterized by an acute angle. The limit lines are sloped for two reasons. One is that the aperture is fixed; as a beam gets larger, the relative size of its aperture shrinks and the average particle density at the aperture increases. The other reason is that the limiting tune shift naturally decreases as the opposing beam gets smaller with constant tune shift. This is due to microscopic effects that cause particle stability to decrease with amplitude relative to the RMS width of the opposing beam.

When the apertures of the two beam systems are finite, they bend the beam-beam limit lines up and close the accessible area of the (r_1, r_2) plane. This is shown figure 3(d). The finite apertures cause lifetimes to decrease when the beam sizes get too large. Note that if the beam currents are unlimited, the beam sizes are typically increased as much as possible to accomodate maximum current. In this case, the open (accessible) space in figure 3(d) may become very small.

Flip-flop Limit

The third beam-beam limitation on luminosity comes from the flip-flop effect. Most electron-positron colliders normally operate with both beams somewhat blown-up by the beam-beam interaction (usually in the vertical direction only). Because the amount of blow-up in one beam is affected by the amount of blow-up in the other, the steady state is characterized by a stable self-consisent equilibrium. When the currents in the two beams increase, a bifurcation can occur which causes the stable equilibrium to become unstable or disappear altogether. The result is a sudden asymmetric change in the sizes of the two beams; commonly, one beam becomes very large while the other collapses to its natural (single beam) size. For historical reasons, this instability is called the flip-flop effect. A flip-flop condition arises when the sensitivity of blow-up to tune-shift exceeds a critical value. For symmetric machines, the bifurcation or "critical point" occurs when the sensitivity of beam size to tuneshift exceeds

$$\left.\frac{\partial r_1}{\partial \xi_1}\right|_{N_2} > \frac{r_2}{2\xi_1} \tag{3.1}$$

To understand this criterion, and the flip-flop phenomena itself, it is necessary to consider the behavior of one beam when the size of the other beam is controlled. This is sometimes called the "weak-strong beam" model of the beam-beam interaction, even though neither beam is necessarily weak. The following discussion is specific to round beams, but is qualitatively valid for flat beams as well.

For a given set of beam parameters (those listed in section 2), the size of each beam can be plotted in the (r_1, r_2) plane as a function of the size of the opposing beam. More specifically, the size r_1 of beam #1 can be plotted as a function of the size r_2 of beam #2

$$r_1 = S(r_2; \mathbf{p}_1)$$

$$\mathbf{p}_1 = \{\nu_1, \nu_{s1}, \beta_1, \eta_1, T_1, r_{o1}, N_2, ..., etc.\} \tag{3.2}$$

and the size r_2 of beam #2 can be plotted as a function of the size r_1 of beam #1

$$r_2 = S(r_1; \mathbf{p}_2)$$

$$\mathbf{p}_2 = \{\nu_2, \nu_{s2}, \beta_2, \eta_2, T_2, r_{o2}, N_1, ..., etc.\} \tag{3.3}$$

The result is two curves that cross each other at one or more points. The intersections of the two curves mark the locations of the self-consistent equilibria, which correspond to the simultaneous solution of equations (3.2) and (3.3). The stability of a particular intersection depends on the orientation of the crossing. It is proved in reference [2] that the intersection is stable if, at the intersection, the product of the slopes of the two functions (3.2) and (3.3) is less than one.

$$\frac{dS(r_2; \mathbf{p}_1)}{dr_2} \frac{dS(r_1; \mathbf{p}_2)}{dr_1} < 1 \tag{3.4}$$

This is a general result, valid for both symmetric and asymmetric beams. There are typically either one or three equilibria associated with each pair of curves. The two cases are shown in figures 4(a) and 4(b). In figure 4(a), the currents in the two beams are relatively low and there is only one intersection. The corresponding equilibrium satisfies (3.4) and is therefore stable. Figure 4(b) shows the situation at higher currents, when the beams are flip-flop unstable. Here there is a single unstable equilibria and two highly asymmetric stable satellite equilibria.

For a fixed set of parameters $\mathbf{p}_1$ and $\mathbf{p}_2$, a certain region of the (r_1, r_2) plane is inaccessible because its points cannot be made stable to the flip-flop effect. To find this region, two families of curves must be graphed using either simulation or direct experiment. In figure 5(a), r_1 is plotted as a function of r_2 for several different values of r_{o1} (the natural size of beam #1). These curves are smoothed simulation results obtained using the parameter list of the preceding section and a damping time of 20000 turns. A similar set of curves is shown in figure 5(b) for beam #2 with a damping time of 1200 turns.

The superposition of figures 5(a) and 5(b) is shown in figure 5(c). Each intersection is labeled with either a white or black dot. The white dots mark equilibria that satisfy (3.4) and are therefore stable. The black dots mark unstable equilibria. A thick curve separates the white dots from the black. The thick curve in figure 5(c) is called the "flip-flop" bump. The area of the (r_1, r_2) plane below the flip-flop bump is unstable and therefore inaccessible. Note that the inaccessible region is the region of highest luminosity. Thus, in the absence of lifetime limitations, the flip-flop effect almost always limits the maximum attainable luminosity.

The flip-flop limit is represented by a single curve in the (r_1, r_2) plane while the blow-up limit and lifetime limit are each represented by two separate curves. This is because the blow-up and lifetime limits are independent for the two beams; the parameters $\mathbf{p}_1$ that determine the radius and lifetime of beam #1 do not affect the radius of beam #2 and vice versa. The flip-flop limit, on the other hand, is a self-consistent effect and is therefore dependent on all the parameters affecting the blow-ups of both beams.

Note that the luminosity limits affected by blow-up and flip-flop are independent of the size of the aperture. One cannot improve these limits by increasing the size of the aperture. These limits are unique to lepton colliders since they are related to the equilibrium sizes of the two beams; the beams in a hadron collider are never in equilibrium and therefore do not exhibit these limits.

Comments

A superposition of the three different beam-beam limits is shown in figure 6. In the case represented, the actual beam-beam limit is determined by particle loss to the aperture. However, if the aperture were increased in size, the lifetime vertex would move in, possibly penetrating the flip-flop bump.

The actual beam-beam limit is determined by either the flip-flop instability or particle loss. It is possible for particles to be lost to the aperture without noticable blow-up of the beam. This is characteristic of round beams and is due to the fact that particles escape to the aperture through a leak in the phase space. The leak is through a specific region of phase space where both the synchrotron and betatron amplitudes are large (two or three sigma), and where the horizontal and vertical oscillations are almost exactly in phase. The motion in this narrow region of phase space is dynamically unstable. The leak provides a channel for particles to move rapidly from a relatively compact (not blown up) core to the aperture. It is difficult to see in simulations because the phase relations between the horizontal and vertical motions are generally hidden.

A similar leak exists in flat-beam machines, though in a more apparent region of phase space. Flat beam colliders typically exhibit beam blow-up long before the lifetime begins to decline. They are therefore more likely to be limited by flip-flop instability than round beams. A comparison of the weak-beam size functions for round and flat beams is shown in figures 7(a) and 7(b). The flat beam size function rises more slowly, at lower tuneshifts, than the round beam size function. This is because the flatbeam vertical tuneshift is inversally proportional to the vertical beam size while the round beam tuneshift is inversally proportional to the square of the beam radius. This means that round beams are quite resistant to beam blow-up but do not, generally speaking, have higher particle loss limits. If the round beam aperture is very large, it is possible to reach the flip-flop limit. The round beams tend to go flip-flop unstable with only 10% - 20% blowup.

Note that the beam-beam limit can be different for the two beams. For example, if the beam apertures are of different sizes, the particle loss limits will also be different. This would be the case even if the blow-up and flip-flop characteristics were perfectly symmetrical. In figure 3, the only asymmetry is the different damping times of the two beams. The limiting tune shifts in figure 3 differ by about 25% and the optimal beam sizes differ by about 10%.

4. Balance Point Sensitivity

If the lifetime vertex is outside the flip-flop bump, as is usually the case with round beams and often the case with flat beams, then lifetime determines the beam-beam limit. In this case, the vertex shown in figure 6(c) is called the "balance point"; it is the operating point with the highest attainable luminosity. The acute opening angle at the vertex creates an illusion of "instability", especially if one is blindly probing the space with a strong-strong simulation or experiment. This pseudo-instability is called "balance point sensitivity". If one starts from a point just inside the vertex and varies the size of either beam in either direction, one of the beams immediately hits its beam-beam limit and its lifetime drops below the acceptable level. Clearly, the closer the operating point is to the vertex, the more sensitive the balance.

If the lifetime vertex is inside the flip-flop bump, then luminosity is limited by the flip-flop instability. The points on the surface of the flip-flop bump correspond to bifurcation points. All but one of them are tangent bifurcation points (first order phase transitions); the single exception is a critical point. The critical point is a pitchfork bifurcation and is analogous to a second order phase transition. When the flip-flop instability is the limiting phenomena, the balance point is associated with the critical point rather than with the lifetime vertex. The critical point does not always have the highest accessible luminosity in the (r_1, r_2) plane; because the flip-flop bump is convex, there may be other (tangent bifurcation) points on its surface that have higher luminosities. However these other points are dangerous. If one operates too close to a tangent bifurcation, fluctuations can momentarily push the system out of the local minima and a "catastrophe" may occur. The catastrophe appears as a sudden irreversible change in the sizes of the two beams.

A catastrophe is less dangerous close to the critical point since the region surrounding the critical point corresponds either to a single (marginally stable) equilibrium or to three separate but very tightly bunched equilibria (for practical purposes they're a single point).

Close to the critical point, the beam sizes are extremely sensitive to small changes in any of the beam-beam parameters $\mathbf{p}$; for example $\partial r_1/\partial r_{o1}$ becomes very large (the sensitivity is infinite at the critical point). This means that one can expect high balance-point sensitivity at the coincident beam-beam limit even when the lifetime vertex is well inside the flip-flop bump.

Other phenomena characteristic of operation close to a critical point include long relaxation times and large fluctuation response functions. The former can be tens or hundreds of times longer than a damping time. Generally, the damping time determines the weak-strong relaxation time: the strong-strong or self-consistent relaxation times are determined by the stability of the equilibrium. Relaxation times that are much longer than the damping time have been observed in experiment[4]. The second effect, high sensitivity to fluctuations, is typically more important in simulation than actual machines. In simulations, super-particle discreteness drives fluctuations that become very large close to a critical point. For this reason, it is almost impossible to investigate flip-flop criticality with a standard strong-strong beam simulation code.

Since the beam-beam balance at both the lifetime vertex and the critical point are extremely sensitive to any change in the value of any beam-beam parameter (any change pushes the system over the beam-beam limit), it is always necessary to back away from the true balance point (ideally, by slightly increasing the natural beam sizes). This reduces both the sensitivity and the luminosity. The loss of luminosity associated with this "backing away" is small, usually less than 10%.

Balance point sensitivity exists in symmetric and asymmetric configurations, and in both one-ring and two-ring accelerators. In one-ring machines, the problem is ameliorated by the fact that virtually all beam-beam parameters, other than the currents, are constrained to be the same for the two beams. Any imbalance can then be attributed to current mismatch and can be removed by adjusting the currents. In addition, because fluctuations in other parameters are constrained to be the same in the two rings, one-ring machines are expected to be less sensitive to fluctuations than two-ring machines.

In a two-ring machine, there is no practical advantage to running the beams in a symmetric configuration; the symmetric and asymmetric configurations have the same balance point sensitivities. Even if a two-ring machine is designed to be symmetric, it will not be balanced when it is first turned on (or following any changes in either lattice). Although imbalance in a two-ring accelerator can come from any of several parameter asymmetries, imbalance in one pair of parameters can be corrected by introducing a compensating imbalance in another pair of parameters.

Being able to compensate in this way means that the beams in a two-ring accelerator can be both asymmetric and balanced.

Summary

The beam-beam limit is determined by either particle loss to the aperture or the flip-flop instability, depending on the aspect ratio of the beam, the size of the aperture, and other parameters such as the damping time. When beams are running close to either the lifetime limit or the flip-flop limit, they exhibit high sensitivity to small changes in parameter values, usually of either sign. Luminosity optimization therefore requires careful balancing of the two beams against one another, especially when the two beams are in different rings. In an asymmetric machine, this balancing is probably most efficiently accomplished by adjusting the emittances of both beams. There is no particular advantage to making a two-ring collider symmetric.

Acknowledgements

The work described in this report was supported by three institutions: the Lawrence Berkeley Laboratory, the California Institute of Technology, and the Stanford Linear Accelerator Center. The simulations were performed on a CRAY Y-MP8/864 at the San Diego Super Computer Center. Machine time was provided by the San Diego Super Computer Center with funds from the National Science Foundation.

Reference

1. J.L. Tennyson, "The Beam-Beam Limit in Asymmetric Colliders: Optimization of the B-Factory Parameter Base", Caltech preprint CALT 68-1632. To be published in AIP Conference Proceedings **The Workshop on Beam Dynamics Issues of High-Luminosity Asymmetric Collider Rings** 1990.

2. J.L. Tennyson, "Flip-Flop Modes in Symmetric and Asymmetric Colliding Beam Storage Rings", LBL preprint LBL-28013. To be published in Physical Review D (1990).

3. J.L. Tennyson, "Luminosity dependence in Round Beam Lepton Colliders: a Simulation Study", in preparation.

4. Private communication; Martin Donald & Ewan Paterson.

Figure captions

Figure 1.

A log-plot of the luminosity L as a function of the current per bunch N shows L proportional to N^2 below the blow-up threshold and proportional to N above it. The curve ends at the "beam-beam limit" due to one of the effects described in section 3.

Figure 2(a).

Weak beam size functions for beam #1. The size of beam #1 is plotted against the size of beam #2 for several natural radii of beam#1. These graphs are representative of actual functions. When the size of beam #2 falls below a certain value, beam #1 begins to blow up. The blow-up makes certain areas of the plane inaccessible.

Figure 2(b).

Weak beam size functions for beam #2. This figure is identical to 4(a) except that the roles of beam #1 and #2 have been interchanged.

Figure 3(a)

Results of simulation runs which measured the lifetimes of beam #1 for several different values of the natural beam sizes. Beam size pairs with lifetimes greater than one minute (assuming a ring size equal to that of PEP, 2200 meters) are marked with triangles; those with lifetimes less than one minute are marked with squares. The aperture is at 500 microns, so the beam-beam limit is fairly straight in this plane. The damping time of this beam (#1)is 20000 turns.

Figure 3(b)

Results of simulation runs which measured the lifetimes of beam #2 for several different values of the natural beam sizes. The aperture is at 500 microns. The damping time of this beam (#2) is 1200 turns. Note that the limiting tune shift on the diagonal is about $\xi = .060$ compared to $\xi = .087$ for the highly damped beam in the previous figure.

Figure 3(c)

Superposition of figures 6(a) and 6(b). The shaded area is the region in which the lifetime of one or both beams is less than one minute. Maximum luminosity is found close to the vertex. Note that the limiting tuneshift for the weakly damped beam is some 25% below that of the strongly damped beam.

Figure 3(d).

The lifetime limits actually bend back and enclose the accessible region when the apertures are finite. The closing point is 1/6–1/5 of the aperture for each beam. This is a representative figure (not derived from simulation). The box is roughly the area shown in figure 3c.

Figure 4(a).

At low and moderate currents, the two "weak-beam size function" curves intersect as a single stable point. This figure corresponds to a fully symmetric collider and a blow-up of about 10% at the equilibrium.

Figure 4(b).

At high currents, the beams are unstable to flip-flop. The weak-beam size functions intersect at three separate points. The center symmetric point is unstable while the two satellite flip-flop equilibria are stable.

Figure 5(a).

Weak beam size functions for beam #1. These curves are derived from numerical simulation using the parameters listed in section 5. The damping time of beam #1 is 20000 turns.

Figure 5(b).

Weak beam size functions for beam #2. The damping time of beam #2 is 1200 turns. The natural radii for each curve is shown. This beam clearly blows up at a lower tune shift than beam #1 (shown in the preceding figure).

Figure 5(c).

Superposition of figures 5a and 5b shows the regions of flip-flop stability and instability. Each crossing of a grey and black line has one of two orientations. Crossings with one orientation correspond to stable equilibria and crossings with the other orientation correspond to unstable equilibria. The regions of stable and unstable equilibria are well separated and their interface is represented here with a thick curve. Because the boundary is characteristically convex, the unstable region is referred to as the "flip-flop bump".

Figure 6.

Superposition of the three limits. In this case, the lifetime vertex is outside the beam-beam bump and therefore determines the actual beam-beam limit.

Figure 7(a).

Typical weak-beam size function for round beams. The "weak" beam remains close to its natural size until the blow-up threshold is reached at relatively high currents. When the threshold is exceeded, the weak beam size rises very rapidly. The rapid rise is due to the fact that the tune shift is inversally proportional to the square of the radius. Round beams become flip-flop unstable when the blow-up is no more than 10%–20%. Although resistance to blow-up is high in comparison to flat beams, round beams do not show significantly better lifetimes.

Figure 7(b).

Typical weak-beam size function for flat beams. Here the tuneshift is inversally proportional to vertical size, so the weak beam size increases more slowly when the size of the strong beam is decreased. Flat beams can blow up more than round beams before they become unstable to flip-flop.

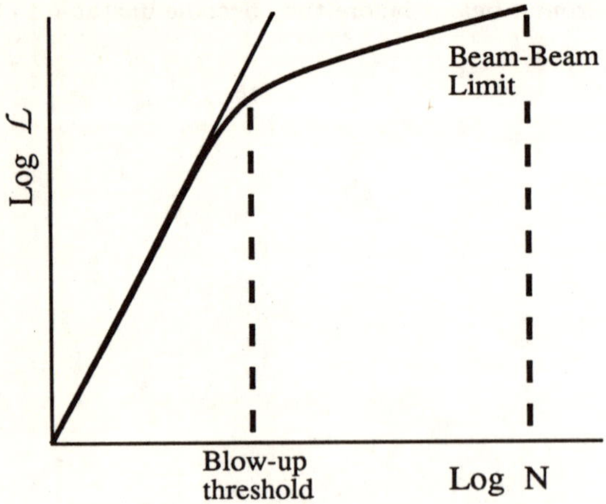

Fig. 1

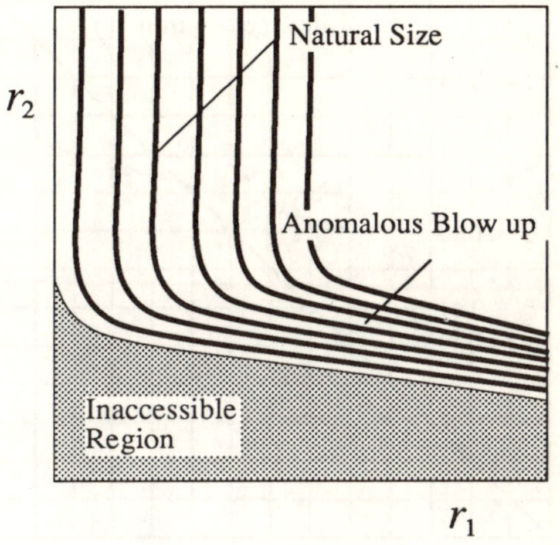
Natural Size
r_2
Anomalous Blow up
Inaccessible
Region
r_1

Fig. 2(a)

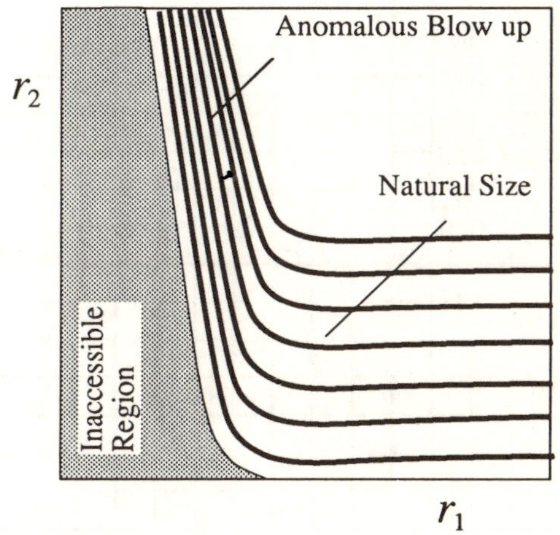
Anomalous Blow up
r_2
Natural Size
Inaccessible
Region
r_1

Fig. 2(b)

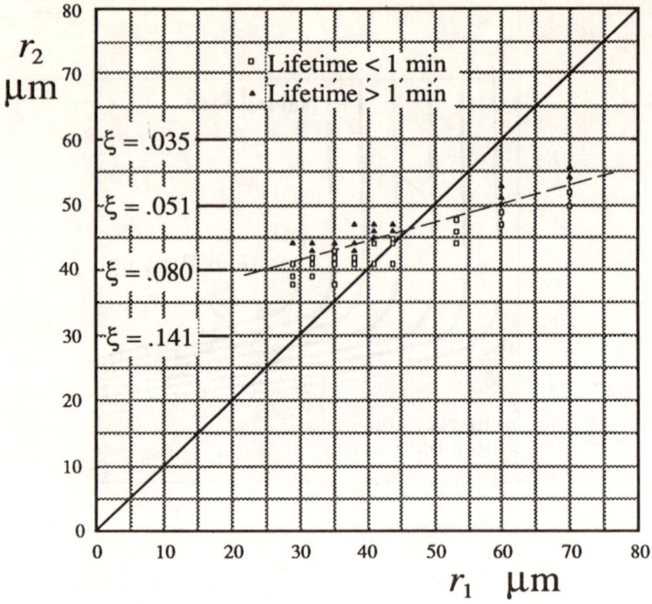
r_2
μm
Lifetime < 1 min
Lifetime > 1 min
ξ = .035
ξ = .051
ξ = .080
ξ = .141
0
10
20
30
40
50
60
70
80
r_1 μm

Fig. 3(a)

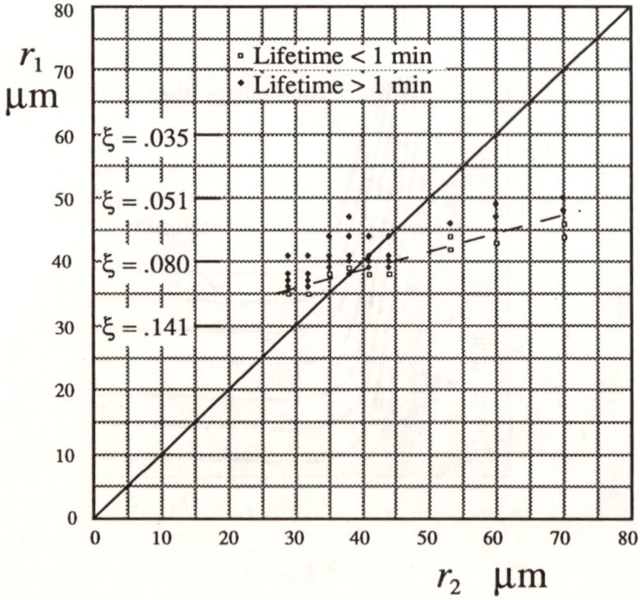
r_1
μm
Lifetime < 1 min
Lifetime > 1 min
ξ = .035
ξ = .051
ξ = .080
ξ = .141
0
10
20
30
40
50
60
70
80
r_2 μm

Fig. 3(b)

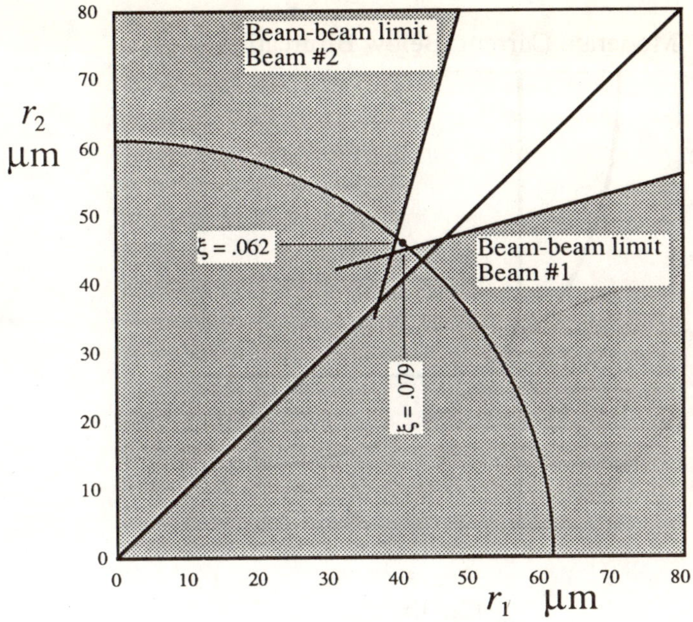
Beam-beam limit
Beam #2
Beam-beam limit
Beam #1
ξ = .062
ξ = .079
r_2
μm
r_1 μm
0
10
20
30
40
50
60
70
80

Fig. 3(c)

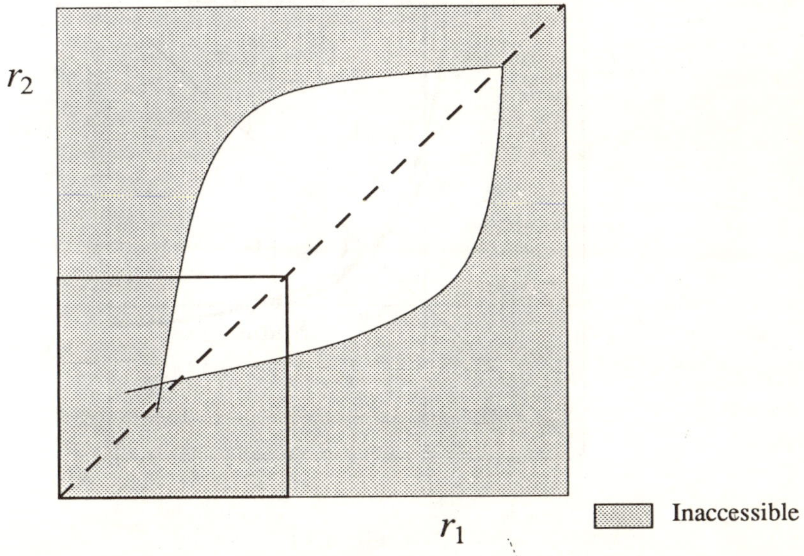
r_2
r_1
Inaccessible

Fig. 3(d)

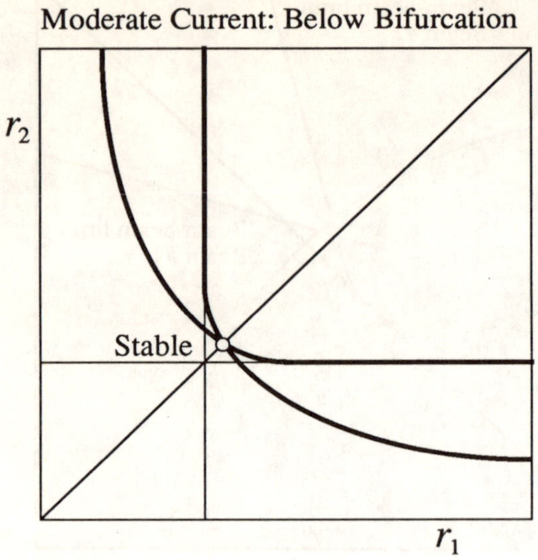
Moderate Current: Below Bifurcation
r_2
Stable
r_1

Fig. 4(a)

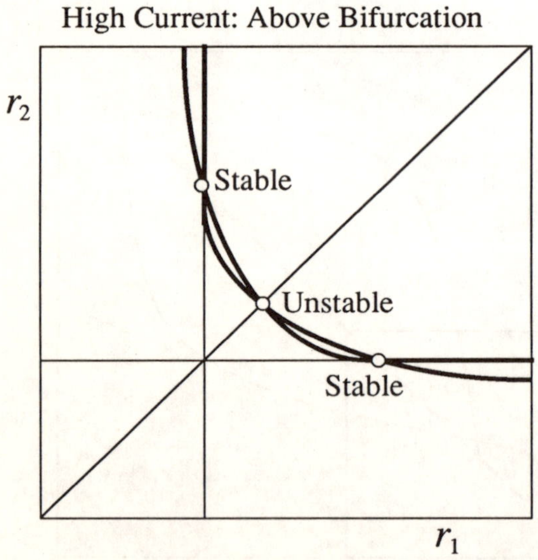
High Current: Above Bifurcation
r_2
Stable
Unstable
Stable
r_1

Fig. 4(b)

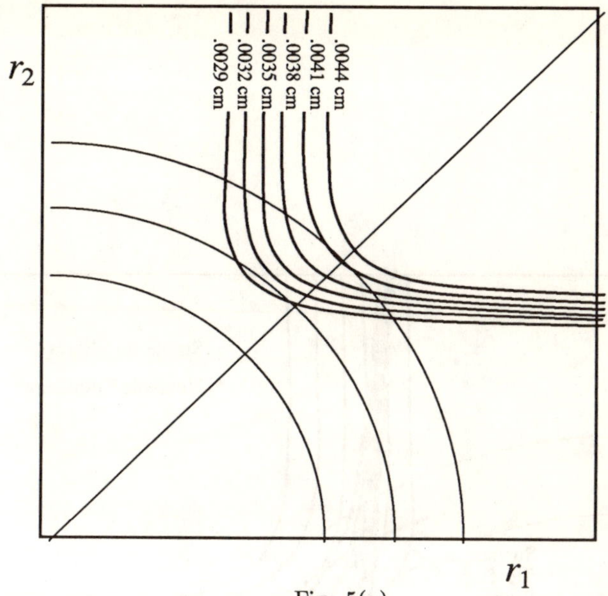

r_2
.0044 cm
.0041 cm
.0038 cm
.0035 cm
.0032 cm
.0029 cm
r_1

Fig. 5(a)

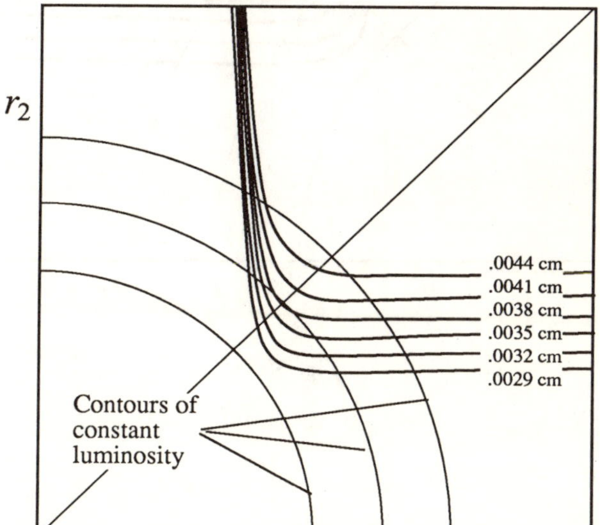

r_2
.0044 cm
.0041 cm
.0038 cm
.0035 cm
.0032 cm
.0029 cm
Contours of constant luminosity
r_1

Fig. 5(b)

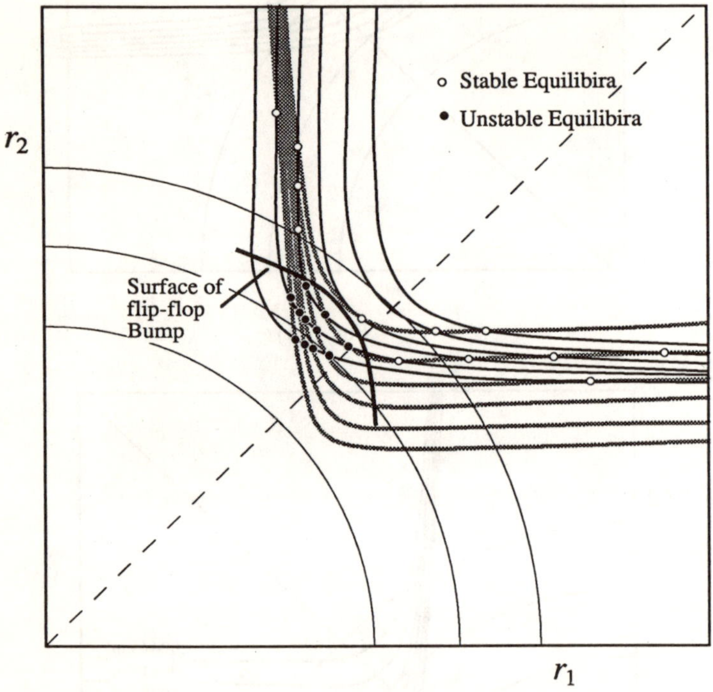

Stable Equilibira
Unstable Equilibira
r_2
Surface of
flip-flop
Bump
r_1

Fig. 5(c)

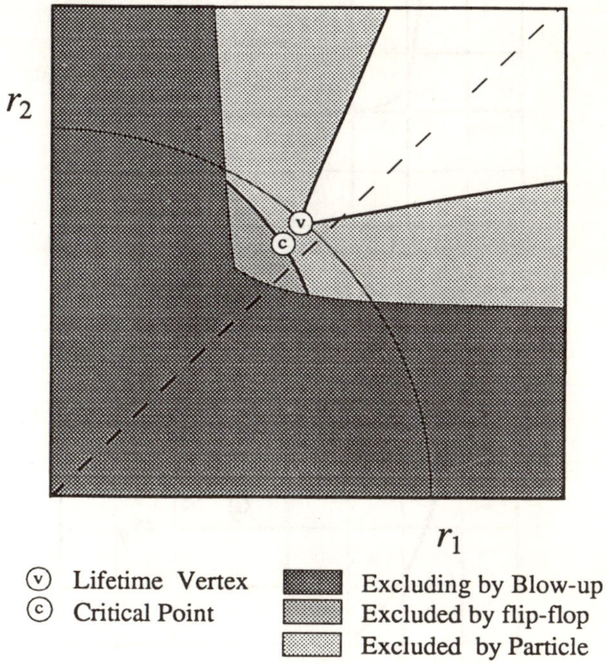
r_2
r_1
v
c
Lifetime Vertex
Critical Point
Excluding by Blow-up
Excluded by flip-flop
Excluded by Particle

Fig. 6

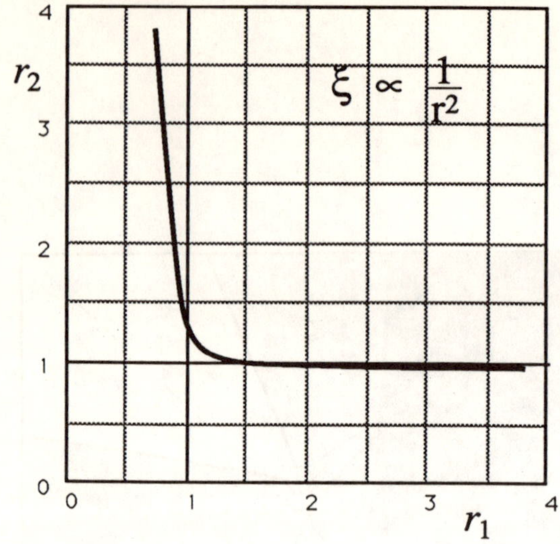
4
r_2
3
2
1
0
$\xi \propto \frac{1}{r^2}$
0
1
2
3
4
r_1

Fig. 7(a)

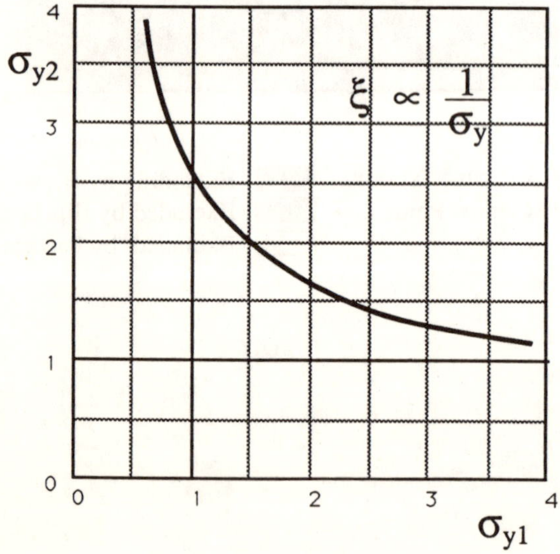
4
σ_{y2}
3
2
1
0
$\xi \propto \frac{1}{\sigma_y}$
0
1
2
3
4
σ_{y1}

Fig. 7(b)

SOME REMARKS ON CONSERVATIVE AND SYMPLECTIC SCHEMES

SALVADOR JIMÉNEZ, *
Courant Institute of Mathematical Sciences, New York University, 251 Mercer Street, New York, N.Y. 10012, U.S.A.

and

LUIS VÁZQUEZ,
Departamento de Física Teórica, Facultad de Ciencias Físicas, Universidad Complutense, 28040–Madrid, Spain.

ABSTRACT
We describe the discretization features of the equation of the motion of a particle in a general potential with eventually a dissipation. A conservative scheme with a possible symplectic structure in the mean is analyzed.

1. One-dimensional motion of a newtonian particle in a conservative field

1.1 *Introduction*

Consider a nonlinear wave equation with dissipation and noise of the form

$$\varphi_{tt} - \varphi_{xx} = F(\varphi) + D(\varphi_t) + S(\varphi, x, t), \tag{1}$$

where F is a nonlinear self-interaction, D represents dissipative terms and S stochastic terms. In the numerical study of Eq. 1, it is difficult to establish reliable numerical schemes that reproduce the long time behavior and do not introduce additional dissipation or noise due to discretization effects. We seek schemes free from "numerical chaos", that is, such that the dynamics of the discrete system is close to the dynamics of the continuous equation. As a first step we can consider numerical schemes that show a good behavior when restricted to the deterministic and/or dissipative cases.

On the other hand, if we perform first a discretization in space and then in time, the numerical study of Eq. 1 can be considered as the study of a system of coupled particles, each one corresponding to a point in the discrete spatial grid. Besides its importance as a tool for low-dimensional nonlinear problems, the study of a single particle is relevant as a preliminary step, and presents an introduction to

---

* On leave from:
Departamento de Física Teórica, Facultad de Ciencias Físicas,
Universidad Complutense, 28040–Madrid (SPAIN).

the basic features to be considered in the more general case. In this sense, we study numerical representations of the motion of a particle in a general potential.

1.2 *The continuous system*

Let us consider the equation

$$m\frac{d^2x}{dt^2} = -\frac{dU}{dx} = F(x). \tag{2}$$

The dynamics of the system is contained in the conservation of the energy:

$$E = \frac{m}{2}\left(\frac{dx}{dt}\right)^2 + U(x), \tag{3}$$

$$\frac{dE}{dt} = 0, \tag{4}$$

the symmetry under time inversion: $t \to -t$, and the stability properties of the equilibrium points of the potential $U(x)$. In the hamiltonian formalism, we have:

$$\frac{dx}{dt} = \frac{p}{m}, \tag{5}$$

$$\frac{dp}{dt} = F(x). \tag{6}$$

The evolution

$$\begin{pmatrix} x(0) \\ p(0) \end{pmatrix} \longrightarrow \begin{pmatrix} x(t) \\ p(t) \end{pmatrix} \tag{7}$$

is governed by a canonical transformation with generatrix function the hamiltonian

$$H(x,p) = \frac{p^2}{2m} + U(x) \tag{8}$$

which is a constant of the motion. The volume of the phase space is preserved under the evolution described by the mapping Eq. 7.

1.3 *The discrete system*

We represent the continuous variables $x(t)$ and $p(t)$ by

$$x^n = x(n\Delta t), \tag{9}$$

$$p^n = p(n\Delta t), \tag{10}$$

with $t = n\Delta t$, Δt being the mesh step of the discretization. We transform the ordinary differential equations Eq. 2, or Eq. 5 and Eq. 6, into finite difference equations or evolution maps. In order to have an accurate discrete representation of the continuous system, the scheme must be convergent (the discrete solution approaches the continuous one as $\Delta t \to 0$) and stable (initial errors are controlled throughout the computations). The stability is in general difficult to prove and some other properties are needed to ensure a good behaviour for long times: the local and global accuracy of the scheme, related to the maximum value of Δt, and the existence of discrete counterparts of the conservation laws and symmetries of the underlying continuous system. Two related points are the preservation of the stability character of the equilibrium points and, in the case of dissipative systems, the discrete analogous of the rate of change of certain quantities (energy, Lyapunov functions, etc.).

Besides "conventional" schemes such as Euler Methods, Predictor-Corrector (Runge-Kutta, etc.), there exist numerical schemes that are constructed around some of the symmetries of the continuous system. Generally it is not possible to reflect exactly all the symmetries in the discrete level (unless the equation is linear). We have for instance *conservative schemes*, that exactly preserve discrete counterparts of the conservation (or variation) laws and *symplectic schemes*, that preserve the canonical structure of the phase space. There are also *variational schemes*, in which adjusting a variational parameter at each step gives either a conservative or a symplectic scheme (in this case, however, the effect of the parameter on the accuracy of the solution is not clear, as well as the relation between its values at different steps). The "conventional" schemes applied to nonlinear equations are in general unstable for large times and may present "ghost" solutions, while this second kind of schemes represent more accurately the continuous solutions.

2. Conservative schemes

2.1 *Symmetries and conservation law*

In the case of Eq. 2, the conservation law is the preservation of the energy. Different conservative schemes that exactly preserve the energy can be defined (see for instance [1] and [2]). We consider here [3]:

$$m\frac{x^{n+1}-2x^n+x^{n-1}}{\Delta t^2}+\frac{U(x^{n+1})-U(x^{n-1})}{x^{n+1}-x^{n-1}}=0. \tag{11}$$

The scheme is time-reversible since this equation is symmetric with respect to n. On the other hand, Eq. 3 and Eq. 4 have as discrete counterparts

$$E^n=\frac{m}{2}\left(\frac{x^{n+1}-x^n}{\Delta t}\right)^2+\frac{1}{2}\left(U(x^{n+1})+U(x^n)\right) \tag{12}$$

and

$$\frac{E^{n+1}-E^n}{\Delta t}=0. \tag{13}$$

The existence of this conserved discrete energy ensures the boundedness of the numerical solutions when the continuous solutions are bounded.

2.2 *Symplectic structure*

We have two consistent discretizations for the Hamilton equations Eq. 5 and Eq. 6:

$$m\,\frac{x^{n+1}-x^n}{\Delta t}=p^n, \tag{14}$$

$$\frac{p^{n+1}-p^n}{\Delta t}=-\frac{U(x^{n+1}+\Delta t p^{n+1})-U(x^n)}{\Delta t(p^{n+1}+p^n)}, \tag{15}$$

and

$$m\,\frac{x^{n+1}-x^n}{\Delta t}=p^{n+1}, \tag{16}$$

$$\frac{p^{n+1}-p^n}{\Delta t}=-\frac{U(x^{n+1})-U(x^n-\Delta t p^n)}{\Delta t(p^{n+1}+p^n)}. \tag{17}$$

Consider now the evolution mapping

$$T:(x^n,p^n)\mapsto(x^{n+1},p^{n+1}). \tag{18}$$

Since we use Eq. 11 to integrate numerically the problem, we suppose the mapping T implicitly defined for any value of n. A one-dimensional mapping is symplectic if and only if the determinant of its jacobian matrix is unity. Let be M^n the jacobian matrix of T at n. We have

$$\det M^n = 1-\Delta t\,\frac{R^n}{Q^n}, \tag{19}$$

where

$$R^n=\frac{x^{n+1}-2x^n+x^{n-1}}{\Delta t^2}+\frac{1}{2}\left(F(x^{n+1})+F(x^{n-1})\right), \tag{20}$$

and

$$Q^n=\frac{x^{n+1}-x^n}{\Delta t}+\frac{\Delta t}{2}F(x^{n+1}). \tag{21}$$

The numerator R^n is in fact a discrete approximation of Eq. 2 with an averaging scheme and since Eq. 11 is $O(\Delta t^2)$, we get $R^n=O(\Delta t^2)$ too. On the other hand the denominator Q^n is $O(1)$ and we have

$$\det M^n = 1+O(\Delta t^3). \tag{22}$$

We see that T can be considered in good approximation a symplectic evolution mapping (in the linear case the scheme is exactly symplectic). Furthermore, since we can expect $\det M^n$ to oscillate with n around the unity (this has been observed por certain potentials in [3]), we have a conservation in the mean of the symplectic structure of the phase space at the discrete level.

2.3 *Maximum step size*

The numerical scheme requires solving the functional equation Eq. 11 for the unknown x^{n+1} at each time step. Given x^n and x^{n-1}, or for instance, the initial values x^o and x^1, some restrictions have to be imposed to Δt in order to have a unique real solution.

Theorem: *Let $U \in C^2(\mathbf{R})$ be a function such that $U(x) \to +\infty$ as $|x| \to \infty$. Let be*

$$\alpha = \inf_{x \in \mathbf{R}} U(x).$$

For any $C > \alpha$ we define $I_C = \{x, U(x) \geq \alpha\}$ *and*

$$M(C) = \frac{1}{2} \sup_{x \in I_C} |U''(x)| > 0.$$

Let x^o and $\Delta t > 0$ be given, $x^1 = x^o + \Delta t\, \dot{x}(0)$. We choose

$$C = U(x^o) + U(x^1) - \alpha + \left(\frac{x^1 - x^o}{\Delta t} \right)^2 .$$

Then if $\Delta t^2 < 1/M(C)$ there exists a unique sequence $(x^n)_{n \geq 0}$ satisfying the discrete equation Eq. 11. (For the exact formulation and demonstration see [4].)

There is a second limitation we have to impose on Δt. The numerical scheme is based on an expansion of Eq. 2 in Δt series. The radius of convergence of the series is determined by the location of the nearest singularity of x^{n+1}, considered as a function of Δt, in the complex-Δt plane [5]. Let be $x^{n+1} = X(\Delta t)$, the maximum value of Δt for which the series converge is τ, such that

$$\lim_{\Delta t \to \tau} \frac{dX}{d(\Delta t)} = \infty. \tag{23}$$

Given x^n for Eq. 11, this corresponds to

$$\frac{X(\tau) - x^n}{\tau^2} + \frac{1}{2} F(X(\tau)) = 0. \tag{24}$$

Let us consider two examples:

$U(x) = x^4/4$. We have $\tau = +\infty$ and $X(\tau) = \left(8E - (x^n)^4\right)^{1/4}$. In this case there is no additional bound on Δt.

$U(x) = -x^4/4$. If we fix the value of x^n and consider τ as a function of E, we have: for $E \leq -(x^n)^4/8$, the value of τ is infinite and there is no additional bound; for $E > -(x^n)^4/8$, the value of τ is finite and decreases as E increases, such that finally: $\tau \to 0$ as $E \to +\infty$. In this last case there is a second bound on Δt that becomes stricter as the energy grows.

2.4 *Application to first order equations*

Let us consider now the first order equation

$$\frac{dx}{dt} = V(x). \tag{25}$$

We can consider this equation as a special case of Eq. 2 with $U(x) = -mV^2(x)/2$ and $E = 0$. We can apply the conservative scheme to this problem: from Eq. 12 we have

$$\frac{x^{n+1} - x^n}{\Delta t} = \varepsilon^n \left(\frac{V^2(x^{n+1}) + V^2(x^n)}{2} \right)^{1/2}, \tag{26}$$

with

$$\varepsilon^n = \begin{cases} 1 & \text{if } V(x^n) > 0 \\ 0 & \text{if } V(x^n) = 0\,. \\ -1 & \text{if } V(x^n) < 0 \end{cases} \tag{27}$$

This scheme preserves the stability character of the equilibrium points of Eq. 25 and has no "ghost" solutions [4]. As an illustrative example let us consider the case $V(x) = x(1-x)$. The solution of Eq. 25 is

$$x(t) = \frac{x_o e^t}{1 - x_o + x_o e^t}, \tag{28}$$

with x_o the initial value. For $x_o > 0$ the solution tends to a stable limit of value 1 as t grows, while for $x_o < 0$ it has a blow-up at $t^* = \log(1 - 1/x_o)$. The conservative scheme Eq. 26 reproduces both these behaviors [4], but if we consider other schemes this is no longer true. For instance the central difference scheme

$$\frac{x^{n+1} - x^{n-1}}{2\Delta t} = x^n(1 - x^n) \tag{29}$$

gives solutions that after a certain time oscillate and show numerical chaos [6].

The Euler-Method (EM) is also not suitable for the study of the long time behavior of the system. The numerical scheme for EM is

$$\frac{x^{n+1} - x^n}{\Delta t} = V(x^n), \tag{30}$$

and considering two steps we have

$$\frac{\frac{x^{n+2} - x^{n+1}}{\Delta t} - \frac{x^{n+1} - x^n}{\Delta t}}{\Delta t} = \frac{V(x^{n+1}) - V(x^n)}{\Delta t}, \tag{31}$$

that is

$$\begin{aligned} \frac{x^{n+2} - 2x^{n+1} + x^n}{\Delta t^2} &= \frac{V(x^{n+1}) - V(x^n)}{x^{n+1} - x^n} \frac{x^{n+1} - x^n}{\Delta t}, \\ &= \frac{V(x^{n+1}) - V(x^n)}{x^{n+1} - x^n} V(x^n). \end{aligned} \tag{32}$$

This scheme, defined by EM for the associated second order equation, is not conservative, not symplectic and not time-reversible.

3. Symplectic schemes

3.1 *Symmetries and conservation law*

The are several ways to define a symplectic scheme, different techniques can be found in [7,8,9]. The important feature is to ensure the preservation of a canonical evolution in the discrete level, as a counterpart of the canonical evolution given by Eq. 7 for the continuous system. Besides the theoretical interest, these schemes can give very accurate solutions, although they may require a considerable work to be constructed [9] and/or high orders in Δt as we shall see. Here we are interested in a symplectic scheme, of similar order and complexity as the previous conservative one, to study Eq. 2. The discrete evolution equation is now

$$m\,\frac{x^{n+1} - 2x^n + x^{n-1}}{\Delta t^2} + F(x^{n+1}) = 0. \tag{33}$$

The scheme is time reversible but in this case there is not a discrete exact counterpart of the conservation law (except in the linear case). The change in the kinetic energy cannot be written as the difference between two potentials at the discrete level (Siegel Theorem):

$$\frac{m}{2}\left(\frac{x^{n+2} - x^{n+1}}{\Delta t}\right)^2 - \frac{m}{2}\left(\frac{x^{n+1} - x^n}{\Delta t}\right)^2 = \frac{F(x^{n+1})}{2}\,\frac{x^{n+2} - x^n}{2\Delta t}. \tag{34}$$

As a result of the nonexistence of a conserved discrete energy, the numerical solution can present an unphysical blow-up while the solutions of the underlying continuous system are bounded. As an example let us consider the quartic potential $U(x) = x^2/2 + x^4/4$ which has only bounded periodic solutions. It has been numerically checked [3] that Eq. 33 has a blowing-up solution in finite time for initial conditions $x(0) > 100.1$ and $\dot{x}(0) = 10$. When $x(0) = 100.458$, for instance, the blow-up time is $t^* = 7.67$ with $\Delta t = 0.01$. In this case, a second order symplectic scheme is not sufficient to ensure good accuracy.

3.2 *Symplectic structure*

As before, two consistent discretizations of Eq. 5 and Eq. 6 are possible, and we have

$$x^{n+1} = x^n + \Delta t\,\frac{p^n}{m}, \tag{35}$$

$$p^{n+1} = p^n + \Delta t F(x^{n+1}), \tag{36}$$

or

$$x^{n+1} = x^n + \Delta t\,\frac{p^{n+1}}{m}, \tag{37}$$

$$p^{n+1} = p^n + \Delta t F(x^n). \tag{38}$$

Both sets of equations correspond to symplectic evolution maps. In fact the second one is the standard map and corresponds to the Poincaré mapping of the impulsive hamiltonian

$$H_\delta = \frac{p^2}{2m} + U(x)\Delta t \sum_{n=0}^{\infty} \delta(t - n\Delta t). \tag{39}$$

The discrete problem corresponds exactly to a hamiltonian different of the original Eq. 8. This fact and the resonance analysis of H_δ explain the unphysical blow-up of the solutions of Eq. 33 mentioned before [10,11].

4. Applications of the conservative scheme

We present some applications of the conservative scheme previously defined to some other related problems of interest in Physics, such as dissipative systems and electric and magnetics fields.

4.1 *Dissipative systems*

Let us consider now Eq. 2 with a dissipative term

$$m\frac{d^2x}{dt^2} = F(x) + g(x)\dot{x}, \tag{40}$$

with:

$$F(x) = -\frac{dU}{dx} \quad \text{and} \quad g(x) = \frac{dG}{dx}.$$

The expression of the energy of the deterministic part is the same as Eq. 3 and its variation is

$$\begin{aligned} \frac{dE}{dt} &= g(x)\dot{x}^2, \\ &= \frac{dG}{dt}\dot{x}. \end{aligned} \tag{41}$$

Depending on the sign of $g(x)$, E increases or decreases and this property should be satisfied by the numerical scheme. The discrete equation is now

$$m\,\frac{x^{n+1} - 2x^n + x^{n-1}}{\Delta t^2} + \frac{U(x^{n+1}) - U(x^{n-1})}{x^{n+1} - x^{n-1}} = \frac{G(x^{n+1}) - G(x^{n-1})}{2\Delta t}, \tag{42}$$

and the energy, defined as in Eq. 12, changes as

$$\begin{aligned} \frac{E^{n+1} - E^n}{\Delta t} &= \frac{G(x^{n+1}) - G(x^{n-1})}{x^{n+1} - x^{n-1}} \left(\frac{x^{n+1} - x^{n-1}}{2\Delta t}\right)^2 \\ &= \frac{G(x^{n+1}) - G(x^{n-1})}{2\Delta t}\,\frac{x^{n+1} - x^{n-1}}{2\Delta t}. \end{aligned} \tag{43}$$

This condition ensures a good representation of Eq 42. As an example, let us consider the Van Der Pol Oscillator: $m = 1$, $F(x) = -x$, $g(x) = -\varepsilon(x^2 - 1)$. Eq. 42 can be rewritten as

$$(1+\frac{\Delta t^2}{2}-\frac{\varepsilon\Delta t}{2})x^{n+1}+\frac{\varepsilon\Delta t}{6}(x^{n+1})^3+(1+\frac{\Delta t^2}{2}+\frac{\varepsilon\Delta t}{2})x^{n-1}-\frac{\varepsilon\Delta t}{6}(x^{n-1})^3-2x^n = 0. \tag{44}$$

The solution of this cubic equation in x^{n+1} is unique if $\Delta t < 2/\varepsilon$ and the scheme reproduces the long time behavior of the continuous system [12]. A similar approach for nonlinear wave equations with dissipation can be found in [13].

4.2 *Motion in an electric field*

Let us consider the non relativistic motion of a charged particle in an electric field $\vec{\mathcal{E}}$. The equation of the motion is

$$m\frac{d^2\vec{x}}{dt^2} = e\vec{\mathcal{E}}(\vec{x},t) \\ = -e\vec{\nabla}\phi(\vec{x},t), \tag{45}$$

where e is the charge of the particle and ϕ the electric potential. The energy E is given by

$$E = \frac{m}{2}\left(\frac{d\vec{x}}{dt}\right)^2 + \phi, \tag{46}$$

and its variation law is

$$\frac{dE}{dt} = e\frac{\partial\phi}{\partial t}. \tag{47}$$

A conservative numerical scheme can be defined such that it has a consistent discretization of both the energy and its variation law. Let be the discrete position vector $\vec{x}^n \equiv (x^n, y^n, z^n)$, the equation for the discrete evolution is

$$m\frac{\vec{x}^{n+1} - 2\vec{x}^n + \vec{x}^{n-1}}{\Delta t^2} = -e\left(\phi(\vec{x}^{n+1},t^n) + \phi(\vec{x}^{n-1},t^n)\right) \\ \times\left(\frac{\vec{\imath}}{x^{n+1}-x^{n-1}} + \frac{\vec{\jmath}}{y^{n+1}-y^{n-1}} + \frac{\vec{k}}{z^{n+1}-z^{n-1}}\right), \tag{48}$$

with $t^n \equiv n\Delta t$. We define the discrete energy as

$$E^n = \frac{m}{2}\left(\frac{\vec{x}^{n+1} - \vec{x}^n}{\Delta t}\right)^2 + \frac{e}{2}\left(\phi(\vec{x}^{n+1},t^{n+1}) + \phi(\vec{x}^n,t^n)\right) \tag{49}$$

and the variation is

$$\frac{E^{n+1} - E^n}{\Delta t} = \frac{e}{2}\left(\frac{\phi(\vec{x}^{n+2},t^{n+2}) - \phi(\vec{x}^{n+2},t^{n+1})}{\Delta t} + \frac{\phi(\vec{x}^n,t^{n+1}) - \phi(\vec{x}^n,t^n)}{\Delta t}\right). \tag{50}$$

For an application of this scheme, see [14].

We consider now the one-dimensional relativistic motion of a charged particle in a constant electric field with the possibility of a dissipation. Let the motion be in the x direction, p be the x-coordinate of the momentum and m_o the mass at rest. Let be a constant electric field $\vec{\mathcal{E}} = (\mathcal{E},0,0)$ such that $\mathcal{E}$ is constant and $\phi(x) = -e\mathcal{E}x$. The equations of the motion are

$$\dot{x} = \frac{p}{\sqrt{m_o{}^2 + p^2}}, \tag{51}$$

$$\dot{p} = e\mathcal{E} - \alpha\dot{x}, \tag{52}$$

where α is the coefficient of the dissipation. The energy is now

$$E = \sqrt{m_o{}^2 + p^2} + \phi(x), \tag{53}$$

and its variation

$$\frac{dE}{dt} = -\alpha\dot{x}^2. \tag{54}$$

We define now the associated conservative numerical scheme

$$\begin{aligned}\frac{x^{n+1} - x^n}{\Delta t} &= \frac{\sqrt{m_o{}^2 + (p^{n+1})^2} - \sqrt{m_o{}^2 + (p^{n+1})^2}}{p^{n+1} - p^n} \\ &= \frac{p^{n+1} + p^n}{\sqrt{m_o{}^2 + (p^{n+1})^2} + \sqrt{m_o{}^2 + (p^n)^2}},\end{aligned} \tag{55}$$

$$\frac{p^{n+1} - p^n}{\Delta t} = -\frac{\phi(x^{n+1}) - \phi(x^n)}{x^{n+1} - x^n} - \alpha\frac{x^{n+1} - x^n}{\Delta t}. \tag{56}$$

The discrete energy is

$$E^n = \phi(x^n) + \sqrt{m_o{}^2 + (p^n)^2} \tag{57}$$

and its variation

$$\frac{E^{n+1} - E^n}{\Delta t} = -\alpha\left(\frac{x^{n+1} - x^n}{\Delta t}\right)^2. \tag{58}$$

In the case of no dissipation, $\alpha = 0$, the solution of Eq. 51 and Eq. 52 is

$$x(t) = x_o + \frac{1}{eE}\left(\sqrt{m_o{}^2 + (eEt)^2} - \sqrt{m_o{}^2 + p_o{}^2}\right), \tag{59}$$

$$p(t) = p_o + eEt, \tag{60}$$

with $x(0) = x_o$ and $p(0) = p_o$. The numerical solution agrees with the analytical one with a relative error less than 10^{-15} for $\Delta t = 10^{-3}$. On the other hand, due to its definition Eq. 55, the numerical velocity is always smaller than unity as the theory requires.

Let us consider now a dissipative case with α depending on the velocity $v \equiv \dot{x}$:

$$\alpha(v) = -\frac{A}{v^3}(B + C\sqrt{1 - v^2}), \tag{61}$$

with A, B and C some constants. There exists a critical value of the velocity, v_c, such that $\dot{p} = 0$. For $v(0) > v_c$, it has been found that an integrating IMSL routine diverges when $v(0) \approx v_c$ and otherwise its convergence is slow, while the solution with the conservative scheme converges in all cases and does it some 20 times faster than the IMSL routine.

4.3 *Motion in a magnetic dipole field*

The equation of the three-dimensional motion of a non relativistic charged particle in a magnetic field is

$$m\frac{d^2\vec{x}}{dt^2} = \frac{e}{c}\dot{\vec{x}} \times \vec{B}, \tag{62}$$

where c is the speed of light. The magnetic dipolar field $\vec{B}$ is given by

$$\vec{B} = \vec{\nabla} \times \vec{A}, \tag{63}$$

$$\vec{A} = A(-y, x, 0), \tag{64}$$

$$A = \frac{-\mu}{(x^2 + y^2 + z^2)}. \tag{65}$$

This system has two conserved quantities, the energy and the azimuthal angular momentum. This permits to reduce the problem to a two-dimensional motion of a particle in the ρ-z plane, $\rho \equiv \sqrt{x^2 + y^2}$. Introducing dimensionless variables, the resulting hamiltonian of the problem is

$$H = \frac{1}{2}(\dot{z}^2 + \dot{\rho}^2) + U(\rho, z), \tag{66}$$

with

$$U(\rho, u) = \frac{1}{2}\left(\frac{1}{\rho} - \frac{\rho}{(z^2 + \rho^2)^{3/2}}\right)^2. \tag{67}$$

This is known as the Størmer problem, related to magnetospheric and radiation belt phenomena (see [15] and references therein). A property of the system is the existence of a threshold value of the energy separating trapped and untrapped orbits. Some numerical schemes have been used to study this problem [15] but the long time behavior of the solutions was not well represented due to cumulative errors for long integration runs. Different reliable conservative schemes can be defined for this problem [16], we consider here:

$$\frac{z^{n+1} - 2z^n + z^{n-1}}{\Delta t^2} + \frac{U(\rho^{n-1}, z^{n+1}) - U(\rho^{n-1}, z^{n-1})}{z^{n+1} - z^{n-1}} = 0, \tag{68}$$

$$\frac{\rho^{n+1} - 2\rho^n + \rho^{n-1}}{\Delta t^2} + \frac{U(\rho^{n+1}, z^{n+1}) - U(\rho^{n-1}, z^{n+1})}{\rho^{n+1} - \rho^{n-1}} = 0. \tag{69}$$

The associated conserved energy is

$$E^n = \frac{1}{2}\left(\frac{z^{n+1} - z^n}{\Delta t}\right)^2 + \frac{1}{2}\left(\frac{\rho^{n+1} - \rho^n}{\Delta t}\right)^2 + \frac{1}{2}\left(U(\rho^{n+1}, z^{n+1}) + U(\rho^n, z^n)\right). \tag{70}$$

5. Conclusions

We have defined a conservative numerical scheme to study the motion of a particle in a general potential, with very accurate features and reliable results. We have compared it with a same order symplectic scheme with a positive result. We have seen some applications to some other related problems and obtained satisfactory results. All this shows the relevance of the numerical schemes with discrete analogous of the conservation or variation laws. It is possible, as a second step, to generalize these properties in order to construct a conservative numerical scheme for nonlinear wave equations with possible dissipative and/or stochastic terms. Some applications can be found in [1,3,17,18,19].

ACKNOWLEDGEMENTS

One of the authors (S.J.) is supported by the Fellowship EX88–02852481 of the Ministerio de Educación y Ciencia. We thank the financial support from the Dirección General de Investigación Científica y Técnica, under grant PB86–005.

REFERENCES

1. W.A.Strauss and L.Vázquez, *J. Comp. Phys.* **28**, (1978) 271.
2. D.Greenspan, *J. Comp. Phys.* **56**, (1984) 28.
3. S. Jiménez and L. Vázquez, *App. Math. Comp.* **35**, (1990) 61.
4. L.Vázquez, *An. Fís.* **83A**, (1987) 254.
5. J.Hagel and B.Zotter, *CERN report LEP-TH/*__88-59__, (1988).
6. M.Yamaguti and S.Ushiki, *Physica* **3D**, (1981) 618.
7. R.D.Ruth, *IEEE Trans. Nuc. Sci.* **NS-35**, (1983) 2669.
8. K.Feng, *J. Comp. Math.* **4**, (1986) 279.
9. P.J.Channell and J.C.Scovel, "Symplectic Integration of Hamiltonian Systems", submitted to *Nonlinearity*, and references therein.
10. G.Guignard and J.Hagel, *CERN report LEP-TH/*__85-3__, (1985).
11. J.Hagel and L.Vázquez, in preparation.
12. J.A.García and L.Vázquez, *Revista Telefónica* **19**, (1988) 73.
13. P.J.Pascual and L.Vázquez, *Phys. Rev.* **32B**, (1985) 8305.
14. A.P.Navarro, L.Rodríguez and L.Vázquez, "A model for runaway dynamics in tokamaks using a conservative numerical scheme", submitted to *J. Comp. Phys.*.
15. A.J.Dragt and J.M.Finn, *J. Geophys. Res.* **81**, (1976) 2327.
16. S.Jiménez and L.Vázquez, *App. Math. Comp.* **25**, (1988) 207.
17. A.Sánchez and L.Vázquez, in *Lecture notes in Physics* **353**, eds. M.Barthes and J.Leon (Springer-Verlag, New York, Heidelberg, Berlin, 1990) 251.
18. Yu. S.Kivshar, S.A.Gredeskul, A.Sánchez and L.Vázquez, *Phys. Rev. Lett.*__64__, (1990) 1693.
19. S. Jiménez, L. Vázquez and W.W.Zachary, "Behavior of solutions of Duffing Equations and Wave Equations with cubic nonlinearities, cubic dissipation and periodic forcing", submitted to *Zeitschrift für Naturfoschung*.

TRANSITION TO INSTABILITY AND CHAOS IN 3-D HAMILTONIANS

G. CONTOPOULOS
Astronomy Department, University of Athens
Athens, Greece

and

Astronomy Department, University of Florida
Gainesville, Fl., USA

SUMMARY

Systems of 3 degrees of freedom present qualitatively new phenomena like complex instability and Arnold diffusion. The appearance of complex instability shows that a system is not integrable. At a transition to complex instability there is no bifurcation of another family of periodic orbits. We found evidence that in many 3-D Hamiltonians there are no infinite Feigenbaum sequences. The sequences of bifurcations terminate in general either at a complex instability or at an inverse bifurcation.

The structure of phase-space has been explored. We found, in particular, the changes of the topology of the complex unstable regions as various parameters change.

The behaviour of nonperiodic orbits was studied by calculating the Lyapunov characteristic number (LCN). The Lyapunov numbers are almost constant in the large stochastic sea, but they are small in regions of complex instability. However after a long time the approximate value of LCN changes abruptly to the value appropriate for the large stochastic sea. This phenomenon is analogous to the crossing of a cantorus. We found the various families of periodic orbits (invariant points) and of invariant lines on the surfaces of section in order to understand the transition from ordered to stochatic motion.

1. Complex Instability

Hamiltonian systems of 3 or more degrees of freedom present new phenomena that do not appear in systems of 2 degrees of freedom. Such phenomena are[1]

(1) Complex instability,

(2) Arnold diffusion, and

(3) Collisions of bifurcations.

In the present review we will discuss recent work concerning the first two topics.

Complex instability appears when th eigenvalues of a periodic orbit are complex conjugate but not on the unit circle. The eigenvalues (λ) are given by an equation of the form

$$(\lambda^2+b_1\lambda+1)(\lambda^2+b_2\lambda+1) = 0 \ , \qquad (1)$$

where b_1,b_2 are the stability parameters of the periodic orbit. In the complex unstable case b_1 and b_2 are complex conjugate. When one of b_1,b_2 becomes equal to -2 or 2 we have a bifurcation of equal or double period respectively. However when complex instability is introduced b_1 is equal to b_2, but not ± 2 in general. Therefore there is no bifurcation of another family of periodic orbits. At such a transition chaos is introduced abruptly, without an infinity of period doubling bifurcations.

There is numerical evidence that in most 3-dimensional cases (i.e. cases that cannot be separated into a 2-dimensional system and an independent 1-dimensional system) there are no infinite period-doubling sequences. The sequences of bifurcations are terminated either at a complex instability or at an inverse bifurcation[2]. The same is true in 4-dimensional symplectic maps[3]. As pointed out by Mao and Greene[4] the symmetric 4-dimensional symplectic maps that are fixed under the period doubling renormalization operator consist of 2 <u>uncoupled</u> 2-D area preserving maps. Therefore such maps do not represent generic 4-D maps.

Thus we can state that in generic 4-D maps the introduction to chaos follows a different pattern than in 2-D maps.

In order to study complex instability in 3-D Hamiltonian systems we introduced the so-called "existence diagrams" [5]. These diagrams give, in a 2-D parameter space, the regions where a particular family of periodic orbits exists, and is stable, or unstable. In particular we found the regions of complex instability, for several families of periodic orbits, in the simple Hamiltonian

$$H \equiv \frac{1}{2}\left(\dot{x}^2+\dot{y}^2+\dot{z}^2+Ax^2+By^2+Cz^2\right) - \varepsilon xz^2 - \eta yz^2 = h \ , \qquad (2)$$

For fixed values of A,B,C and of the energy h the existence diagram in the parameter space (ε,η) has the form of Fig. 1. We notice that the complex unstable region has only one point in common with the ε-axis (η=0). This is a generic sitution. In fact for η=0 the Hamiltonian (2) is reduced to two uncoupled Hamiltonians

$$H_{xz} = \frac{1}{2}(\dot{x}^2+\dot{z}^2+Ax^2+Cz^2)-\varepsilon xz^2, \quad H_y = \frac{1}{2}(\dot{y}^2+By^2), \qquad (3)$$

therefore we cannot have complex instability. Only at certain points of the ε-axis we may have $b_1=b_2$, i.e. the boundary of a complex unstable region. (In an exceptional case one may have $b_1=b_2=\pm 2$ along the whole ε-axis; this axis is then the boundary of the complex unstable region).

The appearance of complex instability proves that the system is nonintegrable. It is not even partially integrable, i.e., it cannot have even 1 integral besides the Hamiltonian itself as in the case of the system (3).

We studied the change of the complex unstable region of Fig. 1 by changing the parameters A,B and C and found the following types of evolution.

(1) The region Δ of Fig.1 increases to the right, and reaches infinity (Fig. 2). In other cases the region Δ reaches infinity upwards or both to the right and upwards (Fig. 3).

(2) The region Δ of Fig. 1 becomes thinner and approaches the η-axis (Fig. 4a). As C decreases the region Δ reaches the η-axis at one point and then splits into two parts (Fig. 4b). One part has only point in common with the η-axis, while the other part has points in common with both axes ε and η. As C decreases further the second part shrinks to one point (the origin) and disappears.

(3) A region Δ similar to Fig. 1 reaches the η-axis at one point (Fig. 5a) and then detaches itself from the ε-axis so that it remains attached only to the η-axis (Fig. 5b).

(4) A region Δ which has originally 1 point on the η-axis (like Fig. 5b) detaches itself from this axis (Fig. 6).

(5) A region Δ similar to Fig. 1 shrinks to one point on the ε-axis and disappears. This happens to the family 1b for B=C=0.4 when A decreases below A=1.

(6) A region Δ similar to Fig. 6 shrinks to one point outside the ε- and η-axes and disappears. This happens to the family 2k for B=0.42, C=0.4 as A increases beyond 1.61.

(7) A region Δ similar to Fig.5b splits into two, at a point outside the axes (Fig.7).

Every family of periodic orbits has its own evolution as the various parameters A,B,C change. In the present review we do not consider the stable regions or the simply and doubly unstable regions. A detailed discussion of the main qualitative changes that appear in the system (2) is given by Contopoulos[6] and Zachilas[7].

Similar existence diagrams have been calculated in realistic models of galaxies[8,9] and in more general 3-D

Hamiltonians and 4-D maps[3]. In particular complex unstable regions are very common in such systems. Therefore the phenomenon of complex instability is quite general and affects appreciably the dynamics of galaxies, as well as the dynamics of systems with quite general Hamiltonians.

2. Arnold diffusion

In systems of 3 degrees of freedom the phase space is 5-dimensional (assuming that the Hamiltonian is time-independent). The KAM tori are tree-dimensional, and they do not separate the phase space. Therefore the stochastic regions close to unstable periodic orbits may communicate. In this way orbits starting close to an unstable periodic orbit may diffuse slowly through these stochastic regions (the Arnold web) and go to large distances. This slow diffusion is called Arnold diffusion.

However the time scale of such a diffusion is of importance. E.g. in galactic dynamics it was found that no appreciable diffusion exists in some cases for more than 10^5 periods, while the age of the Universe is about 100 periods. Therefore in galactic dynamics Arnold diffusion is in general unimportant.

On the other hand in plasma physics we may have cases with 10^8-10^{10} oscillations, therefore the role of Arnold diffusion may be important.

Besides Arnold diffusion we find a fast diffusion in certain regions of phase space, and the question arises what is the relation between these two types of diffusion.

In order to study this problem we calculated the maximal Lyapunov characteristic number (LCN) for a large number of nonperiodic orbits in the Hamiltonian (2)[10]. The LCN was calculated by finding empirically the limit of the quantity

$$x = \frac{|\ln(\xi/\xi_o)|}{t} \tag{4}$$

as $t \to \infty$, where ξ is the solution of the variational equations with intitial conditions ξ_o, and t is the time.

We found that for large regions of initial conditions (Fig. 8) the LCN is practically the same (LCN$\approx 10^{-2}$) (Fig. 9). These regions constitute the "large stochastic sea", where diffusion is very fast. We found also large regions where the LCN is zero (below the level of the noise in the calculations). These regions we call ordered (blank regions in Fig. 8). The boundary between the stochastic and the ordered regions is not well defined, and it may be a fractal.

Close to the complex unstable orbit 1a we found a region where the LCN is nonzero but appreciably lower than the value of LCN in the large stochastic sea (Fig. 9). Our calculations extended to about $t \approx 2\times10^6$, while the period of the orbit 1a is about t=16. Therefore for at least

1.3×10^5 periods the orbits in the stochastic region near 1a do not communicate with the large stochastic sea.

We repeated the calculations for a larger value of ε (ε=0.56), keeping the other constants the same. In this case the stochastic region around the complex unstable orbit 1a is much larger, reaching the large stochastic sea (Fig. 10). However orbits starting in this region seem to have a much smaller Lyapunov number than the LCN in the large stochastic sea (Fig. 11). In fact the value of x seems to stabilize for some time at about $x \approx 0.0005$. But after a long time ($t \approx 220000$) the value of x increases abruptly to about 0.01, and the LCN takes probably the common value of the large stochastic sea (LCN≈0.0014). The transition, which occurs after about 14000 periods, is clearly seen in a projection of the phase space in Figs. 13a and b. We calculate the successive consequents (for z=0) of the orbit that gives the lower curve of Fig. 11. About 14000 consequents are restricted in the dark region of Fig. 12a, which seems to be well limited by an external boundary. However after ~14000 periods the consequents cross this apparent boundary and from then on they stay in the large stochastic sea (Fig. 12b).

This transition occurs also in other orbits of the small sea but the transition time shows a large dispersion.

The explanation of this phenomenon seems to be the following. In both the large and the small stochastic seas there is a fast diffusion, due to the lack of a large set of KAM surfaces. However between the small and the large sea there is a partial barrier composed of a locally concentrated set of KAM surfaces. Although this barrier is not a continuous 4-D surface, separating the large sea from the small sea, nevertheless it does not allow a fast communication between the two seas. The barrier seems to have several openings, that are like the holes of a cantorus. Thus we can have a diffusion through these openings after a long but not excessive time interval.

This situation can be visualized if we reduce the dimensions of the phase space and of the KAM tori by 2. Then we have a 3-D space containing 1-D tori (strings). Although the strings are not dense they provide some (leaky) barriers, if their density is large in certain regions. In the stochastic seas there are few strings, thus there is no obstacle for diffusion. However in the separating regions the measure of the strings is large. Although there are holes everywhere the total measure of such holes may be small. Thus the diffusion rate through them is small. The diffusion is rather abrupt if the thickness of the barriers composed of strings is small.

The situation is quite analogous to the diffusion through a large number of cantori[11] with holes of different sizes.

If the perturbation is small (in our case smaller ε) the thickness of the leaky barrier is very large. E.g. in Fig. 8 the blank region is presumably permeated by thin regions of stochasticity. However these regions are very small, while the distance between the small sea around the orbit 1a and the large sea is so large that the time needed for joining the two seas is extremely large.

On the other hand if ε increases beyond $\varepsilon=0.56$ the barriers between the small and the large sea become more and more leaky, i.e. the set of KAM tori (or strings) becomes smaller and smaller. For $\varepsilon=0.597$ we checked that the diffusion between the region around 1a and the large stochastic sea is very fast, and we cannot define a partially separated small sea any more.

Returning to the question formulated at the beginning of this paragraph it seems that there is no clear distinction between Arnold diffusion and fast diffusion in the stochastic seas. The fast or slow diffusion depends mainly on the relative importance of the KAM tori in various regions of phase space.

A further exploration of the diffusion mechanism and of the importance of the sets of KAM tori in various situations would be highly desirable.

3. Changes in the topology of 3-D orbits

In order to find the topology of an orbit in phase space we use projections of the orbit on particular planes, like (x,y) or $(x,\dot{x})$. Of special interest are the projections of the sets of consequents on a surface of section (e.g. $z=0$). The surface of section is 4-dimensional, therefore such projections do not give all the necessary information. More information can be provided by 3-dimensional stereoscopic projections[12,13]. However if an orbit is ordered then its consequents lie on an invariant surface and two 2-dimensional projections give an understanding of its qualitative characteristics.

Barbanis (private communication) has recently made an extensive study of the topological characteristics of orbits in the Hamiltonian (2), using mainly projections on the planes (x,y) and $(x,\dot{x})$. A few examples are given in Figs. 13(a,b) and 14(a,b,c,d,e), and refer to orbits starting at particular points of Fig. 10.

Of special interest are the exeptional cases in which the invariant surfaces are reduced to invariant lines (Fig. 14d). These are resonant cases in which two frequencies have a rational ratio. In fact an ordered orbit is quasiperiodic with 3 frequencies $\omega_1,\omega_2,\omega_3$, and in general the ratios $\omega_1,/\omega_2/\omega_3$ are irrational. On the other hand if all ratios are exactly rational we have a periodic orbit. The cases in which only one ratio is rational are intermediate. Such orbits were found already by Martinet and Magnenat[13], but no systematic study has been made up to now. Orbits of this type can be stable and define the topology of nearby nonresonant orbits (compare Figs. 14d and 14a), or they can be unstable [compare Fig. 14c (stable) and 14e (stochastic)]. The role of the unstable resonant quasiperiodic orbits is very important in defining the transitions from one topological form of orbits to another, or from ordered to stochastic orbits.

We are presently finding various periodic families and families of invariant lines bifurcating from a "central" family of periodic orbits (e.g the family 1a). In cases of 2 degrees of freedom we found that cantori are formed when all the periodic families with rotation numbers (ω_1/ω_2) close to a noble rotation number become unstable[14,15]. We expect that in a similar way we have a transition to fast diffusion when the periodic and resonant quasiperiodic families with rotation numbers $\omega_1/\omega_2/\omega_3$ close to noble numbers become unstable.

Acknowledgements

I thank Dr. b. Barbanis for providing Figs. 13 and 14, and Dr. L. Zachilas for drawing the other figures.

References

1. Contopoulos, G.: 1986, Particle Accelerators 19, 107.
2. Contopoulos,, G.: 1983, Lett. Nuovo Cimento 38, 257, 1983.
3. Contopoulos, G. and Giorgilli, A.: 1988, Meccanica 23, 19, 1988.
4. Mao, J.-M and Greene, J.M.: 1987, Phys.Rev.A 35, 3911.
5. Contopoulos, G. and Magnenat, P.: 1985, Celest.Mech. 37, 387.
6. Contopoulos, G.: 1986, Astron. Astrophys. 161, 244.
7. Zachilas, L.: 1988, PhD Thesis, Univ. of Athens.
8. Martinet, L. and Pfenniger, D.: 1987, Astron. Astrophys. 173, 81.
9. Patsis, P.A. and Zachilas, L.: 1990, Astron. Astrophys. 227, 37, 1990.
10. Contopoulos, G. and Barbanis, B.: 1989, Astron. Astrophys. 222, 329.
11. Meiss, J.D. and Ott, E.: 1986, Physica 20D, 387.
12. Froeschle, C.: 1970, Astron. Astrophys. 4, 115.
13. Martinet, L. and Magnenat, P.: 1981, Astron. Astrophys. 96, 68.
14. Greene, J.M.: 1979, J. Math. Phys. 20, 1183.
15. Contopoulos, G., Varvolgis, H. and Barbanis, B.: 1987, Astron. Astrophys 172, 55.

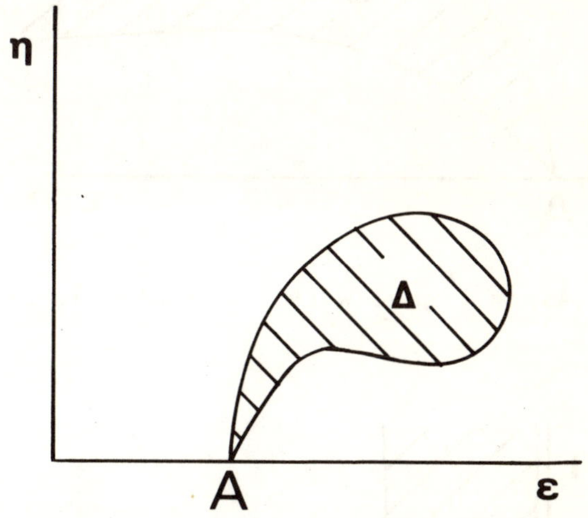

Fig. 1: Schematic "existence diagram" for the simple periodic orbits 1a in the Hamiltonian (2) with h=0.00765, A=0.9, B=0.4, C=0.225. The complex unstable region Δ, reaches the ε-axis at the point A.

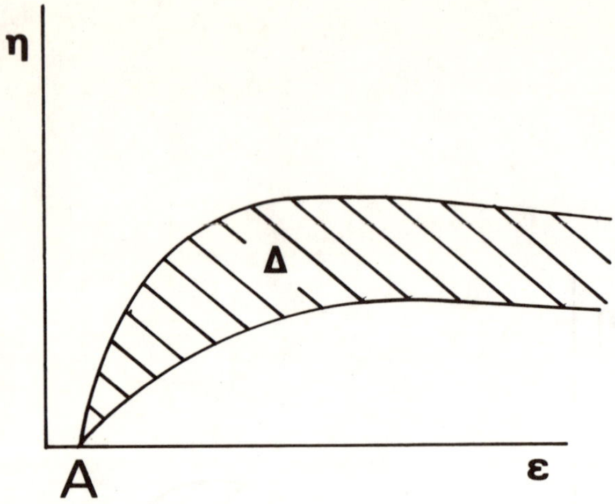

Fig. 2: As in Fig. 1 with A=0.9, B=0.5, C=0.225.

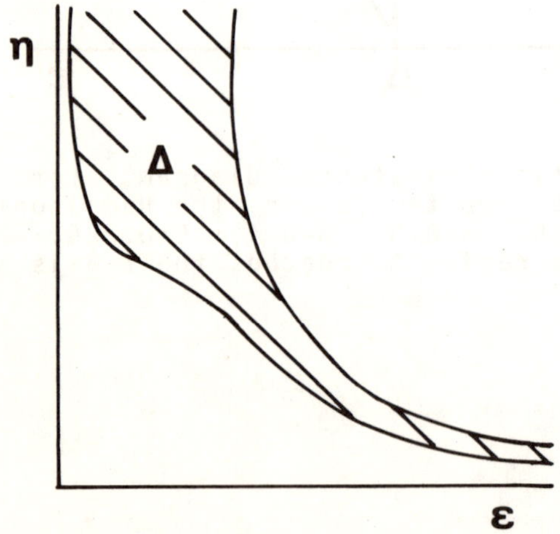

Fig. 3: As in Fig. 1 for the family 1g'.

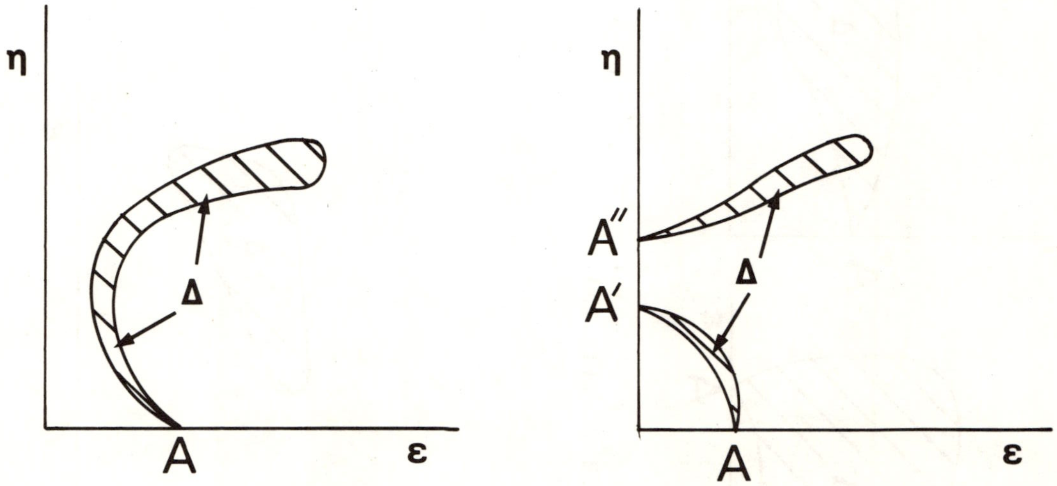

Fig. 4: As in Fig. 1 with A=1.6, B=0.4 and (a) C=0.23, (b) C=0.228.

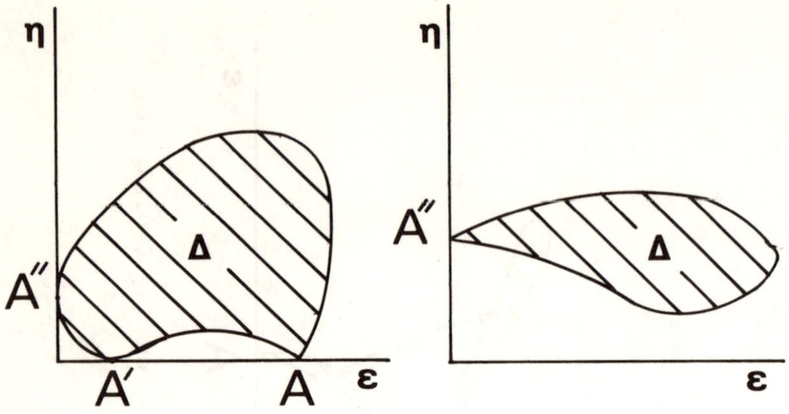

Fig. 5: As in Fig. 1 with A=1.2, B=0.4 and (a) C=0.747, (b)C=0.8 (Family 1b).

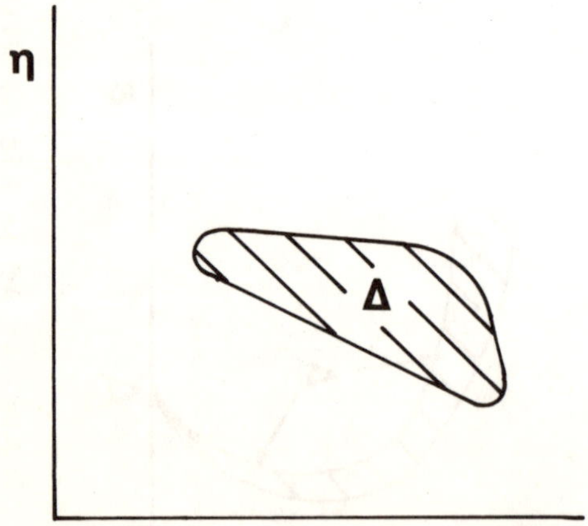

Fig. 6: As in Fig. 1 for the family 1c and A=1.4, B=C=0.4.

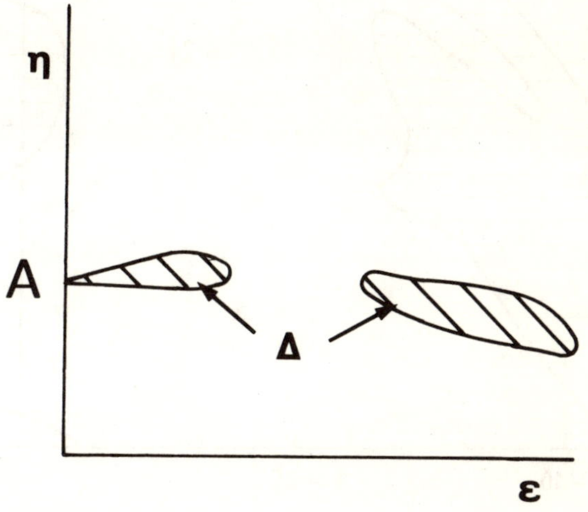

Fig. 7: As in Fig. 1 for the family 1i and A=1.5, B=C=0.4.

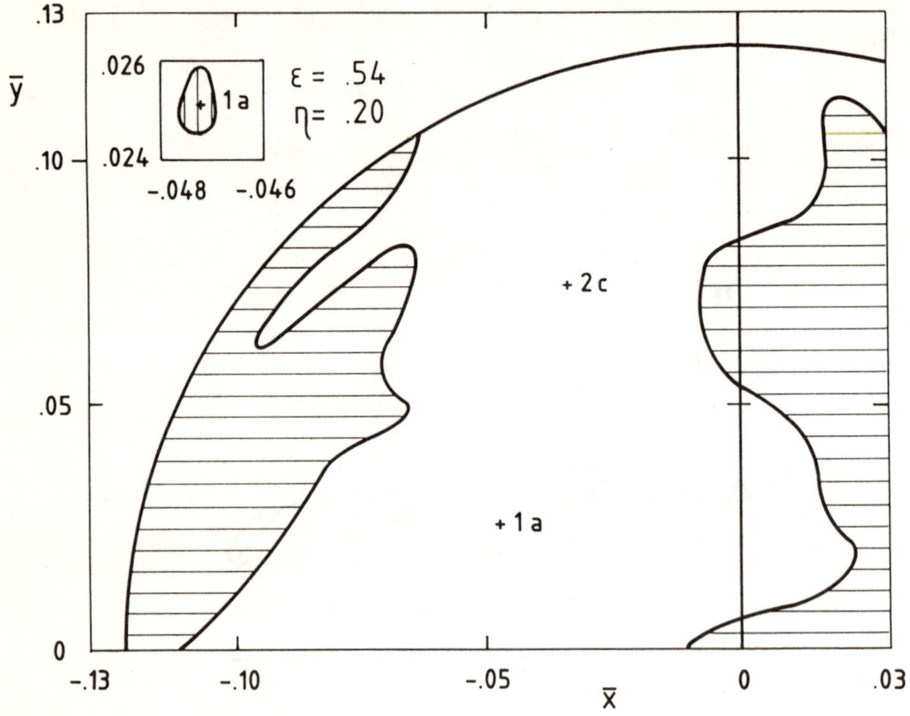

Fig. 8: Ordered (blank) and stochastic (dark) regions in the Hamiltonian (2) with A=0.9, B=0.4, C=0.225, $\varepsilon=0.54$, $\eta=0.2$. The orbits start with $z=\dot{x}=\dot{y}=0$, h=0.00765 upwards ($\dot{z}>0$) at many points $\bar{x}=A^{1/2}x$, $\bar{y}=B^{1/2}y$). The points 1a and 2c represent periodic orbits. The orbit 1a is complex unstable and it has a small neighbourhood (insert), where the LCN is different from zero (small sea).

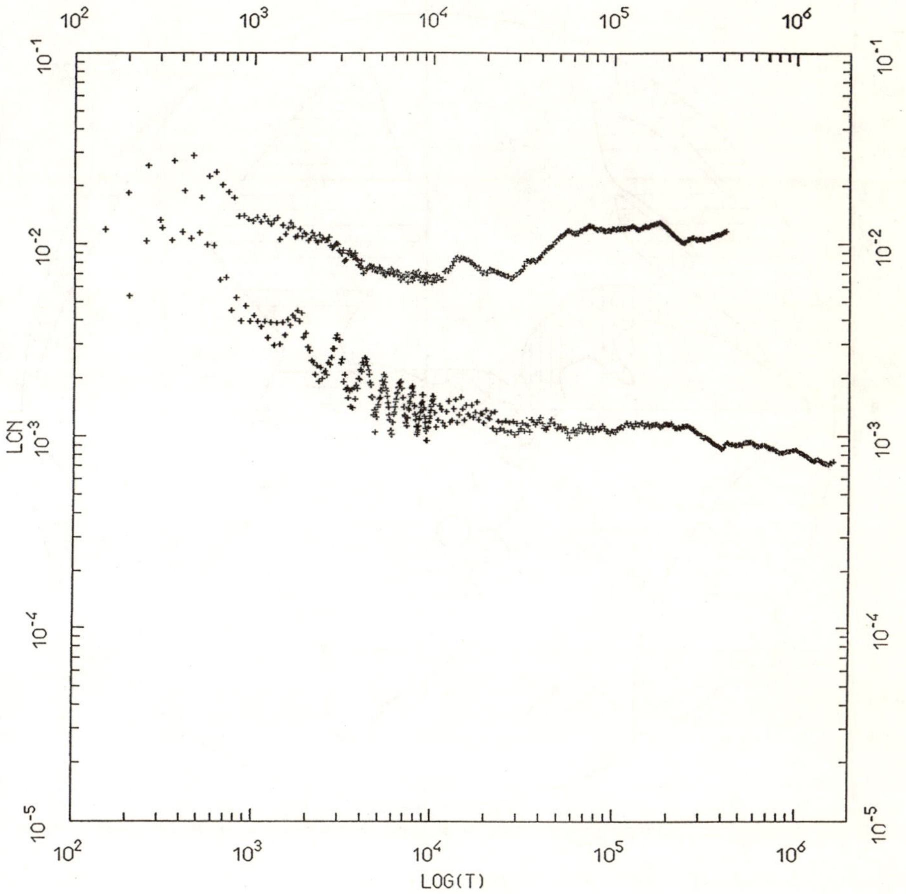

Fig.9: Calculation of the LCN's in the case of Fig. 9. In the large stochastic region the LGCN is about 10^{-2}, while in the neighbourhood of 1a it is close to 7×10^{-4}.

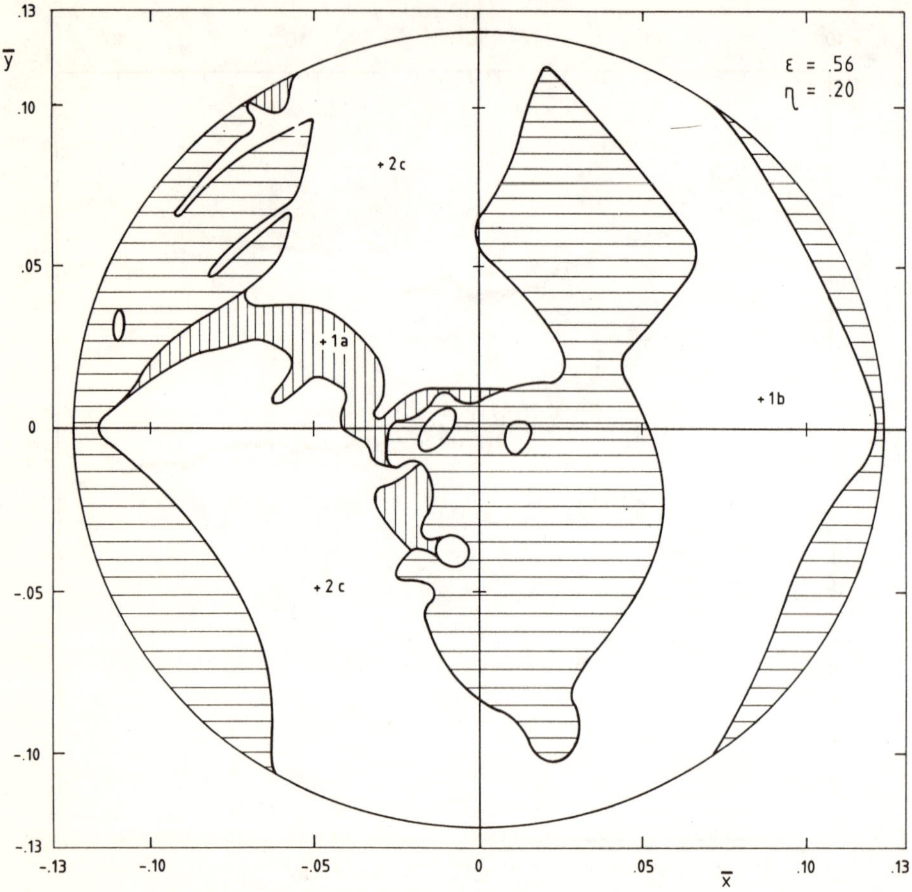

Fig.10: The same as in Fig. 8 for ε=0.56. In the dark regions with vertical lines (small sea) the value of x remains small for a large time and then jumps to the value appropriate for the large stochastic sea (horizontal lines).

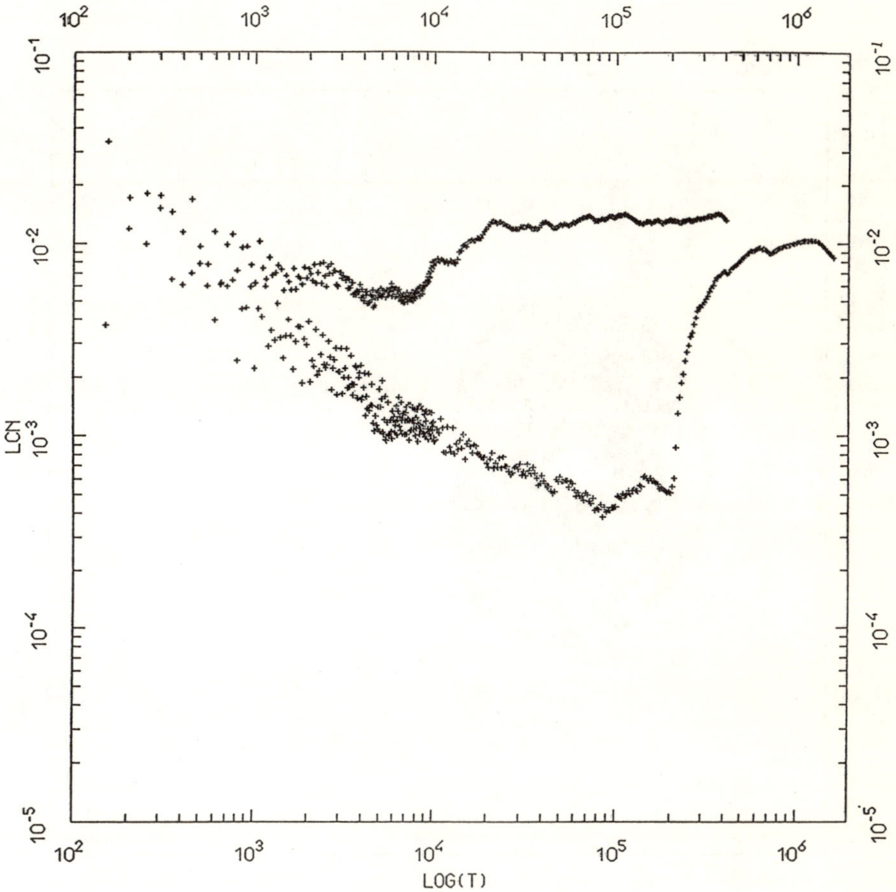

Fig.11: Calculation of the LCN's in the case of Fig.10. In the large stochastic sea LCN≈0.014, while in the small sea the value of x seems to stabilize near 0.0005, but then jumps to ~0.01.

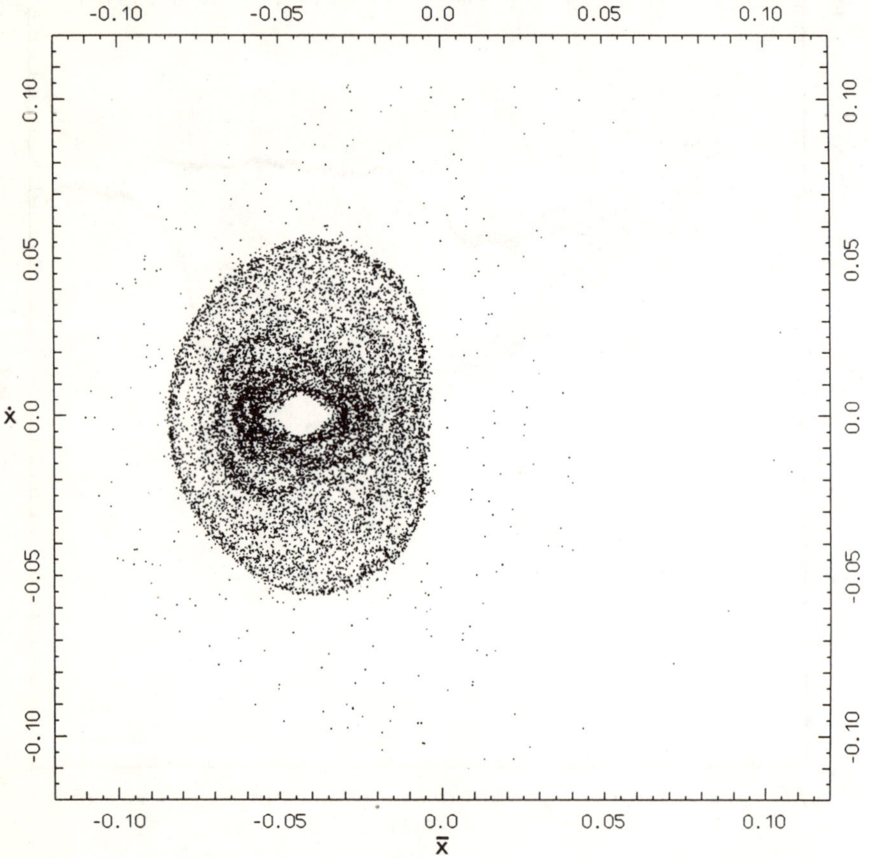

Fig. 12a

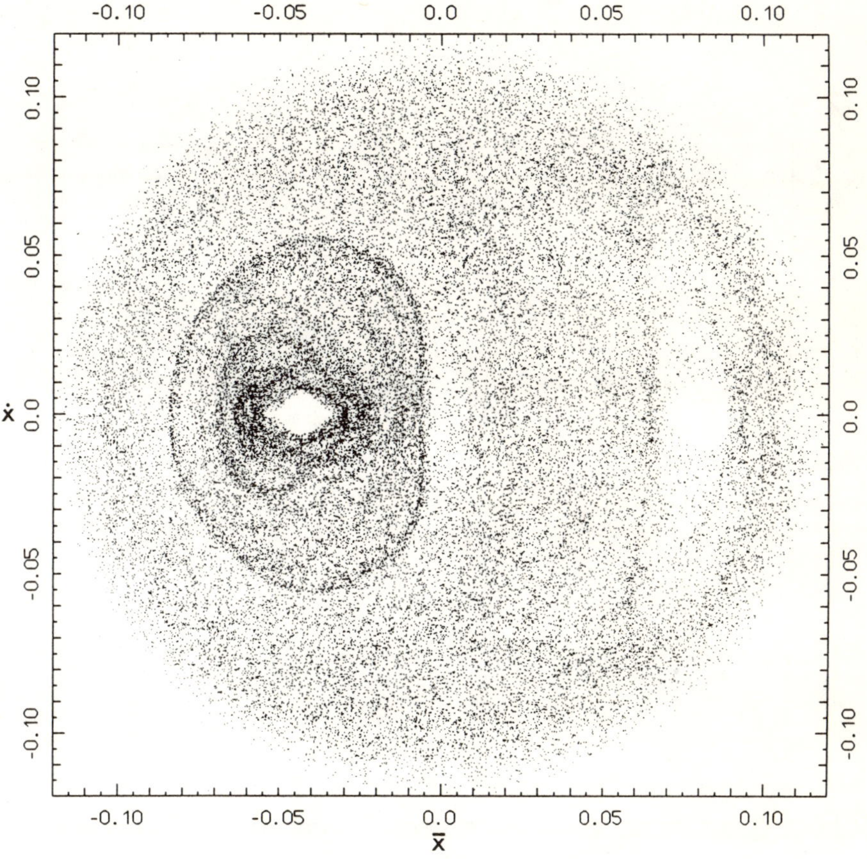

12b

Fig.12a,b: Projection of the consequents of an orbit starting in the small sea on the plane $(\bar{x},\dot{x})$. (a) The first 14000 consequents. (b) The first 50000 consequents.

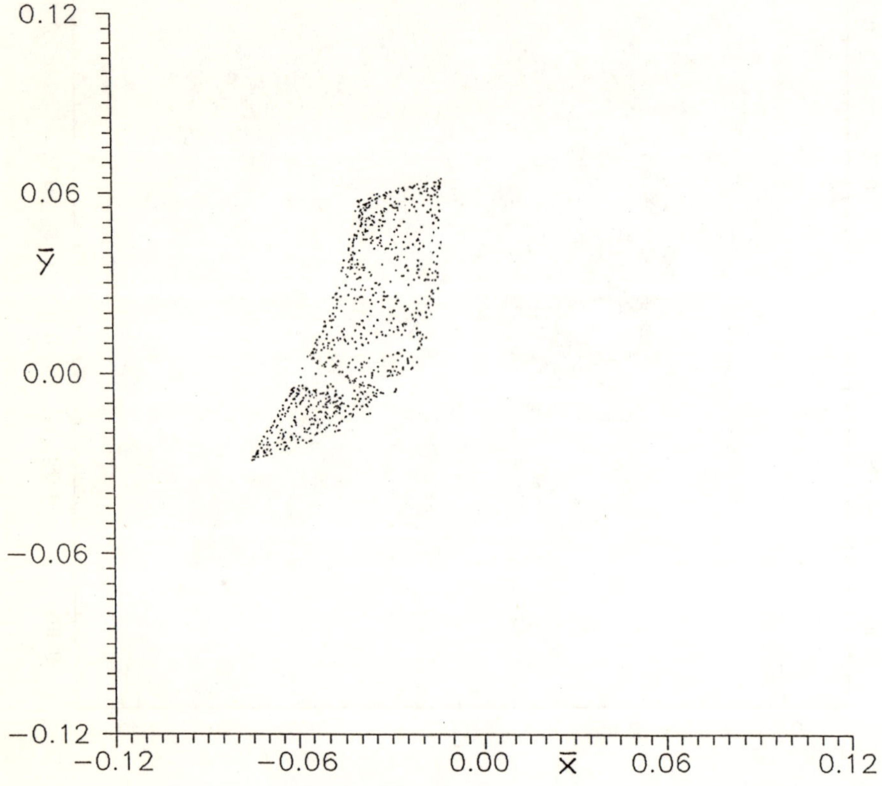

Fig. 13a

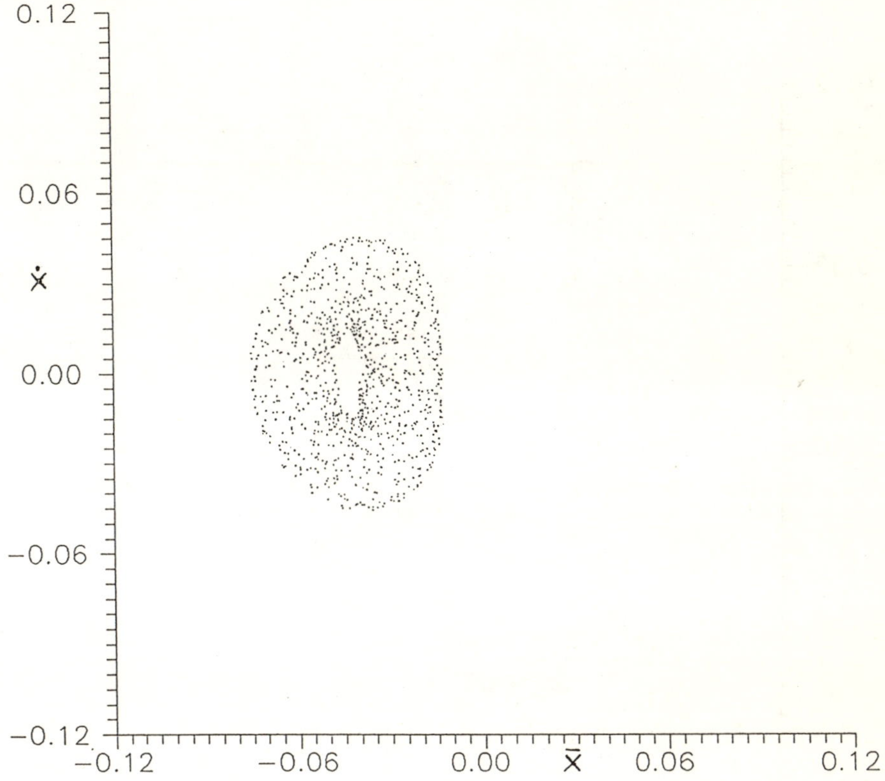

13b

Fig.13a,b: Projections of 2000 successive consequents of an orbit in the model of Fig.10 with initial conditions ($\bar{x}$=0.013, $\bar{y}$=0.065, $z=\dot{x}=\dot{y}=0$, h=0.00765, $\dot{z}>0$) on the planes (a) ($\bar{x},\bar{y}$) and (b) ($\bar{x},\dot{x}$).

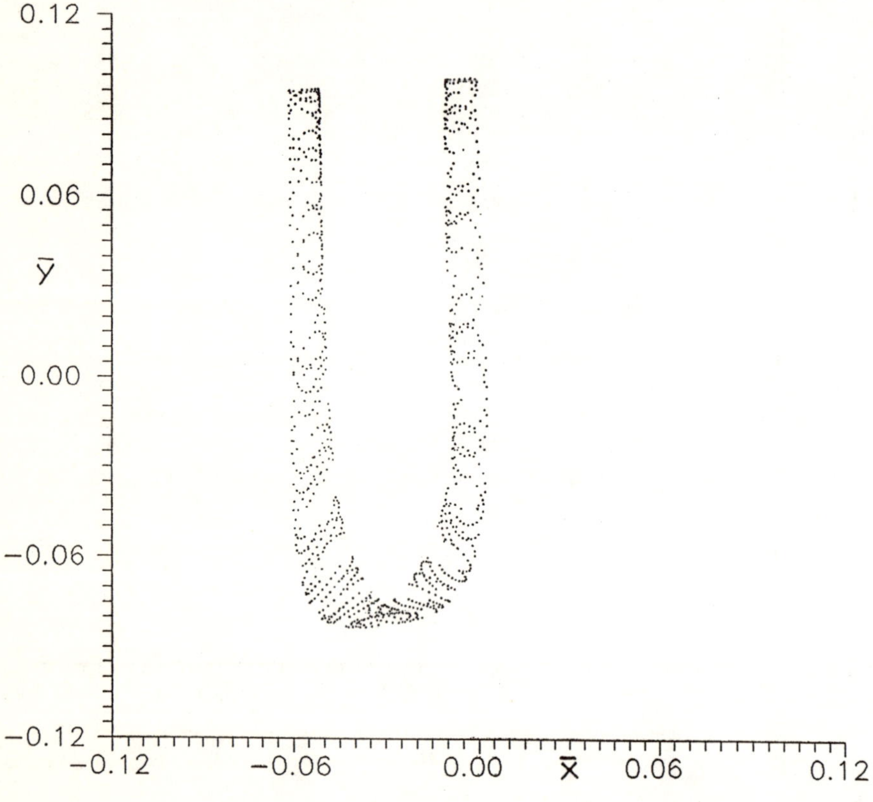

Fig. 14a

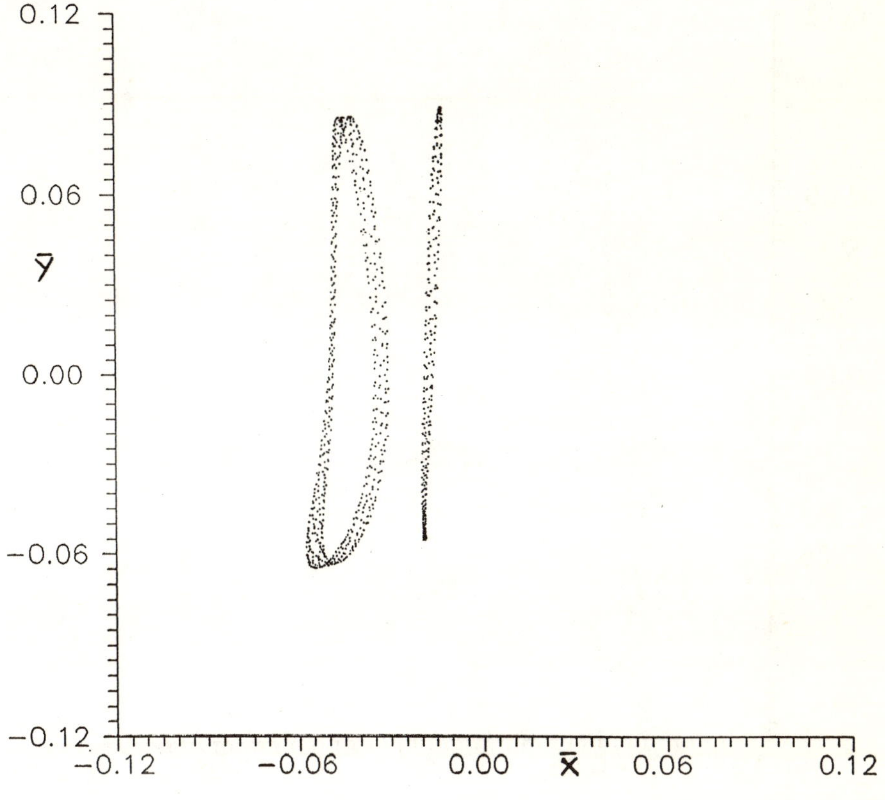

Fig. 14b

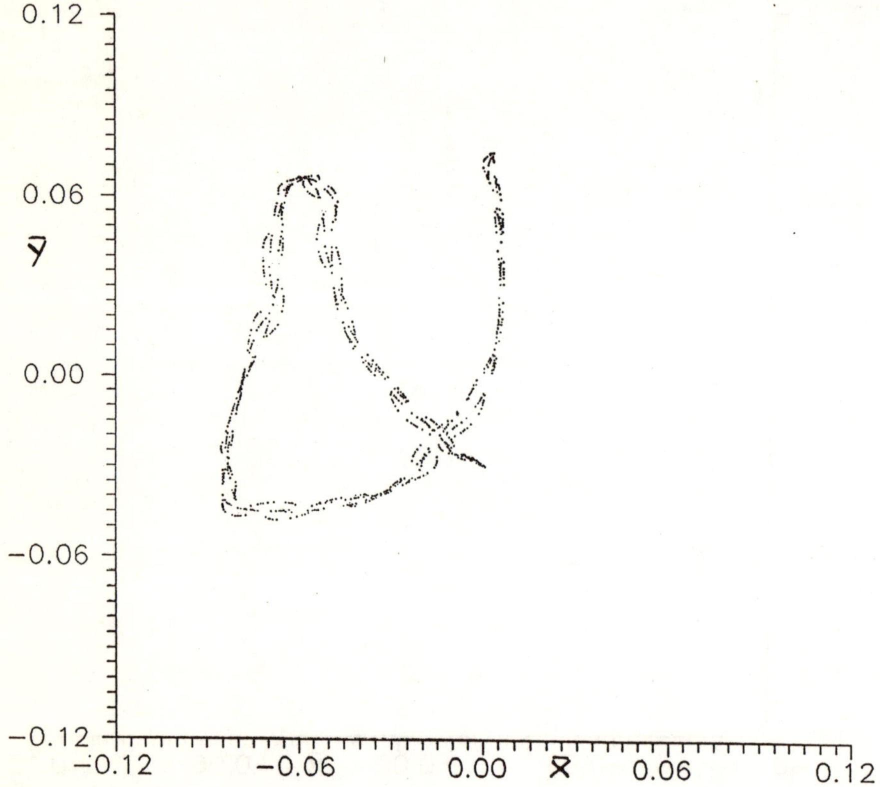

Fig. 14c

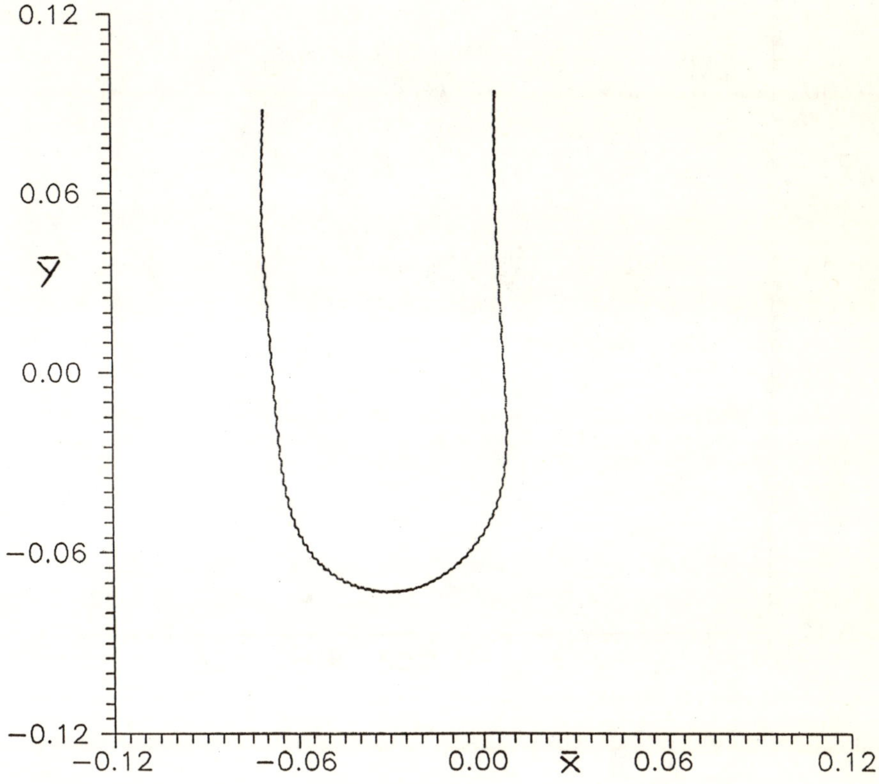

Fig. 14d

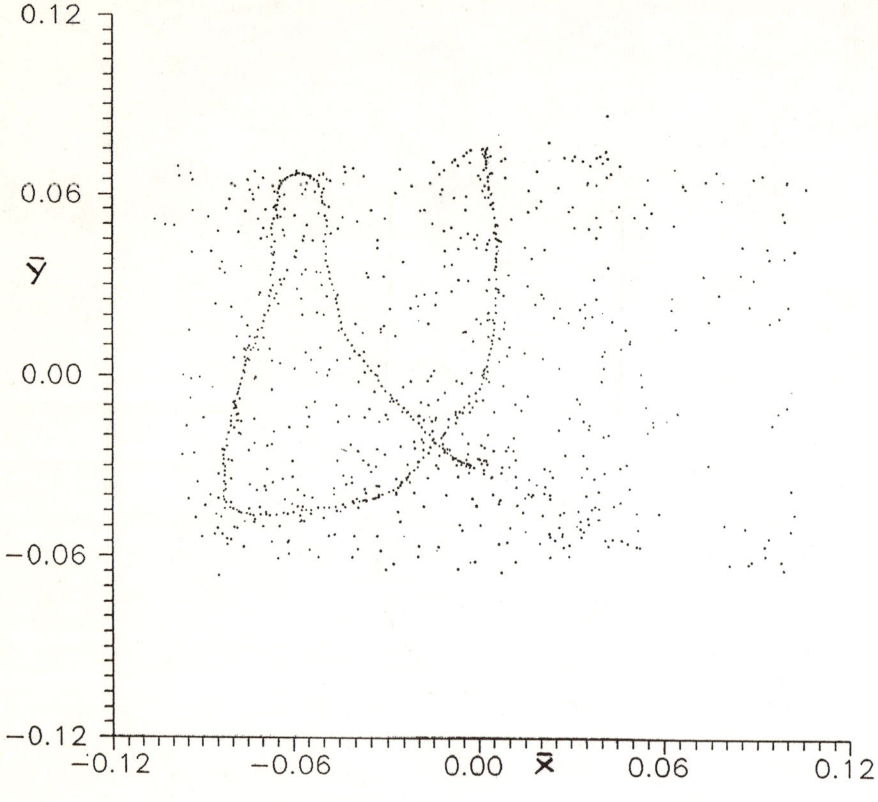

14e

Fig.14a,b,c,d: Projections on the plane ($\bar{x},\bar{y}$) of 2000 consequents of orbits starting at particular points of Fig.10 ($z=\dot{x}=\dot{y}=0$ and $\dot{z}>0$). (a) $\bar{x}=-0.011$, $\bar{y}=0.099$; (b) $\bar{x}=-0.013$, $\bar{y}=0.089$, (c) $\bar{x}=0.004$, $\bar{y}=0.075$; (d) $\bar{x}=0.005$, $\bar{y}=0.094$; (e) $\bar{x}=0.003$, $\bar{y}=0.076$.

RANDOM PERTURBATIONS OF DYNAMICAL SYSTEMS

YURI KIFER
Hebrew University of Jerusalem

ABSTRACT

The most celebrated problem in random perturbations of dynamical systems is to determine a stationary measure of a dynamical system which is most stable to N random perturbations and so has a physical sense. I intend to explain that, under certain circumstances, one can prove, and in a more general situation one can expect, that such measures should maximize certain variational formulas which means that they are equilibrium states for corresponding dynamical systems.

Models:

a) Discrete time:
$F : M \to M$ + random error perturbation with distribution $Q_{F_x} \sim \delta_{F_x}$

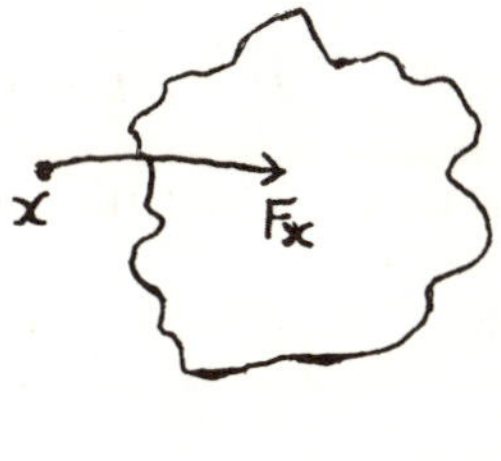

b) Continuous time:
Diffusion X_t^ϵ with generator $L^\epsilon = \epsilon L + (B, \nabla)$ (say $L = \Delta$) where L is a second order elliptic operator (i.e. $P^\epsilon(t, x, \Gamma) = P_x^\epsilon\{X_t^\epsilon \in \Gamma\}$ satisfies $\frac{\partial P^\epsilon}{\partial t} = L^\epsilon P^\epsilon$ and

$dX_t^\epsilon = \sqrt{\epsilon}\sigma dw_t + (\epsilon b + B)dt$; also diffusion is constructed by stochastic differential equations).
Importance: connection with PDE. X_t^ϵ are perturbations of the flow F^t given by the ordinary differential equation

$$\frac{d(F^t x)}{dt} = B(F^t x), \qquad F^t : M \to M \quad t \in (-\infty, +\infty).$$

$M-$ manifold.
X_t^ϵ (with $X_0^\epsilon = x$) is close to $F^t x$ for any time but it turns out that the global param. (as $t \to \infty$) of X_t^ϵ also depends crucially on the behaviour of F^t. Nondegenerate diffusion spreads everywhere $\Longrightarrow$ there are no invariant submanifolds for X_t^ϵ. So problems like in KAM theory (reappearence of invariant tori for perturbed Hamiltonian systems) are not relevant here. (May be for degenerate random perturbations preserving symplectic structure something can be done in this direction). So what one wants to do in random perturbations is to study characteristics of the flow F^t (map F) which can be approximated by some parameters of processes X^ϵ as $\epsilon \to 0$ and so have physical sense.
The only problem in random perturbations I know connected with KAM theory and small denominators is the following. Let

$$\frac{dF^t x}{dt} = B = (\omega_1 \frac{\partial}{\partial \phi_1}, \ldots, \omega_\nu \frac{\partial}{\partial \phi_\nu})$$

$$\omega = (\omega_1, \ldots, \omega_\nu), \qquad \phi = (\phi_1, \ldots, \phi_\nu)$$

be the quasiperiodic motion on T^ν,

$$L = \frac{1}{2} \sum_{1 \le k, \ell \le \nu} a^{k\ell}(\phi) \frac{\partial^2}{\partial \phi_k \partial \phi_\ell} + \sum_{1 \le k \le \nu} b^k(\phi) \frac{\partial}{\partial \phi_k}$$

$a^{k\ell}$ uniformly positive definite, $a^{k\ell}(\phi), b^k(\phi)$ periodic.

$$X_t^\epsilon \quad \text{generated by} \quad L^\epsilon = \epsilon^2 L + B$$

Spectrum of L^ϵ consists of eigenvalues $\{\lambda_m^\epsilon\}$ and the spectrum of B consists of eigenvalues

$$\lambda_m = i(m, \omega), \qquad m = (m_1, \ldots, m_\nu).$$

and the eigenfunctions are

$$r_m(\phi) - \exp[i(m, \phi)]$$

The eigenfunctions of L^ϵ are r_m^ϵ. Let us suppose that there are no resonances i.e. $(\omega_1, \ldots, \omega_\nu)$ are rationally independent.

Theorem (K) For any $m = (m_1, \ldots, m_\nu)$ and $\epsilon > 0$ small enough, there exists an eigenfunction r_m^ϵ and an eigenvalue λ_m^ϵ of L^ϵ, such that

$$|\lambda_m^\epsilon - \lambda_m| < C_m \epsilon^2, \qquad (r_m^\epsilon r_m) = 1, \qquad ||r_m^\epsilon - r_m|| \to 0$$

and such r_m^ϵ and λ_m^ϵ are unique.

One can try to do similar thing in a domain bounded by a torus (accelerator), consider the spectrum of L^ϵ corresponding to *zero* boundary conditions and instead of looking for invariant tori try to find series of eigenvalues close to numbers $i(m,\omega)$ for different frequencies ω.

Ideological problem: probabilistic sense of the spectrum of L^ϵ is not known.

Main problem to study

a) $G \subset M$, $\overline{G}$ compact, exit problems of X_t^ϵ to ∂G, for instance, the distribution of $X_{\tau^\epsilon}^\epsilon, \tau^\epsilon = \inf\{t : X_t^\epsilon \in \partial G\}$ as $\epsilon \to 0$
P.D.E. twin problem — Dirichelet problem

$$L^\epsilon u^\epsilon = 0, \qquad u^\epsilon|_{\partial G} - f$$

Also more general problems

$$L^\epsilon u^\epsilon + V u^\epsilon = -g, \qquad u^\epsilon|_{\partial G} - f$$
$$(u^\epsilon(x) = E^\epsilon \tau^\epsilon \quad \text{satisfies} L^\epsilon u^\epsilon = -1, \quad u^\epsilon|_{\partial G} - 0)$$

Also Newmann boundary conditions.
Nice results can be obtained only when F^t has simple dynamics. Sometimes there is no convergence when $\epsilon \to 0$. (there are examples of nonconvergence of the Dirichelet problem).

b) One can study also for small ϵ the decay of $P_x^\epsilon\{\tau^\epsilon > t\}$ as $t \to \infty$ i.e. the probability of staying in the domain until time t. This may be interesting for accelerators.
It is easy to see that

$$\lim_{t \to \infty} \frac{1}{t} \log P_x^\epsilon\{\tau^\epsilon > t\} = \lambda^\epsilon$$

where λ^ϵ is the principal eigenvalue (i.e. the eigenvalue with the biggest real part) of the operator L^ϵ corresponding to zero boundary conditions . Also if we define the transition operator

$$P_t^\epsilon f(x) = \int P(t, x, dy) f(y)$$

then

$$e^{\lambda_\epsilon} = \lim_{t\to\infty} ||P_t^\epsilon||^{1/t}$$

One can show: $\lambda_\epsilon \to -\infty$ if and only if there is no F^t–invariant set in $\overline{G}$

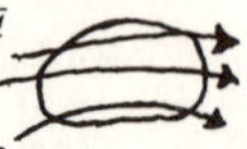

Depending on the kind of invariant set in $\overline{G}$ one gets different behaviour of λ^ϵ as $\epsilon \to 0$

$$P_x^\epsilon\{\tau^\epsilon > t\} = e^{\lambda^\epsilon t}$$

Can be generalized The above problems look differently depending on invariant sets of F^t.

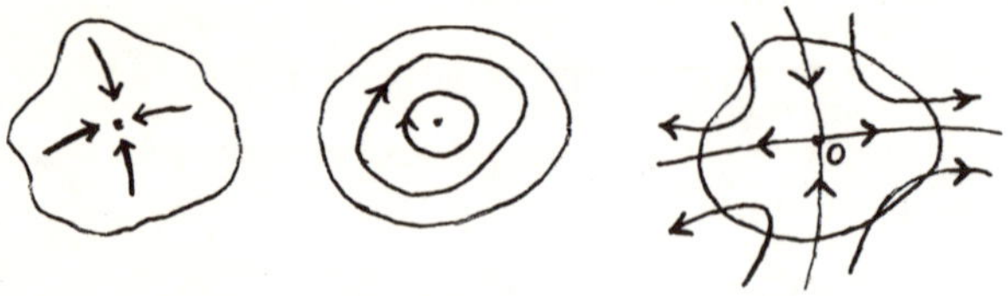

1.	2.	3.
Ventzel Freidlin Day	Hansminskii	Kifer
$\lambda^\epsilon \sim +\exp\frac{-\alpha}{\epsilon^2}$	$\lambda^\epsilon \sim \beta\epsilon$	$\lambda^\epsilon \underset{\epsilon\to 0}{\to} -\sum_i \max(\mathrm{Re}\lambda_i, 0)$
$\tau^\epsilon \sim \exp\frac{\alpha}{\epsilon^2}$	$\tau^\epsilon \sim \frac{c}{\epsilon}$	$\tau_0^\epsilon \sim \lvert\log\epsilon\rvert$ γ_i–eigenvalues of Πs.t.
$\lambda^\epsilon \underset{\epsilon\to 0}{\to} 0$	$\lambda^\epsilon \to 0$	$B(x) = \Pi x + o(\lvert x\rvert)$

The second case treated by Hasminskii in early sixties is closely connected with Hamiltonian systems and is done by averaging. The behaviour of X_t^ϵ is very close to the diffusion, whose coefficients are average of the corresponding coefficients of original diffusion along orbits of F^t. So in fact one has to study the one–dimensional diffusion in the direction transversal to the invariant curves.

This can be done also for higher dimensions, say when orbits of F^t are closed (like in an accelerator). Then after averaging one gets two dimensional diffusion in the cross section. Then also $\lambda^\epsilon \sim \beta\epsilon$ and β can be computed using this averaged diffusion.

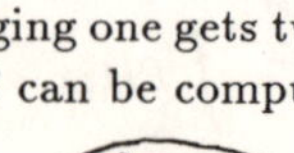

Recently this averaging technique was extended by Wolanski to the case of a random perturbation in the presence of a homoclinic point where there are three types of behaviour. This also occurs in Hamiltonian systems (Duffing equation).

Hasminskii and Kushner developed also stochastic Lyapunov stability (see their books). One can define stochastic Lyapunov function and the study is similar to Lyapunov stability in ODE.

Example $L^\epsilon = \frac{\epsilon}{2}\Delta + (Bx, \nabla)$, eigenvalues of B have negative real part, $X_t^\epsilon = e^{Bt}(X_0^\epsilon) + \epsilon \int e^{\beta s} dw$. c) Another class of problems: (posed by Kolmogorov $\sim$ 40 years ago and also considered by Andronov Pontyiagin et al. 1933)

Behaviour of invariant measures (stationary distributions) μ_ϵ of diffusions X^ϵ as $\epsilon \to 0$ and principal eigenvalues $\lambda^\epsilon(V)$ of operators

$$L_V^\epsilon = L^\epsilon + V, \qquad V\text{–continuous}$$

(subject to zero data on ∂G or without boundary conditions if $G = M$ –compact). These problems are connected, these quantities are stable: no nonconvergence results as $\epsilon \to 0$ are known. Can be studied for flows F^t with complicated (chaotic) dynamics and describe this dynamics nicely

$\lambda^\epsilon(V) \underset{\epsilon \to 0}{\to} -\infty$ iff $\exists$ no F^t–invariant set in $\overline{G} \Longleftrightarrow \exists$ no F^t–invariant measure in $\overline{G}$.

If $F^t = \text{id} \Longrightarrow \lambda^\epsilon(V) \underset{\epsilon \to 0}{\to} \sup_{x \in \overline{G}} V(x)$.

Setup

M manifold, $G \subset M$, G open, $\overline{G}$–compact, ∂G– smooth. Define operators, according to Feynman-Kac formula

$$u(t,x) = T_V^\epsilon(t) g(x) = E_x^\epsilon \chi_{\tau_G^\epsilon > t} g(X_t^\epsilon) \exp\left(\int_0^t V(X_s^\epsilon) ds\right)$$

$$(g|_{\partial G} = 0 \quad \text{if} \quad G \neq M)$$

where $u(t,x)$ satisfies $\partial u / \partial t = L^\epsilon u$ with $u|_{t=0} = g$.

$$\lambda^\epsilon(V) = \lim_{t \to \infty} \frac{1}{t} \log ||T_V^\epsilon(t)||$$

$$V \in C(M), \qquad \lambda^\epsilon(V) \in \text{spec } L_V^\epsilon \text{and if}$$

$$V\text{–Hölder} \Longrightarrow \lambda^\epsilon(V) \text{ –eigenvalue}$$

Dobsker–Varadhan (1975–76). Connected with the theory of large deviations.

$$\lambda^\epsilon(V) = \sup_{\mu\in\mathcal{P}(\overline{G})}\left(\int_{\overline{G}} V\,d\mu - I^\epsilon(\mu)\right),$$

$$I^\epsilon(\mu) = -\inf_{u\in D_+}\int_{\overline{G}}\frac{L^\epsilon u}{u}d\mu$$

$I^\epsilon(\mu)$–lower semicontinuous $\Longrightarrow \exists \mu^\epsilon_V$ s.t. $\lambda^\epsilon(V) = \int V d\mu^\epsilon_V - I^\epsilon(\mu^\epsilon_V)$.
where μ^ϵ–equilibrium state.
V–Hölder $\Longrightarrow \mu^\epsilon_V$ is unique and it is the invariant measure of Markov processes Y^ϵ_t with transition operators

$$\mathcal{T}^\epsilon_V(t) = e^{-\lambda^\epsilon(V)}(r^\epsilon_V(x))^{-1}T^\epsilon_V(t)(r^\epsilon_V g)(x)$$

r^ϵ_V eigenfunctions of L^ϵ_V corresponding to $\lambda^\epsilon(V)$.

Measures μ^ϵ_V generalize invariant measures of X^ϵ_t which are μ^ϵ_V with $V\equiv 0$ and $G = M$–compact.
All weak limits of μ^ϵ_V as $\epsilon\to 0$ are F^t invariant measures!

Corollary If $G = M$ is compact and $\exists$ unique F^t–invariant measure μ then

$$\lim_{\epsilon\to 0}\lambda^\epsilon(V) - \int_M V d\mu$$

Proof.

$$I^\epsilon(\mu)\geq 0 \Longrightarrow \lambda^\epsilon(V) = \int_{\overline{G}} V d\mu^\epsilon_V - I^\epsilon(\mu^\epsilon_V) \leq \int_{\overline{G}} V d\mu^\epsilon_V \underset{\epsilon\to 0}{\to} \int V d\mu \quad \text{as } \epsilon\to 0$$

Now let μ^ϵ be an invariant measure of $X^\epsilon_t \Longleftrightarrow I^\epsilon(\mu^\epsilon) = 0$ (i.e. $\mu^\epsilon - \mu^\epsilon_0$) then

$$\lambda^\epsilon(V) \geq \int_{\overline{G}} V d\mu^\epsilon \longrightarrow \int_{\overline{G}} V d\mu$$

Can be done for an attractor $\overline{G}$ with unique F^t invariant measure.

What happens in general?
Natural to believe that $\lambda^\epsilon(V) \underset{\epsilon\to 0}{\to}$ to a quantity given by a variational formula – in the spirit of convex analysis.
A good bet: the topological pressure (thermodynamical formalism for dynamical systems)

$$P_\Lambda(F,\psi) = \sup_{\substack{\mu-F^t-\text{ invar.}\\ \text{supp}\,\mu\subset\overline{\Lambda}}}\left(h_\mu(F') + \int_{\overline{\Lambda}}\psi d\mu\right)$$

$h_\mu(F')$–Kolmogorov–Sinai entropy.
Another representation:

$$P_\Lambda(F,\psi) = \lim_{\delta\to 0}\lim_{t\to\infty}\frac{1}{t}\log\sum_{x\in S(\delta,t)}\exp\left(\int_0^t \psi(F^s x)ds\right)$$

($S(\delta,t)$–max set s.t. $x,y\in S(\delta,t) \Longrightarrow \exists s\in[0,t]$, $\mathrm{dist}(F^s x, Fsy) > \delta$). (δ,t)–separated.
Similar to the Feynman–Kac formula, $\sum$ in place of averaging E^ϵ_x.
For a large class of flows F^t the entropy $h_\mu(F')$ is upper semicontinuous in μ. In this case if $\exists\phi\in C(\overline{G})$ s.t. $\quad\forall V\in C(\overline{G})$

$$\lim_{\epsilon\to 0}\lambda^\epsilon(V) = P_G(F, V+\phi) \tag{*}$$

then by convexity arguments $\Longrightarrow$ all weak limits of measures μ^ϵ_V satisfy

$$P_G(F,V+\phi) = h_{\mu_V}(F') + \int_{\overline{G}}(V+\phi)d\mu_V$$

Such μ_V are called equilibrium states corresponding to $V+\phi$
If $V\equiv 0 \Longrightarrow \forall$ weak limits of invariant measures μ^ϵ of X^ϵ_t are equilibrium states corresponding to ϕ

Though not simple (*) is usually easier to prove than the convergence of measures. New approach: to prove (*) and get the convergence of measures for free.

Main result

An F^t–invariant $K\subset M$ is hyperbolic, in particular Asonov on M, if $T_K M = \Gamma^s\oplus\Gamma^0\oplus\Gamma^u$ ($\Gamma^0 = 0$ if K is a fixed point), This decomposition is DF^t–invariant, DF^t expands vectors from Γ^u and contracts vectors from Γ^s exponentially fast in t.

$$\mathcal{J}_t(x)\text{– Jacobian of } DF^t:\ \Gamma^u_x\to\Gamma^u_{F^t x}, \qquad \phi^u(x) = -\left.\frac{d\mathcal{J}_t(x)}{dt}\right|_{t=0}.$$

Theorem
Suppose that max invariant set (Axiom A type) of F^t in $\overline{G} = \cup_{i=0}^\kappa K_i$, K_i hyperbolic, $K_i\cap K_j = \emptyset$, $i\neq j$. Then $\forall V\in C(\overline{G})$

$$\lim_{\epsilon\to 0}\lambda^\epsilon(V) = P_G(F, V+\phi^u) = \max_{1\le i\le\kappa} P_{K_i}(F, V+\phi^u)$$

Thus all weak limits of μ^ϵ_V are equilibrium states μ_V corresponding to $V+\phi^u$ (ϕ^u extended continuously to the whole $\overline{G}$). All such μ_V are

$$\mu_V = \sum_{1\le i\le\kappa} p_i\mu_{V,K_i}, \qquad p_i\ge 0, \quad \sum p_i = 1$$

μ_{0,K_i} Sinai–Bowen–Ruelle; μ_V equilibrium measure on K_i corresponding to $V+\phi^u$ and $p_j > 0 \Longrightarrow P_{K_j}(F, \phi^u + V) = \max_i P_{K_i}(,)$.
If $V \equiv 0$ and some K_i are attractors then $\max_i P_{K_i}(,) = 0$ and it is attained on attractors. If it is attained only for one i then μ^{ϵ}_{V,K_j} which is unique if V is Hölder continuous.

Example

Simple (but non trivial) partial case: all K_i are fixed points i.e.

$$B(K_i) = 0, \ \ i = 1, \dots, \kappa \Longrightarrow B(x) = \Pi_i(x - K_i) + o(\mathrm{dist}(x, K_i)).$$

Let $\gamma_1^{(i)}, \dots, \gamma_\nu^{(i)}$ be the eigenvalues of Π_i. Then

$$\lambda^{\epsilon}(V) \underset{\epsilon \to 0}{\longrightarrow} \max_{1 \le i \le \kappa} (V(K_i) - \sum_{1 \le j \le \nu} \max(\mathrm{Re}\gamma_j^{(i)}, 0)$$

where $V(K_i) = P_{K_i}(F, \phi^u + V)$ and $\forall$ w-limits of μ^{ϵ}_V are linear combinations of unit masses sitting on those K_i where the max is attained

REFERENCE BOOKS

[1] On Stochastic Lyapunov stability

R. Hasminskii *Stochastic stability of Differential Equations*, Sijthoff & Noordhoff Alphen aan den Rijn, Netherlands and Rookville, MD, 1980.

H. Kushner *Stochastic stability and control*, Academic Press, New York, 1967.

[2] Large deviations type results
A. Ventzel and M. Freidlin *Random perturbations of dynamical systems*, Springer, New York, 1984.

[3] Random perturbations of chaotic systems
Y. Kifer *Random perturbations of dynamical systems* Birkhäuser, Boston, 1988.

Phase convection: universal mechanism of lifetime limitation.

A.Gerasimov[1*],R.Siemann
Newman Laboratory of Nuclear Studies,Cornell University,
Ithaca NY 14853, USA

Abstract

The problem of lifetime limitation by beam-beam interaction appears as the nonequilibrium statistical mechanics of hamiltonian oscillator. For the case of nearly-integrable dynamics of oscillator, the effect of perturbation (nonlinear resonances) on distribution function in two-dimensional case is exponenetially strong, contrary to the one-dimensional one. Simulation results, verifying the validity of the theory are presented.

1. Distribution Tails and Rare Events in Nonequilibrium Systems

In contrast to single-time, equilibrium, statistical mechanical quantities the general approach to the calculation of time dependent ones is essentially phenomenological. This phenomenology can be based [2 3] on either a relaxation kinetic model of the form

$$\dot{q}_i = -\alpha \frac{\partial V(q)}{\partial q_i} + (2\eta)^{1/2} \xi_i(t) \qquad (1)$$

or on its inertial counterpart in full phase space **q**, **p**

1.*Now at Fermi National Accelerator Laboratory,P.O.Box 500,Batavia IL 60510.
2.H.Risken The Fokker-Planck equation. Springer,Berlin,1983.
3.K.Gardiner Handbook of stochastic methods, Springer,1985.

$$\dot{q}_i = p_i ,$$

$$\dot{p}_i = - \frac{\partial H(p,q)}{\partial q_i} - \alpha p_i + (2\eta)^{1/2} \xi_i(t) . \qquad (2)$$

In these models α is the relaxation rate, η is the diffusion coefficient, and the $\xi_i(t)$ are independent, rapidly varying noise processes that model the interaction of the macroscopic variables with a heat bath. The exact stationary distribution for temperature $T = \eta/\alpha$ is the Hibbs distribution, $\exp(-V(\mathbf{q})/T)$ for system (1) or $\exp(-H(\mathbf{p},\mathbf{q})/T)$ for system (2).

The transition probabilities, the same as the two-time correlation functions, satisfy the Fokker-Planck equation. For the particular case of model (2) with the Hamiltonian $H = \mathbf{p}^2/2 + V(\mathbf{q})$ the Fokker-Planck equation has the form[2,3]

$$\frac{\partial \rho}{\partial t} + \mathbf{p}\frac{\partial \rho}{\partial \mathbf{q}} - \frac{\partial V}{\partial \mathbf{q}}\frac{\partial \rho}{\partial \mathbf{p}} = \frac{\partial}{\partial \mathbf{p}} \left(\alpha \mathbf{p} \rho + \eta \frac{\partial \rho}{\partial \mathbf{p}} \right) . \qquad (3)$$

The kinetic description of thermally nonequilibrium systems naturally follows the same phenomenology. In the most general form, nonequilibrium kinetics is modelled by

$$\dot{q}_i = K_i(\mathbf{q}) + (2\eta)^{1/2} g_i(\mathbf{q}) \xi_i(t) \qquad (4)$$

where $\mathbf{K}$ is the deterministic drift and $\mathbf{g}$ is the diffusion intensity vector. There is no summation over i in eq. (4). The nonequilibrium nature of the system, described either by the absence of detailed balance or deterministic drift that is not derived from a potential[2,3], implies stationary state density fluxes

$$j_i = (K_i + \eta \frac{\partial}{\partial q_i}) \rho_{st}(\mathbf{q}) . \qquad (5)$$

to be circulating in closed loops. Mathematically, such nonequilibrium models appear naturally when the relaxation rate is a vector $\alpha_i(\mathbf{q})$ and/or the diffusion intensity a tensor $\eta_{ij}(\mathbf{q})$. Physically, they appear in both finite dimensional systems and many-body nonequilibrium steady-states; shear flows are an example.[4] Periodically modulated, and therefore nonequilibrium systems, can be fit into category (4) also by treating time as an extra component of **q** so that the stationary distribution in this extended **q'** space is periodic in the primary **q** space.

The stationary distribution satisfying the corresponding stationary Fokker-Plank equation cannot be found explicitly for the general nonequilibrium system (4). This distribution is the main object of interest. The Fokker-Plank equation is a linear, second order partial differential equation that has many resemblances to and analogies with the Schroedinger equation.[5] Thus, weak-noise asymptotics,[6 7] one of the more interesting and insightful approaches to solving the Fokker-Plank equation, parallels the quasiclassical approximation of quantum mechanics. In fact, for equilibrium systems the Schroedinger and Fokker-Planck equations are exactly transformable into each other as are the quasiclassical approximation and weak-noise asymptotics.

The physics of weak-noise asymptotics is the following. A transition probability can be represented by a path integral. In the limit of weak noise, $\eta \sim T \to 0$, one can identify a single trajectory that provides the saddle point of this integral. This trajectory, the so-called extremal, is defined by a classical mechanics variational principle in the same way as in Feynman path-integral representation of quantum mechanics.[8] In this approximation the stationary distribution ρ depends on η as

$$\rho(\eta,\mathbf{q}) = Z(\mathbf{q})\exp\left(-\frac{\Phi(\mathbf{q})}{\eta} + O(\eta)\right) . \qquad (6)$$

4.T. Kirkpatrick, E. Cohen and J. Dorfman, Phys. Rev. **A26**, 950 (1982).
5.R. Graham and T. Tel, Z. Phys. **B60**, 127 (1985).
6.A.Ventzel,M.Freidlin, Random perturbations of dynamical systems, Springer, N.Y.,1984.
7.R.Graham, in: Fluctuations, instabilities and phase transitions, T.Riste ed., Plenum, N.Y. 1975.
8.R.Feynman,A.Hibbs, Quantum mechanics and path integrals, McGrow-Hill, N.Y. 1965

The function $\Phi(q)$ satisfies the Hamilton-Jacobi equation

$$K_i(q)\frac{\partial\Phi}{\partial q_i}+\left(g_i(q)\frac{\partial\Phi}{\partial q_i}\right)^2=0\ . \qquad (7)$$

Weak-noise asymptotics is well-suited for describing the decay of metastable states since it is associated with the tails of distributions. In the simplest case, "deterministic" systems where $\dot{q}_i = K_i(q)$, there are stable points surrounded by regions of attraction. These regions of attraction are limited by absorbing boundaries or boundaries with other regions of attraction. The escape rate r from a region of attraction is exponentially small in η; it is given by $r = \exp(-R/\eta)$, where R is the minimum of Φ on the boundary.

In general the Hamiltonian H' corresponding to eq. (7) (presented as $H'(q,\partial\Phi/\partial q) = 0$) is nonintegrable, and this leads to one of the interesting features of the weak-noise asymptotics of nonequilibrium systems. The true *most probable transition path* is not a single trajectory of the Hamiltonian H', but it consists of pieces of trajectories joined continuously but not differentiably; i.e. the path has cusps. An associated phenomenon is that Φ is not differentiable on the certain surfaces in the space **q**. These cusps appear in nonequilibrium systems only; in equilibrium the Hibbs distribution corresponds to $\Phi = H$ which is always a unique and smooth solution.

Such exotic behavior arises from the fact that the nonintegrable (stochastic) trajectories of the Hamiltonian system provide the saddle-points and not the minimum of the classical mechanics action.[9 10] There was little interest in the trajectories that provided the true minima, since this is an artificial problem in classical mechanics having nothing to do with real dynamics. In the weak-noise asymptotics of nonequilibrium systems these minimum action trajectories are physically relevant and one obtains a new, non-trivial, non-Newtonian

---

9.I.Percival, J.Phys.12A(1979)L57
10.R.Mackay,J.Meiss,I.Percival, Physica 13D(1984)55

mechanics of nonintegrable systems. The fine interrelation of the equilibrium-nonequilibrium statistical and integrable-nonintegrable Hamiltonian systems has a strong fundamental flavour and is not explored fully yet.

A particular nonequilibrium system was considered by one of authors.[11] This was system (2) with an exactly integrable Hamiltonian H_O and a periodic perturbation with period T giving a total Hamiltonian $H = H_O + \epsilon U(\mathbf{q},t)$. As mentioned earlier, the steady-state distribution function is the periodic, asymptotic ($t\to\infty$) distribution that is independent of the initial one; we will refer to it as the *relaxed distribution function*. The Hamiltonian dynamics in the absence of damping and noise for small perturbation ϵ is nearly-integrable and can be described in terms of isolated nonlinear resonances.[12] Nontrivial nonequilibrium effects manifest themselves strongly for two or more spatial dimensions. In this case the resonances are extended localized structures, "tubes", in a space where the axes are the unperturbed action $\mathbf{J}$ and the resonant phase, Ψ. These tubes are narrow, with width of order $\sqrt{\epsilon}$, but protracted. When the damping is added particles outside the separatrix are driven by damping in approximately the same way as in the unperturbed system. Inside the separatrices there is an interplay between the resonant perturbation and the effect of damping, and the dynamics is quite different. If damping is small enough so that it does not destroy the resonance, particles inside the separatrix stay within its limits for arbitrarily long times. They are phase locked to the resonance and move along the resonance line in a direction quite different from what it would be in the absence of the perturbation. This phenomenon is called streaming.[13] Due to this it is often much easier for particles to reach large energies by traveling inside the narrow resonance tubes rather than going by the way they would in the unperturbed case. When the scales of the problem are widely separated

----

11. A.Gerasimov, Physica D41(1990)89
12. B.Chirikov, Phys.Rep. 52(1979)263
13. J.Tennyson, Physica 5D(1982)123

with the unperturbed action being much larger than the resonance width which is much larger than the temperature, $J >> \sqrt{\epsilon} >> T$, the weak noise asymptotics equations for Φ and most probable paths of escape can be explicitly solved in the lowest order in ϵ[11].

The result is that, due to particles travelling along resonances, the deviation of Φ from $\Phi_0 = H_0$ is of order $\Phi_0 - \Phi \sim \Phi_0$. All the characteristic features of Φ described above manifest themselves in this problem and allow a clear physical interpretation. It should be pointed out that Φ was constructed through a direct variational approach and the nonintegrability of the Hamiltonian for the Hamilton-Jacobi equation for Φ was not demonstrated explicitly. This remains a question for future work. In this paper we perform some simulations in order to check the predictions of the theory[11],concentrating on qualitative features.

2. Model Example: Theoretical Prediction

One of the important applications of the theory of phase convection is the problem of beam lifetime in electron-positron storage rings.[14 15 16] There, particles in one beam interact electromagnetically with particles in the opposing bunch. This *beam-beam interaction* is non-linear and periodic with the revolution period. It can cause deterioration of beam lifetimes and be a factor limiting the collider luminosity. Our particular interests lead us to study the beam-beam problem, but, for the purpose of this paper, the beam-beam interaction doesn't possess any special features, and it can be considered a convenient example for studying the general theory.

The transverse motion of particles in e^+e^- storage rings can be modelled approximately by the equations[17]

14.A.Gerasimov,N.Dikansky, to appear in Nucl.Instr.Meth.A, Part I; INP Preprint 88-60 (in Russian)
15.ibid., part2; INP preprint 88-97 (in Russian)
16.ibid., part3; INP preprint 89-3
17.J.Shonfeld, Ann. of Phys. 160(1985)149

$$\frac{d^2x}{dt^2} + \alpha\frac{dx}{dt} + \omega_x^2 x + \frac{\partial}{\partial x}U(x,z)\delta_T(t) = \sigma_x\omega_x(2\alpha)^{1/2}\xi_x(t)$$

$$\frac{d^2z}{dt^2} + \alpha\frac{dz}{dt} + \omega_z^2 z + \frac{\partial}{\partial z}U(x,z)\delta_T(t) = \sigma_z\omega_z(2\alpha)^{1/2}\xi_z(t) \qquad (8)$$

The first three terms on the left-hand-sides represent linear harmonic oscillations with damping; ω_x, ω_z are the oscillation frequencies and α is the damping decrement. The fourth terms on the left-hand-sides correspond to the beam-beam interaction. The potential $U(x,z)$ is nonlinear and is given approximately by eq. (11) below and $\delta_T(t) = \Sigma\ \delta(t-nT)$ where the sum is over $n = -\infty, \ldots, \infty$. The right-hand-sides are the same white noise processes as in eqs. (1) and (2). The quantities σ_x, σ_z are the rms values of coordinates x and z when $U = 0$; in that case the two degrees of freedom are uncoupled and the stationary distribution function is

$$\rho_{st}(x,z) = \exp\left(-\frac{(p_x^2+\omega_x^2x^2)}{2\sigma_x^2\omega_x^2} - \frac{(p_z^2+\omega_z^2z^2)}{2\sigma_z^2\omega_z^2}\right) . \qquad (9)$$

Here and in the following we use p_x, p_z for dx/dt and dz/dt, respectively. In the absence of the thermal terms ($\alpha = 0$, $\sigma_x = \sigma_z = 0$) the system (8) is Hamiltonian with

$$H = \frac{p_x^2}{2} + \frac{p_z^2}{2} + \frac{\omega_x^2x^2}{2} + \frac{\omega_z^2z^2}{2} + U(x,z)\delta_T(t) . \qquad (10)$$

A specific feature of this Hamiltonian, as compared to the general case in eq. (2), is the linearity of the unperturbed motion. At first glance, this could present a difficulty because the stabilizing nonlinearity is proportional to perturbation also. However, from a practical point of view the standard approach to nonlinear resonances can be used by taking a suitably averaged part of the perturbation as the nonlinearity, and low order resonances can be avoided by the proper choice of frequencies ω_x and ω_z because the beam-beam perturbation is sufficiently weak.

The dynamics of the Hamiltonian in eq. (10) can be integrated over the period T, so we can study the exact stroboscopic mapping with the time step T instead of the continuous dynamics(see[18]). The effect of continuous damping and noise can be integrated over period T also, and from eqs. (8) we arrive at a damped, noisy, stroboscopic mapping with the time step T. Thus, eqs. (8) are among the most general stochastic equations that fit the formalism of phase convection[11] and allow for reduction to discrete mappings. The latter is important because this simplifies simulation and saves computer time.

Both the calculations and simulations in this paper are sometimes called weak-strong because the distribution of a "weak" beam is determined assuming it is interacting with a "strong" beam that has a fixed, Gaussian distribution given by eq. (9). The beam-beam potential used is

$$U(x,z)=8\pi S\left[\exp(-x^2/2\sigma_x^2)\left((\sigma_x^2+z^2/2)^{1/2}-(\sigma_z^2+z^2/2)^{1/2}-\sigma_x-\sigma_z\right)+\frac{a^2\sigma_x^3}{4(a\sigma_x^2+x^2)}\right] \tag{11}$$

where a = 5.11 is a parameter that gives good agreement with the exact expression[19] when $\sigma_x >> \sigma_z$. The parameter S determines the strength of the interaction.[20]

A computer program was developed[14] that found the characteristics of all the resonances up to a given order. Graphically, such information can be presented in the form of the resonance libration vectors that describe the separatrices in the plane of unperturbed amplitudes, $A_x \propto (p_x^2 + \omega_x^2 x^2)^{1/2}$, $A_z \propto (p_z^2 + \omega_z^2 z^2)^{1/2}$. A second program was developed that used the information from the first program as input and computed

---

18.F.Izrailev, Physica 1D(1980)243
19.M. Bassetti and G. Erskine, CERN-ISR-TH/80-06 (CERN, 1980).
20.S is related to the beam-beam strength parameter, ξ_z (not to be confused with the ξ_i in (1) and (2)), and the amplitude function at the collision point, β_z^*, by $S = \xi_z\sigma_z/\beta_z^*$.

$\Phi(A_x, A_z)$ through the general theory of phase convection[15,16]. Thus we are able to generate theoretical predictions of Φ and associated quantities.

In the present paper we report the results of simulations and compare them with these predictions. We are particularly interested in the qualitative correspondence between theory and simulations because that tests the validity of the underlying approach of phase convection. As will be clear later this is especially true because some of the cases considered are not trivial.

3. Simulation results and theoretical predictions for particular examples

In Figures 1 and 2, two theoretical predictions of the contours of distribution function tails in the amplitude plane Ax, A_z are given for two particular examples. The predictions are obtained with the help of a specialized computer program as mentioned above. On the same graphs, all the resonances up to the order $N=|l| + |m| \leq 10$ (l and m are harmonic indices) are shown. In both Figures, the contours of Φ for small amplitudes $Ax^2 + A_z^2 \leq 3$ are unperturbed ellipses corresponding to the unperturbed Gaussian distribution (9) in this region. At larger amplitudes, the distribution function is strongly perturbed by the resonances, and the contours have a characteristic bending around the resonance lines.

The importance and not quite self-evident character of phase convection phenomenon from the point of view of accelerator physics can be illustrated in examples of Figs.1 and 2. There, the effect of resonances for the vertical aperture limitation at $A_z = 30$ is observed to be equivalent to the aperture limitations $A_z \simeq 9$ and $A_z \simeq 6$ respectively. The oscillation of A_z under the influence of resonances in the Hamiltonian dynamics (without damping and noise) at the same time never exceeds 6, which is substantially smaller than the effective decrease of aperture.

In the case of Fig.1 , practically only one resonance $6\nu x + 2\nu y = 5$ is perturbing the distribution function, while the second one has a minor effect. In the case of Fig.2 , the situation is more complicated since there are two "active" resonances present. The left one, $4\nu x + 2\nu y = 4$, is "responsible" (as is clear from the bending of the contours) for the perturbation of Φ for vertical amplitudes Az from $A_z \simeq 3$ to $A_z \simeq 19$ and the right one $2\nu x + 6\nu y = 5$ "works" from $A_z \simeq 13$ to the maximum presented amplitudes. The interpretation of this behavior is[11,15] that particles choose these sections of resonance lines as parts of their paths to large vertical amplitudes.

In Figures 3 and 4, the results of direct simulation of distribution function for the cases of Figures 1 and 2 respectively are given in the form of the contours of distribution in the plane of amplitudes Ax, A_z. The comparison with Figures 1 and 2 demonstrates a good agreement between theory and simulation.

In Figures 5 and 6, the data about the phase space density fluxes in the plane of amplitudes Ax, A_z is presented. The fluxes were calculated in the simulation program by counting the number of particles crossing different sides of the boxes in the plane. There were 25 boxes used along each axis of Figures 5 and 6.

The flux lines in Figures 5 and 6 demonstrate a complicated pattern. It's origin is however easily recognizable and fits well into the general approach to the description of such fluxes[11]. Flux lines always form closed loops, parts of which pass along the resonance lines. In the regions between the resonance lines, the flux lines approximately follow the contours of distribution density (cf. Figures 3 and 4). This rule follows from the weak noise asymptotic limit[11], and is therefore rather rough - it ignores the variation of flux lines on the distances of the order of 1σ . It also doesn't apply to the regions of the phase space where the distribution function is nearly unperturbed, the region of small amplitudes in Figures 5 and 6. In weak noise asymptotics, fluxes in these regions are zero, which implies that their relative magnitude (as normalized by the value of distribution function in the points of consideration) is small[11]. However, the absolute magnitude of these fluxes, as compared to those in the regions where the distribution function is strongly perturbed, can be not small. Therefore, taking into account the fluxes in the "asymptotically unperturbed" regions is important for the overall balance of density. This is clear also from Figures 5 and 6, where big loops of flux lines are observed which pass close to the origin $Ax = 0$, $A_z = 0$ and after that going to large amplitudes, where the distribution function is many orders of magnitude smaller.

Another qualitative observation, stemming from the comparison of Figures 5 and 3 is as follows : generically, phase space fluxes appear to be much more sensitive to perturbations by resonances than the distribution function. Thus, the resonance in the left side of Fig. 5 has a minor effect on the distribution function, but its influence on the fluxes is quite appreciable, especially as it changes the topological structure of the loops.

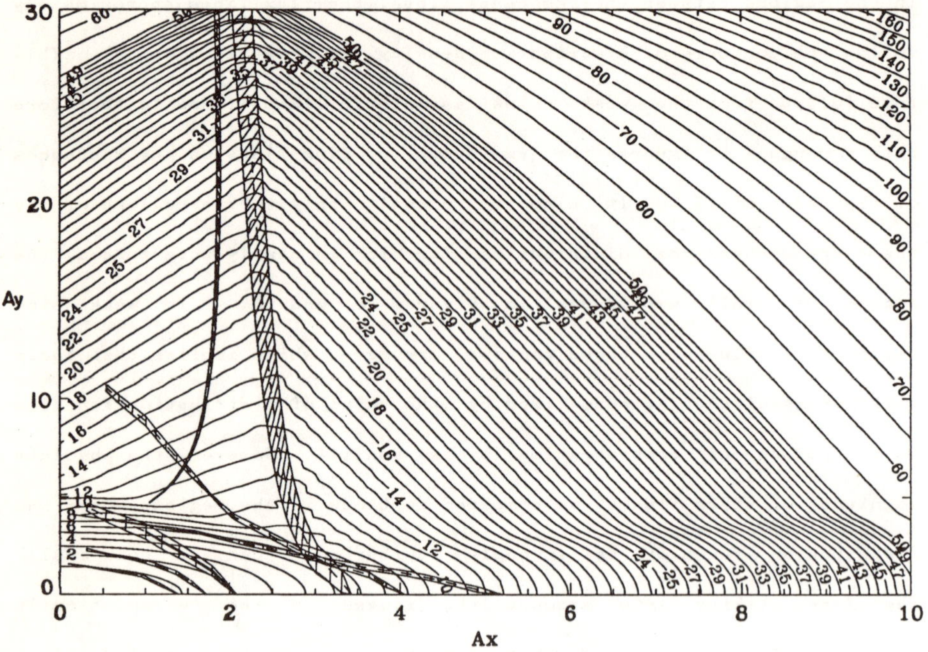

Fig.1 All the resonances up to the order N=|l|+|m|=10 and theoretically predicted (through numerical calculations) contours of relaxed distribution function in the plane of normalized betatron amplitudes A_x, A_z. Contours are marked by the values of the logarithm of distribution function. Storage rings parameters are: linear tune shifts $\xi_x=.03$, $\xi_z=.06$; beam aspect ratio $K=\sigma_x/\sigma_z=100$; tunes $\nu_x=.6008$, $\nu_z=.6669$; damping time (in the number of beam-beam kicks) N=5000.

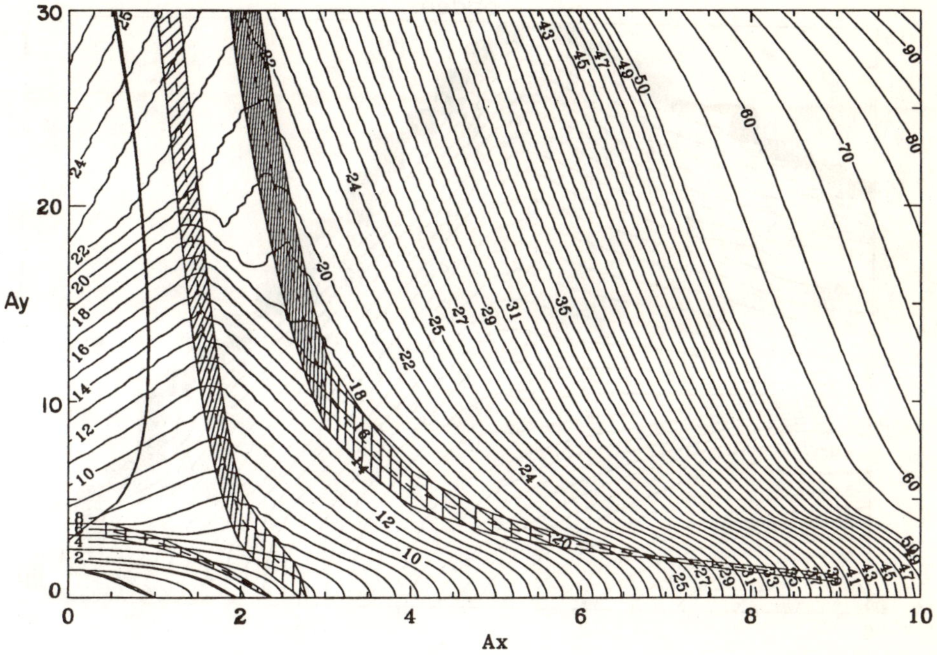

Fig.2 Same as in Fig.1, with the only exception of tunes ν_x=.682, ν_z=.6012.

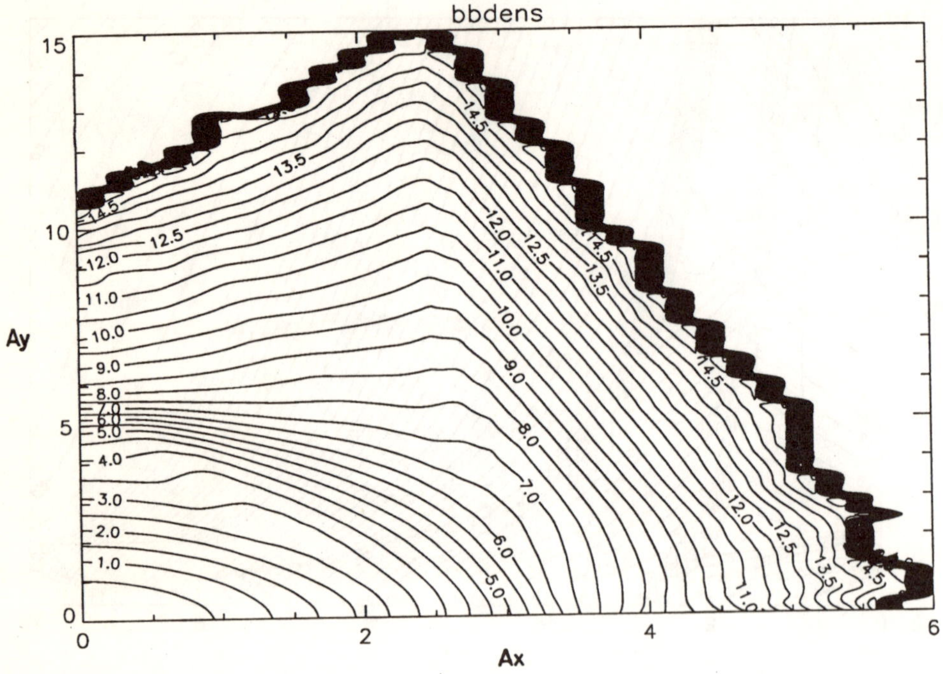

Fig.3 Contours of relaxed distribution function for the case of Fig.1 obtained through numerical simulation.

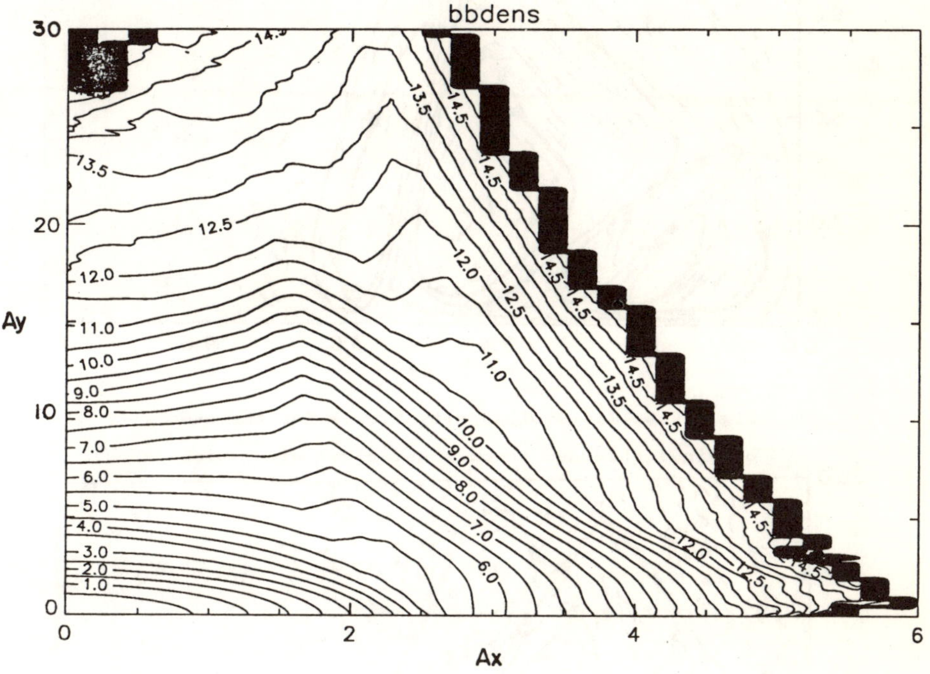

Fig.4 Same as in Fig.3 for the case of Fig.2

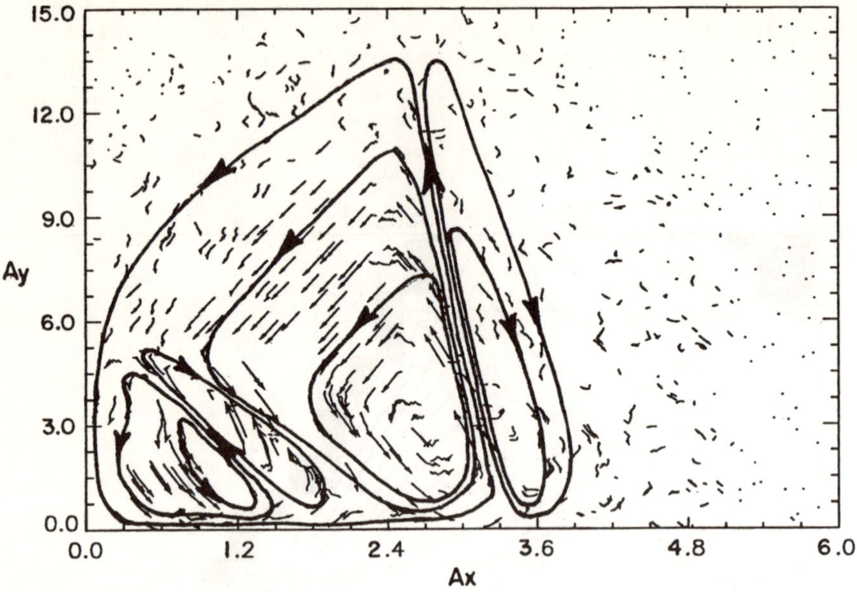

Fig.5 Flux lines of density fluxes for the case of Fig.1 obtained through numerical simulation.

Fig.6 Same as in Fig.5 for the case of Fig.2

CONTRIBUTED PAPERS

LONG TERM PREDICTION AND THE SSC.

Richard Talman

SSC Laboratory, Dallas, TX, 75237.

ABSTRACT

Successful operation of the Superconducting Supercollider (SSC) will depend on the stable circulation of particles for tens of millions of turns around the rings, in the presence of small nonlinear deflecting fields. One design challenge is to set specifications for the maximum allowable field imperfections of this sort, consistent with the required stability. Another challenge is to plan for the inclusion of field compensating elements that will ameliorate the effects of errors. The "tools" available for projecting the long term stability are theoretical, both analytic and numerical, and experimental. These aspects are reviewed.

Introduction. In this paper some of the issues that have arisen in anticipating the long term stability of particles in the SSC will be reviewed. Since much of the work involves a lot off detail that is difficult to include in an overview the discussion here will be rather qualitative. Furthermore it will be somewhat disjointed, amounting to a miscellany of brief discussions of problems, methods, results, conclusions, philosophy and so on. Where possible reference will be made to detailed reports. The topics to be discussed are:

- Operational issues and nonlinear effects that jeopardize performance.
- A progression from complete predictability to complete unpredictability.
- "Proofs" of particle stability.
- "Experimental" mathematics.
- Limitations of "restoring" symplecticity.
- "Exact" tracking.
- Long term tracking results.
- Accelerator modeling and routine error correction procedures in the SSC.
- Beam dynamics experiments.
- Parallel computation.

Operational issues and nonlinear effects that jeopardize performance. The SSC will be used to accelerate protons to an energy of 20 TeV. This will be performed in a sequence of accelerators starting with a linac, followed by boosters (of low, medium, and high energy) and finally the collider itself. The collider consists of two rings, one above the other, with counter-rotating protons. In this sequence there are many questions of dynamics that might be appropriate for study at a workshop on nonlinear dynamics. There are questions of large amplitude particle loss, resonant extraction, particle density dilution due to filamentation, adiabatic invariance, effects of resonances, chaos, turbulence and so on. All of these things relate to all accelerators in the chain. There is, however, one link in the chain that is considered to be the weakest one; that is where studies have mainly been concentrated and it is what will be emphasized in this report.

This most critical phase in the filling cycle occurs in the period during which previously injected protons are circulating in the collider at low energy (2 TeV), waiting for the injection of all the other protons to be completed. At that energy the beam has not yet benefited from the adiabatic reduction of transverse size that will accompany its acceleration to 20 TeV. As a result large amplitude particles, if they are ever to be lost, will be lost at that time. Traveling at the speed of light for, say, an hour, the protons travel 10^{12} meters, circulating some 10^{7} times around the collider. In order not to be lost the particles must remain within the "physical aperture" defined by the vacuum tube, whose bore radius is about 2 centimeters, or by the occasional mask or instrument that may intrude slightly into the chamber. With the a typical transverse amplitude being 1 millimeter this appears not to be very difficult to achieve. However, the coils carrying the currents that generate the dipole (i.e. uniform) magnetic field that bends the protons can, for reasons of cost, be only slightly larger in radius than the bore tube, and that makes it difficult to achieve satisfactory field uniformity. The effect of nonuniform fields is to make a "magnetic aperture" or "dynamic aperture" outside which particles are lost because the motion is unstable. For the SSC, with anticipated field quality, the dynamic aperture may be about one centimeter. Some beam quality degradation can be anticipated for a beam that circulates very long in the outer portion of this region; for that reason an even smaller "linear aperture" can be defined within which the beam circulates without degradation. This will be about 5 millimeters.

In the jargon of the field, the field uniformity is analysed by expanding the field in a (complex) "multipole" power series. After the dipole term comes a quadrupole term that describes linear transverse variation of the field. This component is a nuisance, because it alters the focusing properties, and the horizontal-vertical coupling properties of the lattice (i.e. the totality of elements making up the ring.) It is, however, only a nuisance, as it is reasonably straightforward to compensate for it "operationally". This term is used to describe compensation schemes that rely only on measurements that can be reliably performed using the beam itself, without any knowledge of the magnetic fields of the magnets in the ring. From a theoretical point of view the quadrupole fields can also be regarded as fairly innocuous as they leave the equations of motion linear.

It is higher multipoles, also known as the nonlinear multipoles, that threaten to cause the loss of large amplitude particles. These elements cause the equations describing the transverse motion (betatron motion in the jargon) to be nonlinear. These (relativistic) Newton's Law equations are what make the understanding of accelerators depend on nonlinear dynamics, the subject of the Workshop. It can even be said that accelerators make the best laboratory for the experimental investigation of Hamiltonian Mechanics, for which freedom from loss is the *sine qua non.* For all practical purposes the betatron oscillations are lossless, with damping times of the order of 10^{10} oscillation periods.

The leading nonlinear multipoles are sextupoles and octupoles. These can be present in the lattice both intentionally and unintentionally. Before proceeding to discuss the latter, which are somewhat the more troublesome, we mention two intentional uses as "medicine" to cure various lattice sicknesses. Sextupoles are used intentionally to reduce the chromaticity (i.e. the momentum dependence of the focusing.) Octupoles make the focusing depend on amplitude, which is sometimes exploited to improve current-dependent behavior. These medicines have side-effects, the main one being the loss of large amplitude stability. But since they are present intentionally, the "dose" is known and can be controlled, so that this source of nonlinearity can be handled confidently in a large proton accelerator such as the SSC. For electron machines, in which very strong focusing (leading to large chromaticity) is used to reduce the transverse beam size, this is less true. Even for a large proton accelerator like the SSC there is a kind of design trade-off situation in which stronger focusing can be used to desensitize the lattice to unintentional multipoles; this makes the intentionally present nonlinearities not entirely negligible.

It is unintentional elements that represent the greatest hazard. The strength of each of these multipoles can be expressed as the sum of a *systematic* part, which is the mean strength, a_i or b_i, of the particular multipole for all magnets of the particular type, and a *random* part, centered on zero, that is characterized by an r.m.s. deviation, σ_{ai} or σ_{bi}. These two types of error lead to quite different qualitative behavior. Which of the two is the more serious is largely a quantitative matter, but it is also somewhat subjective as it depends on the credibility of planned compensation schemes. In constructing the accelerator magnets great care is taken trying to make the magnetic fields uniform and identical within the economic constraint that the magnets not be too expensive. When this is done with SSC magnets typical deviations from uniformity are a few parts in ten thousand, at a point one centimeter from the magnet axis.

Systematic field errors can be caused by errors in the design location of the current carrying elements. Since this leads to excitation independent nonuniformity, it is not too serious. Far more serious are the fields caused by persistent currents in the superconducting coils. Not only are these excitation dependent but also they exhibit hysteretic dependence on previous excitation history. The leading effect of systematic field errors is to make the betatron frequencies (they are usually called "tunes") depend on momentum. The degree to which this can be compensated operationally has been studied[2] and continues to be an important issue. The tune shifts caused by systematic multipoles are directly measureable and can be used to set the strengths of compensating elements. This leads to the desirable situation that adjusting the strengths of compensating elements only tends to become difficult when they no longer need adjustment. Unfortunately systematic multipoles also cause the tunes to be amplitude dependent, and the operational procedure mentioned in the previous sentence does not work well in suppressing this effect.[3]

Random multipoles are caused by magnet-to-magnet variation in the manufacturing process. This places a high premium on using uniform coil materials, accurately manufactured parts, reproducible jigging and so on, all tending to increase costs. The leading effect of random multipoles is to violate the invariance of the "Courant Snyder invariant". In a linear lattice this quantity, which could also be called the betatron amplitude, does not vary from turn to turn.[4] Nonlinear elements cause this quantity to vary, with the fractional variation being called the "smear". Empirically it is found that, though smear of a few percent is acceptable, values greater than five or ten percent tend to be accompanied by resonances, chaos, density dilution and particle loss.

A progression from complete predictability to complete unpredictability. As the amplitude of transverse betatron oscillations of protons in an accelerator is increased from small to large values, bands distinguished by qualitatively very different behavior are encountered. At all amplitudes the local motion of an individual particle is governed by an extremely simple equation — an ordinary differential equation, Newton's Law (relativistic). Nevertheless a full spectrum of possibilities has come to be expected for Hamiltonian oscillators. An early work by Hénon,[5] written with no reference whatsoever to accelerators anticipated much of what has by now been abundantly observed. In this section the characteristic features are described.

At small amplitudes the particle behavior is perfectly linear and accurately predictable. The transverse displacement, say x, at a particular point in the lattice varies sinuisoidally in a range $\pm a_x(s)$. Because the focusing strength varies, $a_x(s)$ is a function of the longitudinal coordinate s. When a_x is generalized to account for this "parametric" variation, it becomes the Courant-Snyder invariant.[4] In the small amplitude region that quantity is truly constant.

As a_x is increased, nonlinear forces begin to have an effect. There is a large region in which they can be treated perturbatively. Though a_x is not quite invariant, its variation can be accurately predicted, for example by the introduction of "distortion" functions. The magnitude of the nonlinearity can be quantified by the smear, defined above. If the amplitudes $x, \beta dx/ds$ are recorded, turn-by-turn, in a Poincaré phase space plot taken at a fixed point in the lattice, they lie on a smooth, "regular", if somewhat distorted, closed curve. (The factor β is chosen to make this curve approximately circular.) This is the region in which ordinary accelerator operations live. Particles that start close together, stay close together, or at least diverge reasonably slowly. This makes the motion predictable, at least in the sense that a_x stays constant—tune difference will cause the angle in phase space to diverge eventually.

Already in the region just described, if one looks closely enough, tiny regions can be observed in which the particle tunes exhibit "tune entrainment" onto rational, resonant values. Thin, almost circular, chains of resonance islands are observed. All particles on one of these islands exhibit exactly the same tune. As a_x is increased the the radial extent of these resonance islands become great enough that the islands can not be overlooked. In some areas of this region, if one looks closely enough, chaotic motion is observed. Close-together chaotic trajectories diverge quickly, making them unpredictable.

At larger amplitudes the island chains tend to overlap and the chaotic regions become greater and greater, eventually leaving little area for regular motion. In this region the motion of individual particles is completely unpredictable, and more nearly resembles a random walk in the Poincaré plot. Toward the outside of the region particles are lost; whether they mathematically diverge to infinity, or simply grow to an amplitude at which they strike some obstacle, is equivalent from the point of view of accelerator physics. In this outer region there can still be predictability of the evolution of collections of particles, but only in the sense of diffusion. Such a description would use partial differential equations to govern the time evolution of particle distributions.

All features mentioned in this section have been observed in accelerators. Some of the observations will be described in later sections.

"Proofs" of particle stability. This section and the next will be partly philosophical, partly facetious, partly iconoclastic, but serious nonetheless. They relate precisely to the most important question that can be posed at a workshop on theory and accelerators: to what extent can theory prove in advance that an accelerator will work? In particular, will particles circulate stably?

As a matter of sociology, all that is required of a "proof" of stability is to persuade providers of funds to provide adequate support to enable the accelerator to be built. This means that the "standard of proof" becomes greater and greater as the accelerators become more expensive. Furthermore, as the machines get larger and larger, the apertures get smaller and smaller. In the previously introduced multipole series, higher and higher order terms become important. This make the "proof" harder and harder. Fortunately, the mathematical tools become more powerful, and so do the computers, so the task is not necessarily hopeless.

When Lawrence contemplated building a cyclotron he had only himself to persuade that the particles would circulate stably. He knew that the solution of Newton's differential equation for a charged particle in a uniform magnetic field was a well-bounded curve, namely a circle. QED — an adequate "proof" of stability. He was not troubled by questions of focusing and quadrupoles. It is just as well, as he might have been dissuaded from proceeding by the fact that there was no vertical stability. Because of the large aperture and the small number of revolutions required, this lack of stability was not serious.

Transverse stability was first analysed for the betatron. After appropriate linearization of the equations of motion, they became simple harmonic. "Proof" of stability then amounted to demonstrating that the transverse force had the correct sign to be a "restoring" force. In that case the solutions of the differential equations are sines or cosines. Because those functions are bounded by ± 1 the motion is bounded. QED.

Historically, in the development of accelerators, the terror of nonlinearity first became prevalent just after the discovery of the alternating gradient synchrotron (AGS) principle. This principle, already referred to above,[4] made it possible to focus particles into beams that could be contained in more slender toroidal vacuum chambers than had previously been thought posssible. This permitted the magnets to be much lighter and cheaper. The theory was the first to rely on transfer maps; with only linear forces being contemplated they were called transfer matrices. Evolution of a phase point in the Poincaré plot could be obtained by iterating the once-around transfer matrix. Stability could be related to the eigenvalues of the transfer matrix. Obviously the existence of any eigenvalue of magnitude greater than 1 results in unbounded motion. Symplecticity (intruding into accelerator physics for the first time) assured that the eigenvalues came in reciprocal pairs, and that the trace divided by two gives the cosine of the tune. Stable motion requires that the tune be a real angle. "Proof" of stability amounted then to showing that the trace was bounded by ± 2. QED. Though the AGS principle was invented at Brookhaven, a weak focusing electron synchrotron at Cornell, was the first to prove out the principle experimentally. Already in construction at the time, as a weak focusing machine, it was immediately converted to alternating gradient when the AGS principle was propounded. Shortly thereafter the possibility that nonlinear resonances would make it fail became a real concern. The successful commissioning of the synchrotron laid that fear to rest and (at least at Cornell) removed the incentive to study nonlinear effects for a considerable period of time. The fact that the magnetic aperture was much greater than the physical aperture justified the neglect of nonlinearity.

These examples have been intended to show that proofs of stability in advance of construction have not typically been ironclad. The determination to go ahead has been based more on experience, intuition, and self-confidence than on mathematical rigor. Successful operation has been the only really valid proof. For the next generation of accelerator, SSC or LHC, we are expected to do better, even though the problem has become harder. It may be more realistic to demand that theory veto designs that certainly will not work than that it provide designs guaranteed to work.

At least three theoretical approaches have been taken to proving that particles will be stable for long times. One is true analytic theory such as that of KAM, Arnold and Nekoroshov. This work yields mathematically rigorous results; if the assumed conditions are satisfied the conclusions cannot be doubted. Unfortunately, insufficient progress has been made in weakening the conditions to make them apply to realistic situations, so the results are not yet powerful enough to contribute to the design dialog. The other two appproaches are numerical. Both are conjectural in that they make less than fully substantiated simplifications in order to make the equations tractable. In one approach plausible, but uncontrolled mathematical approximations are introduced as needed to permit solution. In the other, exact mathematics is applied to a simplified model, the idealization of which is at the physics level. In both cases the credibility can be enhanced by showing that weakening the assumptions does not invalidate the conclusions.

It is clear that neither approach can yield a mathematically water-tight proof that protons will circualate stably for 10^7 turns in the as-built accelerator. Different people will attach different credibility to the two approaches. I will shamelessly indicate my bias by attaching the derogatory name "experimental mathematics" (an oxymoron?) to one approach and the congratulatory name "exact tracking" to the other.

These calculations have an influence on design decisions during the accelerator planning phase. Unfortunately errors made can lead either to excessive pessimism (and hence cost extravagance) or excessive optimism (and hence excessive risk). To the extent that computational errors give spurious diffusion they will lead to spurious transverse growth and undue pessimism. The use of too coarse granularity may also enhance damaging resonances and lead to excessive pessimism. On the other hand, calculations are always, to some extent, idealized and this, almost certainly, tends to give too optimistic results. In practice, at least for the SSC, uncertainty in the magnitudes of errors expected for yet-to-be-built magnets results in as much uncertainty as do these theoretical tracking uncertainties. A two or threefold reduction in those errors would render many tracking studies superfluous.

"Experimental mathematics". This section starts with a digression. Much of what is called Physics has developed as follows. With curiosity piqued by some observation, a physical model is formed, based on intuition, guesswork, experience, idealization and so on, and the model is converted to equations. If these equations are neither elegant nor tractible the model is discarded or revised. Otherwise, using valid mathematics and controlled (i.e. known to not change the answer too badly) approximation, the implications of the model are worked out. If the predictions disagree with the observations the model is discarded or revised, or sufficient doubt is cast upon the original observation that it is repeated. If the predictions agree everyone celebrates, new experiments are suggested, new results predicted and the process repeated. If the model has a few successes and no failures, it comes to be accepted as correct, at least in areas close to where it has been tested. There is nothing experimental about the mathematics that enters this sequence.

By experimental mathematics I mean the following. Suppose that, guided by the need to solve them, the equations are transformed, or beautified, or emasculated, or whatever, but simplified sufficiently to be solveable. These manipulations, though possibly guided by considerations of elegance and judgment, are uncontrolled in the sense that their effect on the answer is not known. Suppose also that, in some restricted domain, the equations can be solved by legitimate mathematics. Such a solution in a restricted domain can then play a role much like observations do in physics. If the simplified solution disagrees with the known correct solution then the simplified solution is certainly wrong and must be rejected. If it agrees in the restricted domain, it may well also be valid in a wider domain. As "science" this is just as valid as was the physics of the previous paragraph. But it is not mathematics; or rather, when applied outside the restricted domain it must not be given the same credibility that true mathematics deserves.

A simple, and quite common, instance of this line of reasoning occurs in long time prediction, for example using truncated transfer maps. Here the restricted domain is short times and the extended domain is long times. After a one-turn transfer map has been generated the computer can economically iterate it to predict far into the future, say for 10^7 turns. Based on agreement with a presumably reliable method for short times, say 10^3 turns, one may be tempted to claim that the long time prediction is valid, but that would be fallacious. It is almost as valid to claim, based on the reliable method's having shown stability for 10^3 turns, that the motion is stable forever.

That is not to say that either prediction is wrong; most of accelerator physics has been based on the proposition that stability of equations truncated to linear order assures adequate stability and that has not been found to be seriously wrong yet.

Limitatations of "restoring" symplecticity. Since the particle motion is Hamiltonian, it is necessarily also symplectic. Unfortunately, approximation methods, especially truncation, tend to violate symplecticity. At some level that makes them wrong and unreliable for long time prediction. The simplest example, for one dimensional motion, would have the determinant of the transfer matrix not quite unity, say a bit greater than one. That would lead to inexorable amplitude growth and eventual instability on a sufficiently long time scale. There are ways to restore symplecticity, but they are not unique. Hagel and Zotter[6] give an example in which it is possible to restore symplecticity but only at the cost of failing to conserve energy.

There is however a school of thought[7] saying that long time stability of particles in accelerators and symplecticity are so tightly intertwined that the long time essentials of the motion will be preserved if the description is symplectic, even if it is only approximate. This optimism is perhaps based on the fact that "microscopic" (also known as local) particle densities are preserved at all amplitudes. For linear beam transport systems this has been a powerful guiding principle. It can be carried over to nonlinear systems by restricting the form of nonlinearites. The idea can be explained along with the maxim that "kick codes are automatically symplectic". Here a kick is an infinitely thin element that administers a deflection that depends on the transverse coordinate but not the slope. Phase space coordinates x_-, x'_- before and x_+, x'_+ after such a kick are related by

$$\begin{aligned} x_+ &= x_- \\ x'_+ &= x'_- + f(x_-) \end{aligned} \tag{1}$$

where $f(x_-)$ is any function, linear or otherwise, of x_- but not x'_-. The Jacobean of this transformation is identically equal to 1, and as a result the phase space density is conserved at all amplitudes.

Other quantities are at least as important as local densities for characterizing transverse particle beam distributions: particularly global quantities like emittance (r.m.s. size), and "tails" of the distributions. It has already been shown that symplecticity assures conservation of local densities. A paper by Dragt et al.[8] discusses the conservation of emittance and other moments. For linear transformations there are powerful results assuring the conservation of emittance, but results are sparse in the face of nonlinearity. Actually it is the growth of tails on particle distributions that will lead to loss of particles when the tail particles hit a nearby obstacle. Unfortunately, conservation of microscopic phase space density does not prevent the growth of tails. In the well known process of resonant extraction from an accelerator particle escape to "infinity" along an escaping separatrix, preserving microscopic phase space density all the while. Furthermore the existence of filamentation can cause macroscopic phase space density dilution even with conservation of microscopic density. These effects lead naturally to the growth of tails on the particle distribution function, and those lead naturally to beam loss.

"Exact" tracking. It is clear that the exact representation of every magnetic element of a complicated accelerator lattice is impossible. Nevertheless, for the sake of argument, let us say we have such a lattice description. In that case the numerical method described so far could be called "approximate tracking in an exact lattice". We now describe an alternative method that can be called "exact tracking in an approximate lattice". In this approach the physical model of the lattice is simplified by using only elements for which exact analytical formulas can be derived. This is the approach taken by the accelerator modeling code TEAPOT.[9]

Aside: in this paper issues of round-off error in computers are not addressed. "Exact" and "to machine precision" will be used synonymously, and both will mean that any analytic expressions in the computer code are the same expressions that would be said to give an exact analytic solution of the equations of motion.

The only lattice elements permitted in this code are elements through which particle trajectories can be traced exactly. In practice this restricts the description to thin elements and drifts—the acronym stands for Thin Element Accelerator Program for Optics and Tracking. This does not, however, preclude really quite accurate descriptions, as thick elements can be broken up into "adequately" thin elements. From this point of view the code is just a numerical differential equation solver, or, as it can be called, a symplectic integrator. For small accelerators or for tracking a small number of turns in a large accelerator this is one way the code is in fact used.

For a workshop emphasizing particle stability in future particle accelerators a different emphasis is appropriate. Using thin elements only, one can design an ideal, better-than-true-life accelerator. Given a choice between a zero length and a thick element having the same length-strength product (a physical impossibility unfortunately) the accelerator designer would choose the thin element because it permits simpler, more flexible, design. Using these ideal elements, a lattice that, compared to the true SSC, has the same gross optics, the same dispersion supressors, the same intersection region optics, the same correction elements, and so on is designed. One can then inquire about the long time stability of particles in the better-than-true-life accelerator. This one does by the exact tracking of particle trajectories, which is, by construction, possible. Furthermore it is easy to model realistic deviations from perfection, such as multipole field errors and element misalignment, without giving up the exact tracking capability. If performance of the idealized lattice is unsatisfactory, then it is assumed that the design is insufficiently conservative. If performance is satisfactory, then the design is tentatively accepted, possibly with a safety factor being included, and attention is turned to other potential failure modes. Except for some recent work on long time stability[10] most modeling of the effects of and compensation of imperfections of the SSC have taken this approach.

In the idealized model there are formally only two elements, field-free drift sections and thin element multipoles. The latter are, differentiated into dipoles, quadrupoles, solenoids, markers, etc. on the basis of the numerical values of the multipole coefficients but they are formally equivalent as far as the code is concerned. Propagation through drifts in the computer amounts to solving for the intersection of a straight line and a plane. Though this is entirely elementary, trigonometric functions appear that make even drift regions "nonlinear". Of course, in the limit of paraxial trajectories the equations degenerate into linear equations. Thin elements are also simple though, like drifts, they are not trivial. For example they are not equivalent to the "kicks" described by Eq. (1) . It was emphasized there that the deflection in slope depends on the transverse displacement, but not on the slope itself—that was what seemed to be required to preserve symplecticity. Though this is a good approximation for paraxial trajectories it is not actually correct, even in the limit of zero length elements, as the reader can verify, or look up in reference.[9] Though this might appear to indicate that the exact motion is not symplectic, it really only shows that symplecticity is subtle, as the exact solution is necessarily symplectic.

Long term tracking results. Detailed results are given in reports by reports by Yan[11] and by Ritson[7] in presentations at this workshop, and in many SSC reports, such as the SSC Conceptual Design Report[12] and the SSC Suplementary Conceptual Design Report.[13]

Results tend to have the following qualitative features. Particles launched with large amplitudes are lost after a small number of turns, say less than 100, while small amplitude particle oscillate sinuisoidally, for as long as one has the patience to observe them. There is an intermediate, roughly, but not perfectly, circular band, perhaps ± 20 percent in phase space radius, for which the ultimate fate of any particular particle can only be determined by tracking it for as many as a million turns or more. In more physical terms, it appears to be necessary to track particles for ten or so periods of the longest natural period of the system to achieve selectivity in the determination of whether or not a particle is stable. For the collider, with a synchrotron oscillation period of 1000 turns, this rule of thumb anticipates appreciable evolution continuing for 10,000 turns.

The motion of any particular particle can be bizarre. A particle can oscillate for a million turns with amplitude varying over a fairly narrow band and then, in a small number of turns such as 100, develop wild swings and be lost. It is possible to perform "post mortem" analysis of such a particle, in order to determine the cause of death. In one case where this was done carefully it was clear that a resonance involving all three degrees of freedom was responsible. Energy sloshed between the horizontal and vertical degrees of freedom, with little or no coupling into the longitudinal. However, for several cycles of longitudinal oscillation notable distortion appeared at the same longitudinal phase, and the ultimate blow-up appeared at the same phase. These observations suggest that the damaging resonance is a horizontal-vertical coupling resonance, influenced parametrically by the longitudinal oscillation; finally the slow energy modulation moved the phase space point into a "black hole" from which its loss was almost immediate. It seems to be just a matter of chance whether a particle is launched in a "bad" region, and dies quickly, or in a "good" region, from which it proceeds to the bad region very slowly, if at all. Without intending to imply any particular understanding of the physics, this slow motion can be called "diffusion". If that general picture is correct, then seemingly quite small errors, if they lead to incorrect description of diffusion, can lead to totally wrong predictions of particle loss. Calculational errors are likely to lead to artificially large diffusion; leaving diffusion-causing effects out of the physical model will have the opposite effect.

A paper by Bourianoff, Cole, Talman, Trahern, Yan[16] submitted to this workshop addresses both of those possibilities. It relates to a beam dynamics experiment performed at Fermilab.[15] In that experiment protons were captured on a fifth-order resonance island. They were observed to remain trapped for several minutes—the beam current decayed exponentially with time constant 6.5 minutes, which is about 10^7 turns. This situation was simulated using the program ZTRACK[14] (a vectorized version of TEAPOT, running on a Cray). The observation itself can be interpreted as indicating an extremely small transverse random walk of about 3×10^{-4} meters during 400 seconds. Since there was appreciable nonlinearity present (strong enough, for example, to give the fifth-order resonance) this diffusion could have been due to nonlinear dynamics induced chaos. Since the loss could also have been due to any number of "noise" effects, the measurement has to be regarded as an upper limit on nonlinear dynamics-induced diffusion. In the simulation several particles initially on the resonance island were each tracked for 10^7 turns. All particles remained trapped. The result can be quoted as an upper limit of 1×10^{-4} meters during 400 seconds, for the random walk predicted by the numerical tracking program. If the bulk of the measured loss is due to noise effects, this calculation is consistent with the observation. Diffusion observed in the simulation could be spurious, due to round-off or other errors, or it could be due to nonlinear dynamics. That means that the simulation result can be regarded both as an upper limit on spurious, numerically-induced diffusion, and on nonlinear dynamics-induced diffusion. All of this paragraph has amounted to a convoluted way of saying that no diffusion (above "noise") was observed in the experiment or predicted in the simulation. From the point of view of nonlinear dynamics this is entirely boring. From the point of view of having confidence in future accelerators it is a good result.

Accelerator modeling and routine error correction procedures in the SSC. In the accelerator design phase, no issue is more difficult than assessing how well error correction procedures can be trusted to function. There is tension between two approaches that can be succinctly characterized as "gold-plate, or compensate". If one can rely on in-situ measurements, it will usually be cheaper to compensate for field errors than it is to control the manufacturing process to reduce the error by the corresponding amount. (An example from a more familiar field, may give the idea: it is cheaper to balance an automobile wheel by pounding little lead weights between the tire and the rim, than it is to control the dimensional tolerances of all tires and rims; but a fairly sophisticated wheel balancing instrument is needed to tell where to pound the weights.) Only operationally practical methods are used in simulating the performance of compensation schemes, but there is a large degree of subjectivity in assessing credibility. This subjectivity usually derives legitimately from experience on existing accelerators. To some extent uncertainty can be reduced by controlled experimentation on such accelerators. Some were done in connection with experiments to be described below.

Early simulation codes for the SSC were developed by Schachinger and others. They included closed-orbit smoothing, decoupling, tune and chromaticity control (all built into the TEAPOT code) and are described in various reports referred to in reference[12]. Systematic "homing in" on a satisfactory correction scheme is described by Bintinger et al.[3] Recent work, especially in anticipating the effects of malfunctioning beam position monitors, is due to Bourianoff[17] and others. With the collider good field region being a toroid 86 kilometers in circumference and about 1 centimeter across, it is clear that accurate beam steering is required. Root mean square orbit deviations of 0.5 millimeters away from nominal are expected to be achievable. Deviations exceeding that appreciably will lead to performance degradation; perhaps the worst effect is a conspiracy between two bad effects; feed-down due to random closed-orbit deviations in the presence of systematic field errors yields random field errors. As mentioned previously random field errors cause smear and, perhaps, diffusion-like behavior.

For predicting the detailed performance of an accelerator, accurate description of the accelerator parameters is essential. This may seem mundane, but because of the thousands of elements involved it is as important, and as difficult as any other part of the problem. At the SSC, lattice parameters for all accelerators and transfer lines in the complex are maintained in standard format in a master central commercial database (Sybase) planned and implemented by Peggs, Saltmarsh, and Trahern. Global geometric self-consistency and integrity of parameters as they are used for various purposes can be monitored this way.

Beam dynamics experiments. Various beam dynamics investigations have been performed using the Tevatron at Fermilab, with the intention of contributing to design decisions for the SSC.[15] Being a superconducting accelerator, the Tevatron is the existing facilty that most resembles the SSC. Furthermore it can be considered as essentially linear, so that nonlinear behavior is dominated by the active controlled addition of known nonlinear elements. Here we just categorize the experimental approaches that have been taken, and what has been learned. Mainly the focus has been on studying nonlinear effects.

(i) Phenomenological approach. As has been described above, the level of nonlinearity can be quantified by the quantity called smear which, for a known distribution of nonlinearity, can be detemined without difficulty. What must be determined is how well any accelerator degradation due to nonlinearity correlates with the smear value, and what value of smear can be regarded as tolerable for operations. For these investigations, after "mocking-up" the SSC by turning on nonlinear elements, one can study injection efficiency, damping of injection errors, orbit flattening, storage lifetimes and other operational issues. As smear was raised from the raw machine value somewhat below 1% to about 5%, performance was not greatly impaired, and it was tolerable, though impaired, up to about 10%. this was for a particular distribution of nonlinear elements; the degree to which the degradation is different for different distributions has not, as yet, been determined.

(ii) Engineering approach. It has to be assumed that practical accelerator operations will be restricted almost entirely to the region in which the effects of nonlinearity can be calculated perturbatively. In this region it is possible to make accurate quantitative comparisons between measured and calculated values of smears, tune shifts and decoherence factors. With accuracies at the several percent level agreement is found. This shows that perturbative calculations can be relied upon for predicting most operational procedures. In this region the uncertainty of prediction will be more due to lack of knowledge of the magnetic fields than to calculational uncertainty.

(iii) Pure physics approach. As mentioned before, a functioning proton accelerator can be regarded as a natural laboratory for the experimental study of nonlinear dynamics. Experiments can be conceived of as pure physics without any concern for the degree to which the results will contribute to improved intensity or luminosity or whatever other feature is perceived at the time as being required for improved performance of the device in its primary role as a tool for studying elementary particle interactions. As in the rest of science, one has reasonably high confidence that such investigations will lead to enhanced understanding that will ultimately contribute to the primary goal. Studies at Fermilab[15] in which a "metastable state" of the accelerator was investigated fall into this category. In this state the accelerator acts as a storage ring in which the particles oscillate stably around a fixed pooint that is other than the usual one at the origin. It is one of the stable fixed points accompanying a nonlinear resonance. For a fifth order resonance—the main case studied—the central trajectory closes on itself only after five turns rather than every turn. Used in this way the accelerator is not quite practical for routine operation, in that the lifetime is of order a few minutes instead of a few hours. With effort it could perhaps be made practical, but probably not superior to the reqular operation, and in any case that is not the point of the investigation. One point *is* that particles oscillating at small amplitude relative to the metastable fixed point are oscillating at large amplitude relative to the normal fixed point. A numerical description of the motion that agrees with observations both at the origin and at the amplitude of the nonlinear resonance can probably be relied upon to be correct throughout the perturbative region which is the region that particles visit in normal operations. Some numerical investigations of this sort have been described above.[16]

Many experiments can be designed to study this metastable state. Some have been performed already and some are planned. Its injection efficiency (production probability) and decay rate can be measured under various conditions. Especially apropriate is to study the influence of oscillatory external forces or modulations, as that should probe the oscillatory characteristics of the state, the simplest of which is the frequency of small oscillation around the metastable fixed point.[18]

Another approach that can be taken is to measure, over long times such as half an hour, the evolution of the transverse beam distribution. Ascribing this evolution to diffusion, solving the diffusion equation with empirically adjusted dependence on amplitude of the diffusion constant, a semi-theoretical, semi-phenomenological description results. Qualitatively the results are consistent with a diffusion constant that vanishes for small amplitudes and becomes large at large amplitudes. Pencil beams remain invariant because all particles have such small amplitudes that diffusion is negligible. Somewhat wider beams *spread* slowly, developing tails, while remaining constant in intensity because no particles are at amplitudes large enough to hit any obstacles and be lost. Beams that are wider yet lose intensity and *shrink* in width; that is because tails grow quickly, allowing particles to reach obstacles and be lost; this removes large amplitude particles from the distribution.

(iv) Diagnostics approach. Another experimental activity appropriate to planning for future accelerators is the testing of diagnostic procedures. It has been mentioned above that superconducting magnets have hysteretic magnetic field errors. To control the beam in the presence of such time varying, and somewhat unpredictable forces it will be necessary to have beam property sensors supplying feedback to correctors for stabilization. A loop of this sort that has been successfully tested uses digital Fourier transforms of beam position monitor signals to measure the tunes at two values of the momentum; from this the chromaticity is measured and hence, feeding back to sextupole elements, it is corrected.

Parallel computation. For simulating the performance of particle beams in a large accelerator it is necessary to track a large number of particles through a large number of elements. Provided collective effects are not being considered the evolution of each particle proceeds independent of all the others. This makes it natural to exploit multiple computation processors, with one processor per particle. The previously mentioned code ZTRACK was written to exploit the vector capabilities of the Cray to achieve this parallelism. We[19] have now also revised the tracking code TEAPOT to run on a multiple instruction set, multiple data set computer called the Intel Hypercube. Each of many (say 63) processors (also called nodes) has a complete record of the data describing the lattice, and each tracks a single particle. Another node is dedicated to book-keeping and to directing the efforts of the other 63, noting down intermediate results, launching a new particle whenever a particle is lost or finished and so on. Because the individual calculations are so nearly independent, very little communication is necessary, and the individual nodes calculate with about 98% efficiency, compared to their performance on single processor code. For this particular code then, the 64 node Hypercube calculates at about twice the rate of the Cray Y/MP optimized for the same code. These computations have been performed using a Hypercube computer situated at Oak Ridge National Laboratory. Starting in the fall of 1990 a 64 node Hypercube will be available for accelerator simulations at the SSC Laboratory in Dallas.

REFERENCES

1. G. Bourianoff, B. Cole, and R. Talman, *SSC Aperture Studies Using the Hypercube,* Report in preparation.
2. L. Schachinger and R. Talman, *Simulation of Chromaticity Control in the SSC, SSC-167, March 1988.*
3. D. Bintinger et al. *Compensation of SSC Lattice Optics in the Presence of Dipole Field Errors*, SSC-SR-1038, February, 1989. Updated in B. Cole and R. Talman, *Correction of Systematic Dipole Errors With 90 Meter Half-Cells*, SSC-N-672, October, 1989.
4. E. Courant and H. Snyder, *Annals of Physics*, **3**, 1 (1958).
5. M. Q. Hénon, Appl. Math. **3**, 291 (1969).
6. J. Hagel and B. Zotter, *One Dimensional Motion in a Thick Sextupole, A Comparison of Tracking Methods With the Exact Solution*, CERN/LEP-TH/89-12, March 1989.
7. D. Ritson, *SSCTRK: A Particle Tracking Code For The SSC.*
8. A.Dragt et al. Theory of Emittance Invariants, in Frontiers of Particle Beams; Observation, Diagnosis and Correction, Joint US-CERN School on Particle Accelerators, Capri, Italy, 1988.
9. L. Schachinger and R. Talman, *Particle Accelerators,* **22**, 35 (1987). See also current manual for the program TEAPOT.
10. D. Ritson, report to this workshop.
11. Y. Yan, report to this workshop.
12. SSC Central Design Group, Superconducting Super Collider Conceptual Design, March, 1986.
13. SSC Laboratory Staff, Site-Specific Conceptual Design of the Superconducting Super Collider, July, 1990.
14. L. Schachinger and Y. Yan, *Recent SSC Dynamic Aperture Measurement From Simulations*, SSC-N-664, 1989.
15. A. Chao et al., Phys. Rev. Lett. **61**, 2752 (1988).

16. R. Talman, B. Cole, C.G. Trahern, and G. Bourianoff, *Numerical Calculation of Diffusive Particle Loss From Resonance Islands in the Tevatron.* 1990 Washington Meeting of the A.P.S.

17. G. Bourianoff, J. Peterson, and R. Talman, *Closed Orbit Correction Of The Ssc Collider,* SSC report in preparation.

18. G. Tsironis, S. Peggs, and T. Chen, *Tune Modulation, Mathieu Stability, and the Driven Pendulum,* E.P.A.C., Nice, 1990.

19. G. Bourianoff, B. Cole, and R. Talman, *SSC Aperture Studies Using the Hypercube,* Report in preparation.

PADÉ APPROXIMANTS OF NORMAL FORMS FOR AREA PRESERVING MAPS

GRAZIANO SERVIZI
Dipartimento di Fisica, Università di Bologna
INFN – Sezione di Bologna

ABSTRACT

Rational approximations of Padé type are applied to the radial Fourier amplitudes of the conjugation function and the normal form of a quadratic area–preserving map on the plane, modelling the motion of a particle beam in an accelerator subject to nonlinear sextupole interactions.

Characteristic functions such as tune–shift and smear are computed in this way and a good accuracy is found with respect to the values given by direct iteration of the map. Some phase plot of the direct map and its "Padé approximated" normal form are presented and discussed.

1. Introduction

The study of the motion of a particle in an accelerator can be associated to that of a nonlinear map in the plane, as in the works made since 1982 by the Bologna Mathematical Physics group.[1–8]

Among these maps the simplest one is the quadratic map, taking into account only the nonlinear sextupole effects on the particle beam motion, nevertheless featuring all the relevant properties of a general nonlinear map.

We write the map in the standard way using complex rescaled variables z and z^* as

$$z' = e^{2\pi i\omega}\left[z + i\left(\frac{z+z^*}{2}\right)^2\right] \equiv F(z, z^*),$$

where ω is a non–rational number, and solve the functional equation

$$\Phi \circ \mathcal{N} = F \circ \Phi \tag{1}$$

to find a function $\Phi(\zeta,\zeta^*)$ which conjugates F with its normal form $\mathcal{N}$, namely

$$\zeta' = \exp[2\pi i \Omega(|\zeta|^2)]\,\zeta \equiv \mathcal{N}(\zeta,\zeta^*)$$

where Ω is a real function of its argument.

As in [1] we expand Φ and Ω in formal power series

$$\Phi = \zeta + \sum_{n=2}\sum_{k=0}^{n} a_k^{(n)}\, \zeta^k \zeta^{*n-k} \tag{2}$$

$$\Omega = \omega + \sum_{n=1} \Omega_{2n}\, |\zeta|^{2n} \tag{3}$$

and calculate the coefficients $a_k^{(n)}$ and Ω_{2n} by means of the functional equation Eq.1 and the area–preserving condition $\partial(\Phi,\Phi^*)/\partial(\zeta,\zeta^*) = 1$.

2. Fourier amplitudes and rational approximations

In the present work we are looking for rational approximations of the Padé type (P.A.) to the formal power series (2) and (3) and to the inverse of Φ, Ψ, given by

$$\Psi \equiv \Phi^{-1} = z + \sum_{n=2}\sum_{k=0}^{n} b_k^{(n)}\, z^k z^{*n-k} \tag{4}$$

which closes the commutative diagram

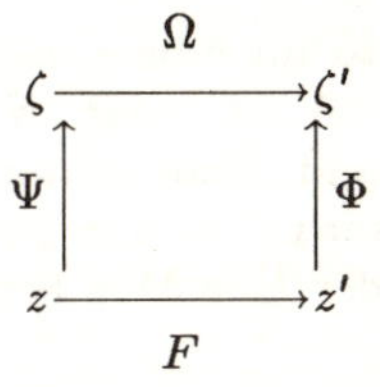

For this purpose we introduce first of all polar coordinates r, θ so that

$$\zeta = r\, e^{i\theta}$$

and the series (2) and (4) become formal Fourier series

$$\Phi = r\, e^{i\theta} + \sum_{n=2} r^n \sum_{k=0}^{n} a_k^{(n)} e^{i(2k-n)\theta}$$

$$\Phi^{-1} = r\, e^{i\theta} + \sum_{n=2} r^n \sum_{k=0}^{n} b_k^{(n)} e^{i(2k-n)\theta}$$

while Ω becomes

$$\Omega = \omega + \sum_{n=1} \Omega_{2n} r^{2n}.$$

When we fix the truncation order N, reverse the sums over k and n, and use the index of summation $m \equiv 2k - n$, we can write Φ (and Φ^{-1} too) as

$$\Phi \equiv \Phi(r,\theta) = \sum_{m=-N}^{N} e^{im\theta} A_m(r)$$

where, after some straightforward calculations, the "Fourier amplitudes" $A_m(r)$ can be written in the following way:

$$A_m(r) = r^{\overline{m}} \sum_{j=0}^{[\frac{N-\overline{m}}{2}]} (r^2)^j a_{\frac{m+\overline{m}}{2}+j}^{(\overline{m}+2j)} = r^{\overline{m}} \tilde{A}_m(r^2).$$

In this formula $\overline{m} = |m|$ for all m except $m = 0$; in this case $\overline{m} = 2$.

The polynomials $\tilde{A}_m$ are functions of r^2, just as the series Ω, and their degree increases by 1 as the Fourier index decreases by 2, starting from the last two polynomials $\tilde{A}_{\pm N}$, which are simple constants, and ending with $\tilde{A}_0$ which has always the same degree as $\tilde{A}_{\pm 2}$.

3. Padé Approximants short review

At this point we are going to introduce the Padé Approximants to a power series $f(z) = \sum f_n z^n$; we call P.A. of order $[N/M]$ to $f(z)$ a rational function $P_N(z)/Q_M(z)$ whose numerator and denominator P_N and Q_M are polynomials of degrees N and M respectively such that the power series expansion of P_N/Q_M coincides with that of f up to order $N + M + 1$, namely

$$Q_M(z)f(z) - P_N(z) = O(z^{N+M+1}). \tag{5}$$

The previous formula gives also the method of finding the coefficients of the two polynomials P_N and Q_M: it has indeed the form of a linear system of equations in $N + M + 1$ unknowns, just as the number of terms which have to vanish on the right hand side.

Since one can always set equal to 1 one of the coefficients of Q_M (or P_N), Eq.5 allows the computation of the P.A. without ambiguity. We choose to set equal to 1 the coefficient of the power z^0 in Q_M: in this way we avoid a pole at the origin, which we know to be a stable equilibrium point in our problem.

The P.A. feature several interesting properties and we are not going to mention all of them here. We only recall thet they are an efficient tool to:

1) accelerate the convergence of a power series (for example the series of e^z on the negative real axis);
2) make analytic continuation of functions beyond the radius of convergence of their power series;
3) reproduce the analytic structure of a function, giving the approximate location of zeros and poles with their residues;
4) for Stieltjes functions $f(z)$, i.e. functions that can be written as integral transforms of positive measures, the sequences of diagonal P.A. $[N/N]$ and sub–diagonal P.A. $[N-1/N]$ are upper and lower bounds respectively to $f(z)$ and moreover the poles of the two sequences interlaces with those of $f(z)$.

We now apply P.A. to the polynomials $\tilde{A}_m(r^2)$, and decide to use diagonal or sub–diagonal P.A. according to the degree of the polynomial $\tilde{A}_m$ itself.

Starting from $\tilde{A}_{\pm N}$ we have the following table for the P.A. associated to each of the $\tilde{A}$'s

$$\begin{array}{cc} \tilde{A}_{\pm N}, \tilde{A}_{\pm(N-1)} & [0/0] \\ \tilde{A}_{\pm(N-2)}, \tilde{A}_{\pm(N-3)} & [0/1] \\ \tilde{A}_{\pm(N-4)}, \tilde{A}_{\pm(N-5)} & [1/1] \\ \vdots & \end{array}$$

and so on increasing alternatively the degree of Q and the degree of P.

4. Results

We show in Figures 1 and 2 the phase plot of the quadratic map near the $\omega = 1/5$ resonance, compared with the images of circles through $\Phi(r, \theta)$ where the $\tilde{A}$'s have been replaced by their P.A. as defined above. On the right the direct iteration of the map is plotted twice to make easier the comparison with the images of circles plotted on the left. We kept N (the truncation order of Φ) equal to 24, but in theory we can easily reach higher orders, provided the number of significant digits during the numerical calculations is large enough.

We observe that, below the resonance, a very high accuracy is reached for all orbits until the so-called dynamical aperture.

The same picture, drawn with the original $\Phi(r, \theta)$, is plotted on the bottom left corner of the figures: it shows regular orbits only in a smaller region around the origin and exhibits five loops on the last closed orbits before losing any predictability.

Above the resonance the series $\Phi(r, \theta)$ again fails to reproduce the phase plot when the island region is reached; on the contrary the Padé approximated Φ is

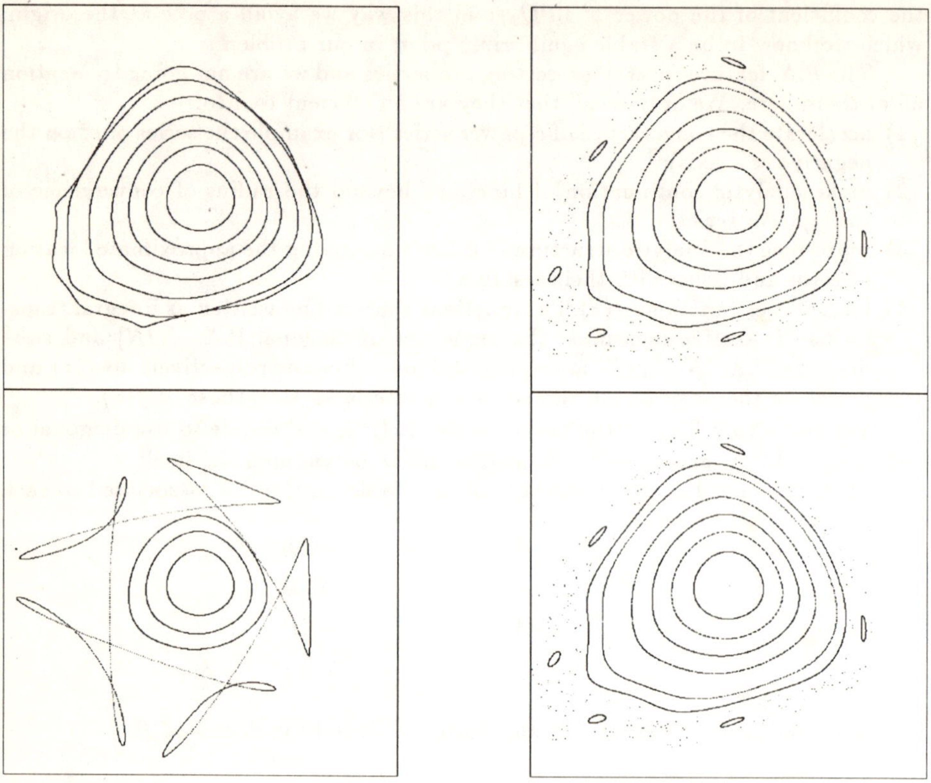

Figure 1: Phase plots for $\omega = 0.195$

able to recover the regular orbits beyond the strip where the resonance islands are located.

In a work of 1988 with Turchetti, Bazzani and Mazzanti[5] the non–resonant normal forms were used to calculate tune shift and smear for a model of linear hadron collider; now we make the same calculations using P.A. and we show in Figures 3 and 4 the behaviour of the tune shift compared with their "true" values, obtained by direct iteration of the map, which are plotted as continuous lines. The values computed through the series expansion and the P.A. are plotted as dashed and dotted lines respectively.

We find, according to the phase plot shown before, that near the 1/5 resonance the P.A. allow to extend the validity of the calculations almost up to the dynamical aperture, while the original power series didn't.

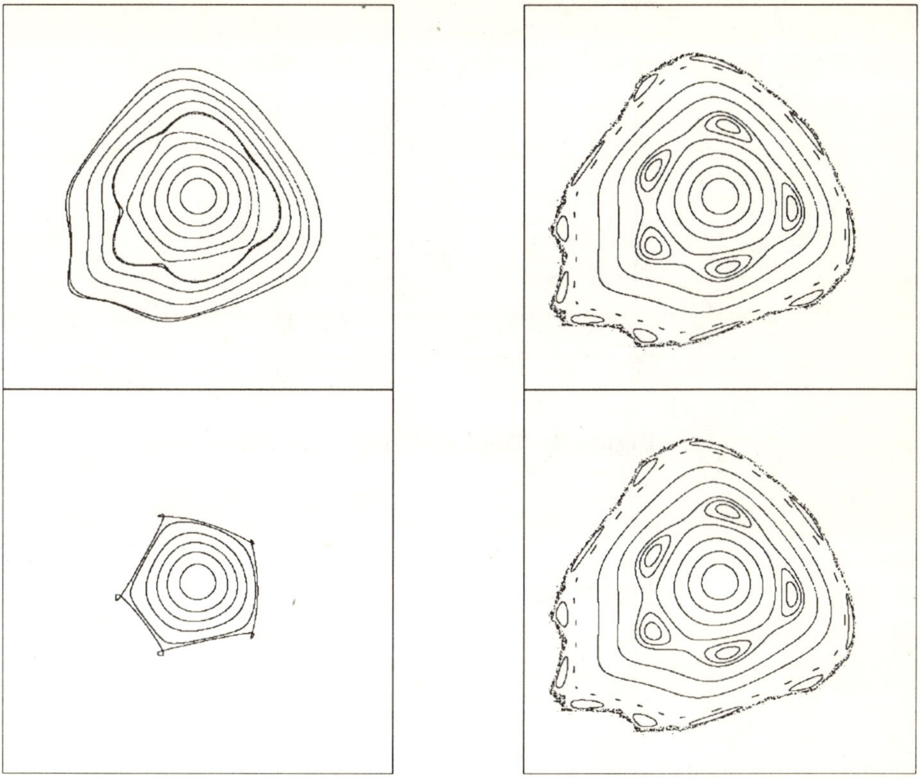

Figure 2: Phase plots for $\omega = 0.205$

5. Conclusions

At this point an accurate analysis of the poles of the P.A. would be required. We conjecture that the P.A. have poles in the islands region with residues proportional to some increasing function of its "width"; this would explain why P.A. can cross regions of irregular orbits without losing the track of the following ones. It will be shown in a future work that indeed the poles occur for values of the distance from the origin which are strictly correlated with the continued fraction expansion of the linear tune ω, namely for

$$r_j^2 \simeq -\frac{\varepsilon_j}{\Omega_2} \tag{6}$$

where Ω_2 is defined in Eq.3 and ε is the absolute value of the difference between ω and its j-th continued fraction approximation. The poles are real or imaginary

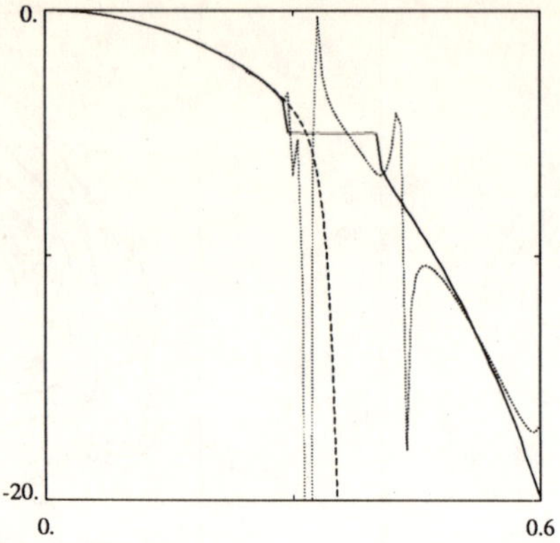

Figure 3: Tune shift for $\omega = 0.205$

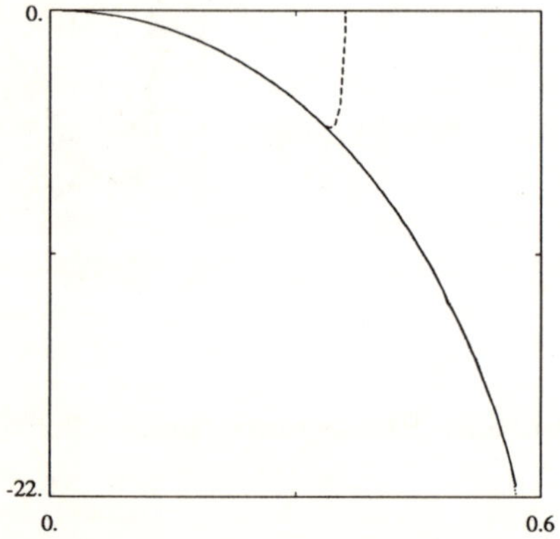

Figure 4: Tune shift for $\omega = 0.195$

according to the sign of Ω_2 or, equivalently, depending on whether the tune is above or below a resonant value.

We plot in Figure 5 the logarithm of the pseudo-radius of convergence of Φ versus the perturbation order for ω equal to the golden mean, whose continued

fraction expansion is given by the Fibonacci numbers. The computation was already done in [1], but now we go up to order 200 using a CRAY machine. The plateaus are exactly located at values given by Eq.6, in agreement with that was previously claimed.

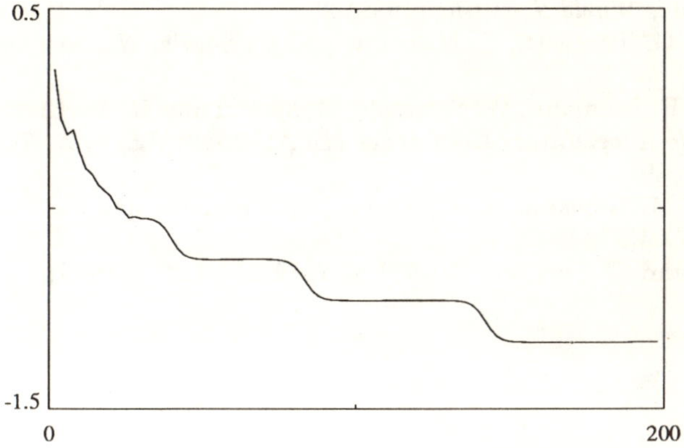

Figure 5: $\log r(N)$ **for** $\omega = \frac{\sqrt{5}-1}{2}$

Another open question is the possibility of a so-called multistep rational approximation meaning that a power series would be approximated with rational functions up to a truncation order N_1 and then up to an higher order N_2 and so on, each time applying P.A. to a "rescaled power series" built with the coefficients of orders $N_k + 1$, $N_k + 2$, ..., $N_{k+1} - 1$, N_{k+1} of the initial series.

While not hard to set up on a theoretical basis, this kind of approximation however would require much more efficient computational tools, such as computer programs which can handle floating point arithmetics with extended precision.

Acknowledgements

I wish to thank Professor Giorgio Turchetti for his help in this work and all members of the Mathematical Physics group of the Bologna University for their friendship. A special thank to Miss Patrizia Nicoli, of the INFN, for her valuable and persevering cooperation.

References

1. G. Benettin, A. Giorgilli, G. Servizi and G. Turchetti, *Phys. Letters* **A95**, (1983) 11.
2. G. Servizi and G. Turchetti, *Computer Physics Communications* **32**, (1984) 201.
3. G. Servizi and G. Turchetti, *Nuovo Cimento* **B95** (1986) 121.
4. A. Bazzani and G. Turchetti *from Nonlinear Dynamics and stochastic processes II Ed. by Livi, World Scientific* (1988) 25.
5. A. Bazzani, G. Turchetti, P. Mazzanti and G. Servizi *Nuovo Cimento* **B102**, (1988) 51.
6. A. Bazzani, B. Leemann, W. Scandale, G. Servizi and G. Turchetti *from European Particle Accelerators Conference (EPAC 1988) Ed. by S. Tazzari, World Scientific* (1989).
7. A. Bazzani, M. Giovannozzi, G. Servizi, G. Turchetti and E. Todesco *CERN SPS/89-24 (AMS)* (1989).
8. A. Bazzani and G. Turchetti *CERN SPS/89-25 (AMS)* (1989).

VARIATIONAL METHOD FOR APPROXIMATE INTEGRALS OF MOTION OF SYMPLECTIC MAPS

ARMANDO BAZZANI

Dipartimento di Fisica, Università di Bologna, ITALY
INFN – Sezione di Bologna, ITALY

ABSTRACT

The problem of constructing approximate integrals of motion for a symplectic map is revisited in terms of a variational principle. We show that the variational principle recovers the perturbative solution if we consider a small neighbourhood of a fixed point and that it may be also used where the perturbative theory fails.
An application to the Henon map illustrates the performances and the limits of the perturbative approach and shows the possibility of the variational method to describe the non linear resonances.
This technique may be useful in the problems of non linear beam dynamics in particle accelerators.

1. Introduction

The non linear effects due to the multipolar magnetic fields on the stability analysis of the transverse oscillations near the reference orbit in hadron particle accelerators (the betatronic motion) is usually reconducted to the study of the dynamics of a symplectic map in $\mathbf{R}^4$ near an elliptic fixed point[1]. The relevant quantities for the accelerators physics are the "tune-shift", the stability estimates for long time (10^{7-9} iterations), the topology of the orbits at large amplitude in a neighbourhood of the elliptic fixed point and information on the measure of the set of the invariant KAM tori or the measure of the stochastic layers near the non linear resonances. Usually

the study of the betatronic motion is carried out by a direct ispection of a great number of iterations of the map. An alternative approach uses the perturbative normal forms method[2] which provides information on both the tune-shift and the topology of the orbits and gives a Nekhoroshev estimate for the stability time[3]. Unfortunatly the normal for approach implies the solution of a non-linear functional equation which can be solved only in a perturbative way. On the contrary if one directly looks for the integrals of motion, the corresponding functional equation is linear.
The knowledge of a set of approximate integrals of motion provides information for the long time stability and the topology of the orbits but actually we do not have any information on the motion along the orbits (i.e. we do not know the tune-shift). In this note we propose a variational method to calculate approximate integrals of motion for a symplectic map in $\mathbf{R}^{2d}$. We work out the variational approach for polynomial integrals of analytic maps with an elliptic fixed point at the origin, but the extension to more general cases is straightforward.
We briefly present the perturbative approach to the integrals of motion as a particular case of the variational method when we introduce the perturbative norm on the space of analytic real functions. By using a perturbative method it is possible in the non resonant case to provide a rigorous estimate of Nekhoroshev type for the stability. This may be of physical relevance as a scaling low of the stability time with respect to the perturbation strength[4]. However this result applies only in a set A where the effects of the non linear resonances of order lower than a suitable order N_{opt} are not present and the perturbative integrals of motion imply an approximated Arnold-Liouville integrability[5] for our map up to term of order N_{opt}.
On the other hand if we use the usual L^2−norm, the variational method may still provide an approximate integral of motion in a subset $A \subset \mathbf{R}^{2d}$ near the dynamical aperture. Moreover we show that we can recover the perturbative result as we make the set A tends to the origin.
We recall that recently other variational principles have been formulated for the Lagrangian and the Hamiltonian mechanics: the nonexixtence of uniform quasiperiodic solutions for quasi integrable hamiltonian systems lead Percival [6] to formulate a fixed frequency variational principle whose solution gives the motion on individual tori and an analogous principle has been formulated by Aubry [7] and Mather [8] in the case of twist area preserving maps. Indeed for a fixed diophantine frequency and for sufficiently small perturbation these variational approaches are equivalent to the usual perturbative method, which can also be used for computational purposes, because the solution is analytic and the perturbative series converge. However the variational principle allowed Mather to obtain weak solutions corresponding to orbits whose closure is no longer a torus but a closed Cantor-like subset of a torus currently named cantorus.
In our case we hope that the variational principle may be used in future to construct weak solutions for the integrals of motion which could give further information on the dynamics; indeed from the KAM theory we know the existence of a C^∞

interpolation of the KAM tori[9] which is in some sense a weak solution to our problem.
From a numerical point of view we remark that the variational principle is very easy to implement since it involves the numerical evaluation of a certain number of integrals and the solution of a linear system; some problems on the numerical stability of our approach are pointed out and the possible solutions are outlined.
This method can be used to analyse directly the tracking data, which give a local information on the dynamics, in order to obtain a global description of the topology of the orbits.
We have applied the perturbative and the variational approach to the study of the dynamincs of the quadratic Henon map[10] in the non resonant case, i.e. a model for the effect of a sextupole on the radial betatronic motion. We show how the variational approach allows to recover non perturbative structures as chains of islands at large amplitudes.
This result suggests that the variational method for the integrals of motion may be usefull in order to understand the dynamics of very complex symplectic maps as in the case of the lattice for an hadron accelerator.

2. The variational principle

Let $\mathcal{M} : \mathbf{R}^{2d} \to \mathbf{R}^{2d}$ be an analytic symplectic map and A a bounded measurable (with respect to the Lebesgue measure) subset of $\mathbf{R}^{2d}$; we say that a non constant real function $I : A \to \mathbf{R}$ is a first integral of motion for the map $\mathcal{M}$ on A if it satisfies the equation

$$I \circ \mathcal{M}(\mathbf{x}) - I(\mathbf{x}) = 0 \quad \forall \quad \mathbf{x} \in A \text{ s.t. } \mathcal{M}(\mathbf{x}) \in A \tag{2.1}$$

where $\mathbf{x} = (q_1, p_1, ..., q_d, p_d)$ and q_i, p_i are the standard symplectic coordinates in $\mathbf{R}^{2d}$. Following an argument due to Siegel[11] it is possible to prove that for a generic map $\mathcal{M}(\mathbf{x})$ the equation (2.1) does not admit any analytical solution if A is a set of positive measure (the area-preserving map with an hyperbolic fixed point are exceptional in this context[12]). Therefore if we introduce the operator $\Delta_{\mathcal{M}}$ according to the definition

$$\Delta_{\mathcal{M}} f(\mathbf{x}) = f \circ \mathcal{M}(\mathbf{x}) - f(\mathbf{x}) \tag{2.2}$$

where $f(\mathbf{x})$ is a real analytic function on $\mathbf{R}^{2d}$, the Siegel theorem assures that for a generic symplectic map $\mathcal{M}$ the kernel of $\Delta_{\mathcal{M}}$ reduces to the constant functions.
Even if an exact integral of motion does not exists, the knowledge of an approximate integral of motion is of physical relevance since it may provide information on the

topology of the orbits and on the stability of the motion for finite (but may be sufficiently long) times. If we introduce the functional

$$\mathcal{F}(I) = \|\Delta_{\mathcal{M}} I\|^2 \tag{2.3}$$

where $\| \quad \|$ is a norm in the space of real functions on $\mathbf{R}^{2d}$, the variational principle

$$\delta \mathcal{F}(I) = 0. \tag{2.4}$$

provides a minimum for the functional $\mathcal{F}(I)$ which may be regarded as an approximate solution of eq. (2.1).
In the following we will consider the case of analytic maps with a fixed point at the origin

$$\mathcal{M}(\mathbf{x}) = \mathcal{M}_1 \circ \hat{\mathcal{M}}(\mathbf{x}) = \mathcal{M}_1 \circ \left(\mathbf{x} + \sum_{n \geq 2} \mathcal{M}_n(\mathbf{x}) \right) \tag{2.5}$$

where $\mathcal{M}_1$ is a linear symplectic map and $\mathcal{M}_n$ are homogeneus polynomial maps of order n and we will discuss the variational principle for analytic approximate integrals of motion

$$I(\mathbf{x}) = I_2(\mathbf{x}) + \sum_{n \geq 3} I_n(\mathbf{x}) \tag{2.6}$$

where $I_2(\mathbf{x})$ is one of the d quadratic invariants of the linear map $\mathcal{M}_1$ which correspond to the Courant-Snyder invariants in the accelerator physics[13]

$$I_2 \circ \mathcal{M}_1(\mathbf{x}) - I_2(\mathbf{x}) = 0. \tag{2.7}$$

In practice cases we have to introduce a cut-off N in the expansion (2.6) and to look for a minimum of (2.3) by varying the coefficients of the polynomials $I_n(\mathbf{x})$ where $3 \leq n \leq N$. Indeed the quadratic part $I_2(\mathbf{x})$ has to be fixed in order to avoid the trivial constant solutions.
We consider two different norms for the funtional $\mathcal{F}(I)$ in the space of polynomials: the usual L^2 norm on the set A

$$\|I\| = \left[\int_A |I|^2 d\mu \right]^{\frac{1}{2}} \tag{2.8}$$

and the "perturbative norm" up to order N which is indeed a pseudonorm

$$\|I\|_{D(N)} = \left[\sum_{n=1}^{N} |I_n|^2 \right]^{\frac{1}{2}} \tag{2.9}$$

where $|I_n|$ is the sum of the absolute value of the coefficients of $I_n(\mathbf{x})$

$$I_n(\mathbf{x}) = \sum_{|k|=n} a_k \mathbf{x}^k \Rightarrow |I_n| = \sum_{|k|=n} |I_k|^2 \tag{2.10}$$

where $k \in \mathbf{N}^{2d}$ and $\mathbf{x}^k = q_1^{k_1} p_1^{k_2} ... q_d^{k_{2d-1}} p_d^{k_{2d}}$. If we introduce the expansion

$$I(\mathbf{x}) = I_2(\mathbf{x}) + \sum_{|k|\geq 3}^{N} a_k \mathbf{x}^k \tag{2.11}$$

the functional $\mathcal{F}(I)$ becomes a quadratic form in the coefficients a_k for both the norms (2.8) and (2.9), consequently the variational condition (2.4) which is made explicit by calculating the derivatives with respect to each a_k, is a linear system in the unknown coefficients a_k. Indeed by using the L^2 norm the functional $\mathcal{F}(I)$ reads

$$\begin{aligned}\mathcal{F}(I) = &\sum_{|k|\geq 3}^{N} \sum_{|h|\geq 3}^{N} a_k a_h \int_A (\Delta_{\mathcal{M}} \mathbf{x}^k)(\Delta_{\mathcal{M}} \mathbf{x}^h) d\mu \\ &+ 2 \sum_{|k|\geq 3}^{N} a_k \int_A (\Delta_{\mathcal{M}} \mathbf{x}^k)(\Delta_{\mathcal{M}} I_2) d\mu + \int_A (\Delta_{\mathcal{M}} I_2)^2 d\mu\end{aligned} \tag{2.12}$$

so that the variational principle (2.4) reads

$$\sum_{|h|\geq 3}^{N} a_h F_{k,h} + B_k = 0 \quad \forall k \text{ s.t. } N \leq |k| \leq 3 \tag{2.13}$$

where we have introduced the quadratic matrix

$$F_{k,h} = \int_A (\Delta_{\mathcal{M}} \mathbf{x}^k)(\Delta_{\mathcal{M}} \mathbf{x}^h) d\mu \tag{2.14}$$

and the vector

$$B_k = \int_A (\Delta_{\mathcal{M}} \mathbf{x}^k)(\Delta_{\mathcal{M}} I_2) d\mu. \tag{2.15}$$

An analogous result holds for the perturbative pseudonorm (2.9) if we define the matrix $F_{k,h}$ and the vector B_k according to

$$\hat{F}_{k,h} = \sum_{|l|\geq 3}^{N} \left[\Delta_{\mathcal{M}} \mathbf{x}^k\right]_j \left[\Delta_{\mathcal{M}} \mathbf{x}^h\right]_j \tag{2.16}$$

and

$$\hat{B}_k = \sum_{|l|\geq 3}^{N} \left[\Delta_{\mathcal{M}}\mathbf{x}^k\right]_j [\Delta_{\mathcal{M}} I_2]_j \tag{2.17}$$

where $j \in \mathbf{N}^{2d}$ and the symbol $[\quad]_j$ indicates the coefficients corresponding to the monomial $\mathbf{x}^j$. We remark that according to the definition (2.9) the perturbative pseudonorm cannot take into account the contribution of the terms of order $> N$ in the expansion of $\Delta_{\mathcal{M}}\mathbf{x}^k$ because

$$\|\mathbf{x}^j\|_{D(N)} = 0 \quad \text{if} \quad |j| > N. \tag{2.18}$$

But this is just what happens when we consider a perturbative approach to the functional equation (2.1) by using the distance from the origin as a perturbative parameter. Indeed it is not difficult to see that the variational principle (2.4) in the case of the perturbative pseudonorm provides exactly the perturbative solution. Let us substitute the expansion (2.11) directly in the equation (2.1) and solve it for each perturbative order n up to order N

$$\sum_{|h|\geq 3}^{N} a_h \left[\Delta_{\mathcal{M}}\mathbf{x}^h\right]_j + [\Delta_{\mathcal{M}} I_2]_j = 0 \quad \forall\, 3 \leq |j| \leq N. \tag{2.19}$$

If we compare the system (2.19) with the system (2.13) in the case of the perturbative pseudonorm which reads

$$\begin{aligned} &\sum_{|h|\geq 3}^{N} a_h \sum_{|l|\geq 3}^{N} \left[\Delta_{\mathcal{M}}\mathbf{x}^k\right]_j \left[\Delta_{\mathcal{M}}\mathbf{x}^h\right]_j + \sum_{|l|\geq 3}^{N} \left[\Delta_{\mathcal{M}}\mathbf{x}^k\right]_j [\Delta_{\mathcal{M}} I_2]_j = \\ &\qquad \sum_{|l|\geq 3}^{N} \left[\Delta_{\mathcal{M}}\mathbf{x}^k\right]_j \left\{ \sum_{|h|\geq 3}^{N} a_h \left[\Delta_{\mathcal{M}}\mathbf{x}^h\right]_j + [\Delta_{\mathcal{M}} I_2]_j \right\} = 0 \end{aligned} \tag{2.20}$$

we can easily see the perfect equivalence of the two systems (2.19) and (2.20).
We discuss briefly the main properties of the perturbative approach. We observe that the system (2.19) can be immediately solved since the matrix $[\Delta_{\mathcal{M}}\mathbf{x}^k]_j$ is just in a triangular form because we actually have

$$\left[\Delta_{\mathcal{M}}\mathbf{x}^k\right]_j = 0 \quad \text{if}\, |j| < |k|. \tag{2.21}$$

Indeed let us consider the case of a symplectic map with an elliptic fixed point at the origin, which is the interesting case for accelerators physics, then we can reduce the linear part $\mathcal{M}_1$ (see eq, (2.5)) in a diagonal form by introducing a set of complex conjugated coordinates[2] in $\mathbf{R}^{2d}$; moreover we suppose that the linear eigenvalues $e^{i\omega_1}...e^{i\omega_d}$ do not satisfy any resonant relation

$$e^{i(m_1\omega_1+...m_d\omega_d)} - 1 = 0 \Leftrightarrow m_1 = ... = m_d = 0 \tag{2.22}$$

where $m_1,....,m_d \in \mathbf{Z}$. With a little abuse of notation we still indicate with $\{\mathbf{x}\}$ the complex coordinates which diagonalize the linear part. The monomials $\mathbf{x}^k$ have the following property

$$\left[\Delta_{\mathcal{M}}\mathbf{x}^k\right]_k = \left(e^{i((k_1-k_2)\omega_1+.....+(k_{2d-1}-k_{2d})\omega_d)} - 1\right) \tag{2.23}$$

and no other contribution of order $|k|$ is present. Incidentally we recall that the quantities in the r.h.s. of eq. (2.23) are the well known small denominators. In this case the system (2.19) reads

$$a_k\left(e^{i((k_1-k_2)\omega_1+...+(k_{2d-1}-k_{2d})\omega_d)} - 1\right) + \sum_{|h|\geq 3}^{N-1} a_h\left[\Delta_{\mathcal{M}}\mathbf{x}^k\right]_j + \left[\Delta_{\mathcal{M}}I_2\right]_k = 0. \tag{2.24}$$

$\forall\, 3 \leq |k| \leq N$ so that if $k_{2i-1} - k_{2i} \neq 0$ $i = 1...d$ we can recursively calculate a_k as a function of the coefficients a_h with $|h| < |k|$. On the contrary if $k_{2i-1} - k_{2i} = 0$ the small denominator is zero and eq. (2.24) reduces to a relation among the coefficients a_h which have just been calculated. Indeed the matrix $\hat{F}_{k,h}$ defined in eq. (2.16) is the Grassmann matrix of the functions system $[\Delta_{\mathcal{M}}\mathbf{x}^k]$ with $3 \leq |k| \leq N$ with respect the scalar product associated to the pseudonorm (2.9). But when we use the perturbative pseudonorm, the operator $\Delta_{\mathcal{M}}$ is equivalent to the operator $\Delta_{\mathcal{M}}^{(N)}$ defined according to

$$\Delta_{\mathcal{M}}^{(N)} f(\mathbf{x}) = \sum_{|j|\geq 0}^{N} \left[\Delta_{\mathcal{M}}\mathbf{x}^k\right]_j \mathbf{x}^j \tag{2.25}$$

which neglects the terms of order $> N$. As one can easily see, the polynomials $\Delta_{\mathcal{M}}^{(N)}\mathbf{x}^k$ do not form a linear independent system for the presence of resonance monomials $\mathbf{x}^k$ with $k_{2i-1} - k_{2i} = 0$ $i = 1...d$. This means that the matrix $\hat{F}_{h,k}$ is singular and the system (2.20) (or equivalently the sytem (2.19)) can be solved if and only if the rank of the complete matrix $\hat{F}_{k,h} \oplus \hat{B}_k$ is equal to the rank of the incomplete matrix $\hat{F}_{k,h}$. Unfortunatly this assert cannot be proved directly but we have to use the Birkhoff normal forms theory. The following proposition holds[4]: let $\mathcal{M}$ be an analytic symplectic map in $\mathbf{R}^{2d}$ with an elliptic fixed point, if the eigenvalues of the linear part $\mathcal{M}_1$ do not satisfy any resonant relation (2.22) and the initial condition $I_2(\mathbf{x})$ for the integral (2.6) is one of the d independent quadratic integrals of motion of the linear part then the system (2.24) is solvable and the resonant coefficients a_k with $k_{2i-1} - k_{2i} = 0$ $i = 1...d$ can be choosen arbitrarily.

This implies that the rank of the matrix $F_{k,h}$ is exactly equal to the number of the resonant coefficients.
In the perturbative case it is also possible to state some properties of the series which define the integrals of motion when $N \to \infty$ and to obtain a stability result of Nekhoroshev type for the dynamics of the map. Indeed if the linear frequencies ω satisfy a diophantine condition

$$\left| e^{i((k_1-k_2)\omega_1+.....+(k_{2d-1}-k_{2d})\omega_d)} - 1 \right| \geq \gamma^{-1}|m|^{-\eta} \tag{2.26}$$

where $\gamma > 0, \eta \geq d$ we have the following estimate for the norm of the homogeneus order I_n

$$|I_n| \leq \gamma^{n-2}(n!)^\eta (n-1)! \left(\frac{3}{2}\right)^{n-2} \qquad n \geq 3$$

where we have supposed without loss of generality that the initial map satisfies the condition $|\mathcal{M}_n| \leq 1 \;\forall\; n \geq 2$. This implies the asymptotic character of the series for the perturbative integrals which diverge in every finite neighbourhood of the origin. The topological reason of such a divergence is the presence of the non linear resonances in the phase space which accumulate at the origin as we increase the order of resonance; the perturbative integral of motion has a pole for each resonance[15] so that the divergent behavour follows since we are expanding in a Taylor series at the origin, which is an accumulation point for the poles of the integrals. Anyway if we fix a certain neighbourhood A of the origin of radius r, it is possible to stop the perturbative calculation at a finite optimum order at which the error is minimum

$$N_{opt} = \exp\left\{-1 - \frac{1}{\eta+1}\log\frac{r}{r^*}\right\} \tag{2.27}$$

where $r^* = \frac{8}{15\gamma}$ and to obtain the following stability result:
if $r \leq r^*(3e)^{-(1+\eta)}/2$ then $|\mathbf{x}_t| \leq 2r \;\forall\, t \leq T$ where $\mathbf{x}_t = \mathcal{M}^t(\mathbf{x})$ $t \in \mathbf{N}$ and

$$T = \frac{128e}{189}\frac{r}{r^{*2}}\left(\frac{r}{r^*}\right)^{\frac{1}{\eta+1}} \exp\left\{\frac{\eta+1}{e}\left(\frac{r^*}{2r}\right)^{\frac{1}{1+\eta}}\right\}. \tag{2.28}$$

The estimate (2.28) may be not good for a physical application but the interesting thing is the dependence of the stability time on the perturbation strength (the radius of the neighbourhood A) which may be an universal low.
The result (2.28) gives the limits of the perturbative approach which is very powerful if the set A is a strict neighbourhood of the origin and it can provide only rough information when we consider a set A at large amplitudes near the dynamical aperture. This is a consequence of the fact that the perturbative pseudonorm (2.9) takes into account only information on a function and its derivatives at the origin. Conversely if we use the L^2-norm (2.8) in our variational principle we take

into account in the matrix (2.14) the global behaviour of the map $\mathcal{M}$ in the set A and consequently it is possible to recover global information on the topology of the orbits for a generic set A. Obviously our results will depend crucially on the choice of the initial ansatz $I_2(\mathbf{x})$ which has to be adapted to the set A. In the case of polynomial map with a non resonant elliptic fixed point it seems, from numerical results (see next section), that the choice of $I_2(\mathbf{x})$ as one of the quadratic invariants of the linear gives good results also at large amplitudes. Moreover the matrix (2.14) has the structure of a Grassmann matrix for the usual scalar product associated to the L^2-norm and we can prove in this case

$$\det|F_{k,h}| \neq 0 \tag{2.29}$$

since the operator $\Delta_{\mathcal{M}}$ (see def. (2.2)) is non singular. Indeed let us suppose that the functions system $\Delta_{\mathcal{M}}\mathbf{x}^k$ would not be linearly independent then it exists a non trivial linear combination such that

$$\sum_{|k|=3}^{N} c_k \Delta_{\mathcal{M}} \mathbf{x}^k = 0 \tag{2.30}$$

where the coefficients c_k are not all zero. But if we recall the definition (2.2) we see immediately that the function $\sum_{|k|=3}^{N} c_k \Delta_{\mathcal{M}} \mathbf{x}^k$ would be an analytic first integral of motion for the map $\mathcal{M}$; this is in general not possible since the Siegel theorem holds. However the determinant of the matrix $F_{k,h}$ may be very near to zero: i.e. the matrix $F_{k,h}$ is a ill conditioned matrix and may be difficult from a numerical point of view to solve the system (2.13). In order to overcome this difficulty we suggest the following procedure: we can use different base systems and different initial conditions for the integrals, we can divide the set A in smaller subsets and calculate an integral of motion for each of them and/or we can use preconditioning technics together with a Newton method. We will not discuss here the reability of these procedures.

Finally we can see that the variational method with the L^2-norm recover the perturbative results as we make the radius of the neighbourhood A tends to zero[14]. Indeed it is sufficient to see that the two norms (2.8) and (2.9) are equivalent when A tends to the origin. This is actually the case because the contribution of the term of order $n > N$ in the entries of the matrix (2.14) becomes negligeable with respect to the contribution of the terms of order $\leq N$ when the radius $r \to 0$ so that also in the L^2-norm the operator $\Delta_{\mathcal{M}}$ becomes equivalent to the operator $\Delta_{\mathcal{M}}^{(N)}$ (see def. (2.15)). However we observe that since the perturbative solution is degenerate while the variational one in L^2-norm is not, the perturbative limit is singular.

We do not know rigorous results like (2.27) and (2.28) on the properties of the series which define the integrals of motion and on the stability time in the case of L^2-norm, but we argue from the perturbative results and the estimate of the measure of the chaotic layers[16,17] that a Nekhoroshev estimate holds in also this case.

We observe that in principle the variational method with the L^2−norm may converge to weak solutions (i.e. C^n solutions or even discontinous solutions[6,7]) for which the perturbative results do not hold any more and which can give better information on the topology of the orbits.
Anyway we remark that the variational approach is very simple to apply from a numerical point of view since it involves the numerical evaluation of the integrals (2.14) and (2.15) and the solution of a linear system. Moreover it can be immediately generalized to a generic base system of functions while the perturbative method may be very combersome in a base system different from the monomials $\mathbf{x}^k$.
The variational approach can be also applied to the problem of calculating an approximate integral of motion in a neighbourhood of an invariant KAM torus of a symplectic map; the only difference from the previous presentation is that we have to use Fourier expansions instead of Taylor expansions.

Numerical results

We have applied the perturbative and the variational method to the quadratic Henon map[10] in $\mathbf{R}^2$

$$\begin{pmatrix} q' \\ p' \end{pmatrix} = \mathcal{R}(\omega) \begin{pmatrix} q \\ p + q^2 \end{pmatrix} \qquad (q,p) \in \mathbf{R}^2 \tag{3.1}$$

where $\mathcal{R}(\omega)$is a rotation matrix of an angle ω. As one can easily see the Henon map can describe the radial betatronic motion for a flat ream when a thin len sextupole is switched on in the magnetic lattice. The quadratic invariant for the linear part $\mathcal{R}(\omega)$ reads

$$I_2(q,p) = \frac{p^2 + q^2}{2} \tag{3.2}$$

ans it will give the initial condition for the integral of motion according to the equation (2.6). We have choosen a linear tune $\frac{\omega}{2\pi} = .21$ so that a non linear resonance of order 5 causes the presence of a chain of 5 islands (see fig. 1) at large amplitude near the border of the stability basin of the origin. In this case the non linear resonances do not allow to obtain realiable results from the perturbative approach. As a consequence the variational method extents the perturbative integral also to the region occupied by the chain of islands.
In order to apply the perturbative method we have to introduce the complex conjugate variables $(z, \bar{z})$ that diagonalize the linear matrix $\mathcal{R}(\omega)$

$$\begin{cases} z = q + ip \\ \bar{z} = q - ip \end{cases} \tag{3.3}$$

so that the map (3.1) reads

$$z = e^{i\omega}\left(z + \frac{i}{4}(z+\bar{z})^2\right) \tag{3.4}$$

and the quadratic invariant (3.2) is given by

$$I_2(z,\bar{z}) = z\bar{z}. \tag{3.5}$$

We look for a polynomial integral of motion of the form

$$I(z,\bar{z}) = z\bar{z} + \sum_{n\geq 3}^{N} I_n(z,\bar{z}) \tag{3.6}$$

by solving the linear system (2.19) which reads

$$a_{j_1,j_2}\left(e^{i\omega(j_1-j_2)} - 1\right) + \left[\left(z + \frac{i}{4}(z+\bar{z})^2\right)\left(\bar{z} - \frac{i}{4}(z+\bar{z})^2\right)\right]_{j_1,j_2}$$
$$+ \left[\sum_{|h|=3}^{N} a_{h_1,h_2} e^{i\omega(h_1-h_2)}\left(z + \frac{i}{4}(z+\bar{z})^2\right)^{h_1}\left(\bar{z} - \frac{i}{4}(z+\bar{z})^2\right)^{h_2}\right]_{j_1,j_2} = 0 \tag{3.7}$$

where the operator $[\ \ ]_{j_1,j_2}$ gives the cofficient of the monomial $z^{j_1}\bar{z}^{j_2}$. The system (3.7) is easily implemented in a computer code which allows to calculate the pertubative integral up to very high orders ($N = 100$ on a VAXSTATION 3100). But since the non linear resonances prevent the integral of motion to converge in any neighbourhood of the origin[15] it is useless to consider very high perturbative orders. In the case of the Henon map it is sufficient to stop the perturbative expansion to a relative low order ($N = 9$) in order to compute quite a good integral of motion up to the chain of 5 islands. Indeed if we take into account the terms up to order N in the expansion (3.6) then our integral is realiable up to the amplitude where the effects of the N order resonance appears. In fig. 3 we report the apparent radius of convergence for the perturbative integral up to order 50 according to the ratio criterium

$$\rho_n = \left(\frac{|I_n|}{|I_{n+2}|}\right)^{\frac{1}{2}} \qquad n \geq 2. \tag{3.8}$$

The stepwise behaviour is a typical feature of the perturbative theory for polynomial symplectic maps. The apparent radius of convergence (3.8) is simply the amplitude at which the islands associated to the non linear resonances are situated; therefore we have a stepwise variation each time we reach a perturbative order $\hat{N}$ at which corresponds a non linear resonance nearer to the origin then the ones of lower order [15]. As a consequence the chains of islands are true barriers for the application of the non resonant perturbative theory as one can clearly see in fig. 4 where we report the relative variation of the integral of motion at order $N = 9$ as a function of the amplitude r

$$\Delta I = \left\{ \frac{1}{M} \sum_{m=0}^{M} \left(\frac{I(\mathbf{x}_m)}{\bar{I}} - 1 \right)^2 \right\}^{\frac{1}{2}} \tag{3.9}$$

where $\mathbf{x}_m$ $m = 0, .., M$ are the points of one orbit $\mathbf{x}_m = \mathcal{M}(\mathbf{x}_{m-1})$ with $\mathbf{x}_0 = (r, 0)$ and $\bar{I}$ is the mean value of the integral along the orbit.
On the other hand the variational approach in the L^2 norm allows to recover an approximate integral of motion also in presence of a non linear resonance. We have evaluated the entries of the matrix (2.14) and the vector (2.15) by means of the sums

$$\sum_{m=1}^{M} (\Delta_{\mathcal{M}} \mathbf{x}_m^k)\, (\Delta_{\mathcal{M}} \mathbf{x}_m^h) \tag{3.10}$$

and

$$\sum_{m=1}^{M} (\Delta_{\mathcal{M}} \mathbf{x}_m^k)\, (\Delta_{\mathcal{M}} I_2(\mathbf{x}_m)) \tag{3.11}$$

where $\mathbf{x}_m$ is a distribution of M point in the set A which is a neighbourhood of the 5 islands. Fixing the order of the integral of motion at $N = 9$ a uniform distribution of 2000 points results to be sufficient. We have checked that our results are stable when we change our sampling. In fig. 2 we show the level curves of the variational integral to be compared with the orbits in fig. 1 that are the results of a direct tracking of the map. The scale of the two figures is the same. Note the quite good agreement between the orbits and the levels curves; the integral of motion properly reproduces the two different topogical structures: the closed orbits around the origin and the islands structure at larger distance.

4. Conclusions

The tracking data are essentially a local information on the dynamics of a symplectic map. An approximate integral of motion provides a global description of the orbits which allows to draw conclusions on the long time behaviour of the orbits and on their stability properties.
Moreover it compresses the information in a small number of parameters. As a consequence the application to the analysis of the geometry of the orbits and to the stability properties of the beam dynamics, seems to be particulary promising.

Acknowledgements

Most of the results I will present have been stated during several discussions with prof. E.Remiddi and prof. G.Turchetti which I thank for their teachings and their constant encouragement. I wish to thank also dr. G.Servizi for his fundamental help in the numerical applications.

Phase plot of the Henon map

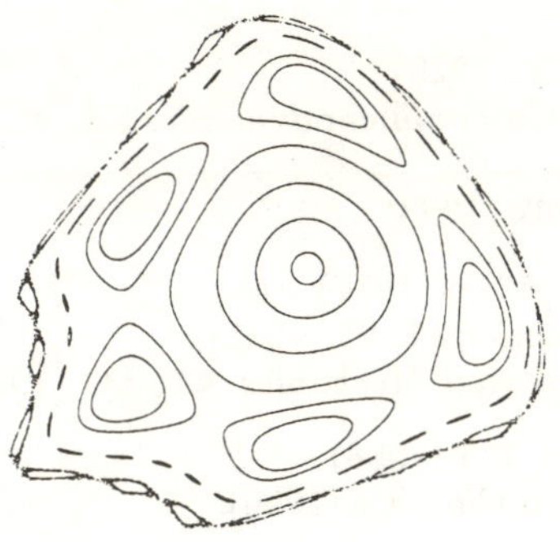

Fig. 1

Level curves of the variational integral

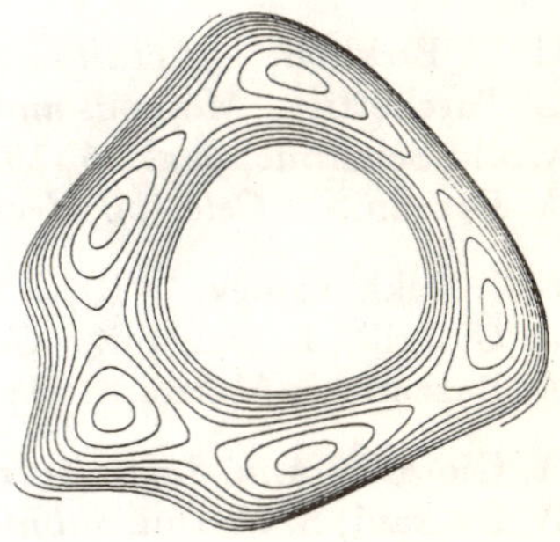

Fig. 2

Apparent radius of convergence

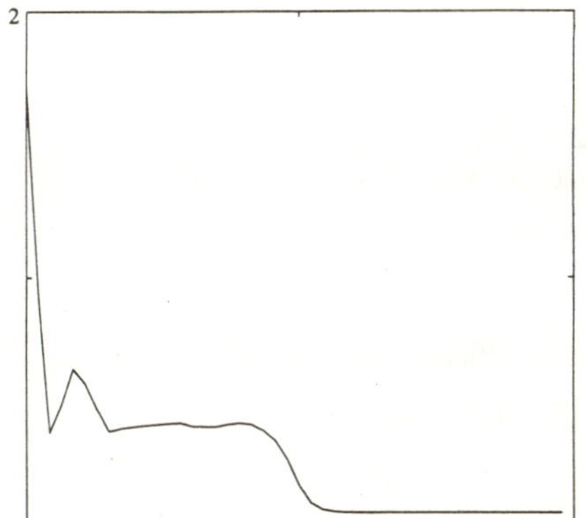

Fig. 3

Perturbative integral error

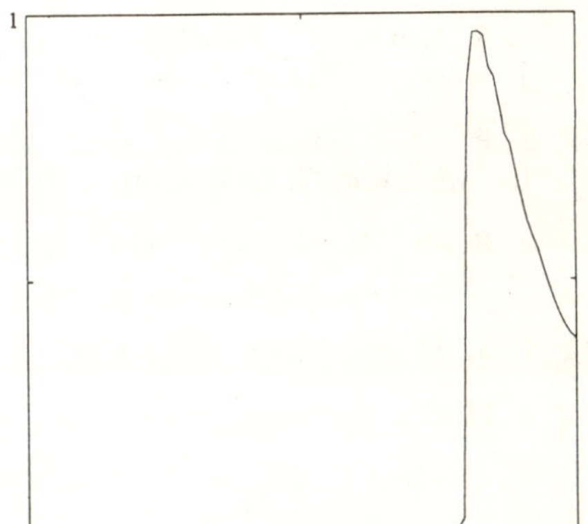

Fig.4

REFERENCES

[1] A.J. Dragt, *Lectures in nonlinear orbit dynamics*, A.I.P. Conf. Proc. **87** (1982)
A.J. Dragt, *Lecture Notes in Physics* **247** (1986)
A. Bazzani, A. Pisent, G. Turchetti, *internal note of SPS division* **38**(AMF) (1987)

[2] G. D. Birkhoff, *Acta Mathematica* **43**, 1 (1920)
G. Turchetti, *Methods and applications in nonlinear dynamics* Ed. A. Saenz, World Scientific, pag. 95 (1987)
A. Bazzani, *Celestial Mechanics* **42**, 107 (1988)

[3] N.N. Nekhoroshev, Russ. Math. Surv. **32**, 6 (1977)
A. Giorgilli, L. Galgani, *Celestial Mechanics* **37**, 95 (1985)
A. Bazzani, S. Marmi, G. Turchetti, *Celestial Mechanics*, **47**, 333 (1990)

[4] A. Giorgilli, *Ann. Inst. Henri Poincaré* **48** 423 (1988)
A. Bazzani, S. Marmi, *submitted to Nuovo Cimento* (1990)

[5] V. I. Arnold, *Methodes Mathématiques de la Méchanique classique* E. MIR, Moscow (1976)

[6] I. C. Percival, *J. Phys.* **A12**, L57 (1979)

[7] S. Aubry, P. Y. La Daeron, *Physica* **8D**, 381 (1983)

[8] J. N. Mather, *Topology* **21**, 457 (1982)
J. Moser, *SIAM Review*, **28** (1986)

[9] J. Poeshel, *Comm. Pure Appl. Math.* **35**, 653 (1989)
L. Chierchia, G. Gallavotti, Il Nuovo Cimento **67 B**, 277 (1982)

[10] J. Roels, M. Hénon, *Bull Astr. Soc.* **2**, 267 (1967)
M. Hénon, *Quart. Appl. Math.* **27**, 291 (1969)

[11] C. L. Siegel, *Annals of Mathematics* **42**, 806 (1941)

[12] J. Moser, *Comm. Pure Appl. Math.* **9**, 673 (1956)

[13] E. Courant,H. Snyder, *Annals of Physics* **3**, 1 (1958)

[14] A. Bazzani,E. Remiddi,G. Turchetti, *preprint Univ. of Bologna* (1990)

[15] G. Turchetti, *Number Theory and Physics* J.–M. Luck, P. Moussa and M. Waldschmidt editors, Springer Proceedings in Physics **47**, 223 (1990)
G. Servizi, G.Turchetti, *submitted to Physics Letters*
G. Servizi, *These proceedings*

Particle Acceleration in Extremely Strong Electromagnetic Wave Fields

K. O. Thielheim
University of Kiel
Institut für Reine und Angewandte Kernphysik
Abteilung Mathematische Physik
FRG

Abstract

In this paper I will discuss certain aspects of charged particle dynamics in extremely strong electromagnetic plane wave fields. The intention is to throw some light on certain features of accelerating mechanisms under such extreme conditions rather than to evaluate specific properties of given accelerator devices.

First I will review solutions of the Lorentz equation for plane wave fields, that is relativistic particle dynamics in plane wave fields without taking into account radiation reaction. As is well known particle motion then is periodic in velocity space so that for given initial conditions there is an upper limit of particle energy.

It is interesting to note that one can on the basis of such solutions obtained for plane wave fields infer certain important characteristics of non - linear particle dynamics in spherical wave fields, although the existence of such characteristics is essentially based on the spherical character of the wave field. I will mention two of these characteristics: One is the acceleration boundary r_B describing the transition from monotonic to oscillatory energy development of particles initally at rest within a spherical wave field. The other is the plasma border r_P defined as the radial distance from the source of the spherical wave field at which particles of given energy approaching from outside are reflected.

Then I will describe solutions of the Lorentz - Dirac - Landau equation for plane wave fields, that is relativistic particle dynamics in plane wave fields taking into account radiation reaction. Linearization of the non - linear equations of motion is achieved through substitution of eigentime by phase. In this case particle motion in general is non - periodic in velocity space and there is no upper limit of particle energy (at least in principle).

1 Relativistic Particle Motion in Plane Wave Fields without Radiation Reaction

In what follows $u^\mu = dx^\mu/d\tau$ are the contra - variant components of the velocity vector $\underline{u} = (u^\mu)$, while $u_\nu = g^{(0)}{}_{\nu\mu}\, u^\mu$ are its co - variant components. $(1,-1,-1,-1)$ is the signature of the metric tensor $g^{(0)}{}_{\mu\nu}$. Consequently in this notation the square of the norm of the four - vector of velocity is $||\underline{u}||^2 = u^\mu\, u_\mu = c^2$. $d\tau = (1/\gamma)dt$ is the differential of eigentime and $ds = c\, d\tau$.

A dimensionless notation will be used for the vector of the differential of the coordinates $d\underline{\mathbf{x}} = (d\mathbf{x}^\mu) = (1/r_L)(dx^\mu)$ and of the vector of velocity $\underline{\mathbf{u}} = (\mathbf{u}^\mu) = (1/c)\,(u^\mu)$. r_L is an appropiately chosen unit of length in the given frame of reference. So in this dimensionless notation $\boldsymbol{ds} = ds/r_L$. Dimensionless notations are also introduced for the field tensor $\underline{\underline{\mathbf{F}}} = (e/m\, c^2)\, r_L\, (F^\mu{}_\nu)$ and for the electric and magnetic field vector, $\vec{f} = (e/m\, c^2)\, r_L\, \vec{E}$ and $\vec{h} = (e/m\, c^2)\, r_L\, \vec{H}$, respectively, so that

$$\underline{\underline{\mathbf{F}}} = \begin{pmatrix} 0 & f^1 & f^2 & f^3 \\ f^1 & 0 & h^3 & -h^2 \\ f^2 & -h^3 & 0 & h^1 \\ f^3 & h^2 & -h^1 & 0 \end{pmatrix}. \tag{1}$$

The Lorentz equation

$$\dot{\underline{\mathbf{u}}} = \underline{\underline{\mathbf{F}}}\, \underline{\mathbf{u}} \tag{2}$$

is the appropriate equation of motion if effects of radiation reaction are neglected. $\dot{\underline{\mathbf{u}}} = d\underline{\mathbf{u}}/\boldsymbol{ds} = r_L\, d\underline{u}/c\, ds = r_L\, d\underline{u}/c^2\, d\tau$ is the dimensionless notation of the vector of acceleration.

For a plane vacuum wave field propagating in x^1 - direction the field tensor is represented by a 'wave - like matrix'

$$\underline{\underline{\mathbf{F}}} = \begin{pmatrix} 0 & 0 & f^2 & f^3 \\ 0 & 0 & f^2 & f^3 \\ f^2 & -f^2 & 0 & 0 \\ f^3 & -f^3 & 0 & 0 \end{pmatrix}, \tag{3}$$

in which the components of the electric vector $\vec{f}$ are functions of the phase $\Phi = \mathbf{x}^0 - \mathbf{x}^1$.

Wave - like matrices have the following algebraic properties: If

$$\underline{\underline{\mathbf{G}}} = \begin{pmatrix} 0 & 0 & g^2 & g^3 \\ 0 & 0 & g^2 & g^3 \\ g^2 & -g^2 & 0 & 0 \\ g^3 & -g^3 & 0 & 0 \end{pmatrix} \tag{4}$$

is another (or the same) wave - like matrix and $\vec{g} = (0, g^2, g^3)^T$ the corresponding vector, their product is

$$\underline{\underline{\mathbf{F}}} \cdot \underline{\underline{\mathbf{G}}} = (\vec{f}, \vec{g})\, \underline{\underline{\mathbf{Q}}} \tag{5}$$

where

$$\underline{\underline{\mathbf{Q}}} = \begin{pmatrix} 1 & -1 & 0 & 0 \\ 1 & -1 & 0 & 0 \\ 0 & 0 & 0 & 0 \\ 0 & 0 & 0 & 0 \end{pmatrix} \tag{6}$$

is a constant matrix. Also:

$$\underline{\underline{\mathbf{F}}} \cdot \underline{\underline{\mathbf{Q}}} = 0. \tag{7}$$

Substraction of the first from the zeroth component of the Lorentz equation leads to

$$\ddot{\Phi} = \ddot{\mathbf{x}}^0 - \ddot{\mathbf{x}}^1 = \dot{\mathbf{u}}^0 - \dot{\mathbf{u}}^1 = 0, \tag{8}$$

and then to the relation between eigentime and phase

$$\mathbf{s} = \frac{1}{v_0}\,(\Phi - \Phi_0), \tag{9}$$

where v_0 and Φ_0 are constants of integration of which v_0 has to be determined through the requirement $||\underline{\mathbf{u}}||^2 = 1$ while Φ_0 results from setting the scale of eigentime relative to the scale of phase. It is then possible to substitute eigentime by phase in the Lorentz equation

$$\underline{\mathbf{u}}'(\Phi) = (1/v_0)\, \underline{\underline{\mathbf{F}}}(\Phi)\, \underline{\mathbf{u}}(\Phi) \tag{10}$$

with $\underline{\mathbf{u}}' = d\underline{\mathbf{u}}/d\Phi$. The components of the Lorentz equation then are

$$\mathbf{u}^{0'} = \mathbf{u}^{1'} \tag{11}$$

$$\mathbf{u}^{1'} = (1/v_0)\,\{\mathbf{u}^2 f^2 + \mathbf{u}^3 f^3\} = (1/v_0)(\vec{\mathbf{u}}, \vec{f}) \tag{12}$$

$$\mathbf{u}^{2'} = f^2 \tag{13}$$

$$\mathbf{u}^{3'} = f^3. \tag{14}$$

For plane, monochromatic vacuum wave fields the components of the electric vector are

$$\vec{f}(\Phi) = f_0 \cdot (0, \cos\alpha\, \sin\Phi, \sin\alpha\, \cos\Phi)^T. \tag{15}$$

where $-\pi/4 \leq \alpha \leq \pi/4$ with $\alpha = 0$ for linear polarization and $\alpha = \pm\pi/4$ for circular polarization. Obviously the mean values of these components over one period vanish: $\overline{f^2} = \overline{f^3} = 0$.

It is useful to introduce

$$\vec{a} = \int_{\Phi_0}^{\Phi} d\Phi' \, \vec{f}(\Phi') + \vec{a}(\Phi_0) \tag{16}$$

with constants of integration $\vec{a}(\Phi_0)$, so that

$$a^1 = 0 \tag{17}$$
$$a^2 = -f_0 \cos\alpha \cos\Phi \tag{18}$$
$$a^3 = f_0 \sin\alpha \sin\Phi \tag{19}$$

and consequently $\overline{\vec{a}} = 0$, and

$$|\vec{a}|^2 = (f_0{}^2/2)\,\{1 + \cos(2\,\alpha)\,\cos(2\,\Phi)\} \tag{20}$$

with $\overline{|\vec{a}|^2} = f_0{}^2/2$. The zeroth component of the velocity vector with the help of (9) is expressed in the form

$$\mathbf{u}^0 = \mathbf{u}^1 + v_0. \tag{21}$$

Integration of (13) and (14) leads to the second and third components of the velocity vector

$$\mathbf{u}^2 = a^2 + \overline{\mathbf{u}^2} \tag{22}$$
$$\mathbf{u}^3 = a^3 + \overline{\mathbf{u}^3}. \tag{23}$$

Insertion of these expressions into (12) results in

$$\mathbf{u}^{1\prime} = (1/v_0)\,\{(\vec{a},\vec{f}) + (\overline{\vec{\mathbf{u}}},\vec{f})\}$$

and integration over the phase Φ in the first component of the velocity vector in

$$\mathbf{u}^1 = (1/v_0)\,\{(|\vec{a}|^2 - \overline{|\vec{a}|^2})/2 + (\overline{\vec{\mathbf{u}}},\vec{a})\} + \overline{\mathbf{u}^1}. \tag{24}$$

Equations (21) to (23) deliver the following relation

$$\begin{aligned} ||\underline{\mathbf{u}}||^2 &= v_0{}^2 + 2\,\overline{\mathbf{u}^1}\,v_0 - \overline{\mathbf{u}^2}^2 - \overline{\mathbf{u}^3}^2 - \overline{|\vec{a}|^2} \\ &= (v_0 + \overline{\mathbf{u}^1})^2 - \overline{\vec{\mathbf{u}}}^2 - \overline{|\vec{a}|^2} \end{aligned}$$

so that with $||\underline{\mathbf{u}}||^2 = 1$ and $\overline{|\vec{a}|^2} = f_0{}^2/2$,

$$v_0 = -\overline{\mathbf{u}^1} + \sqrt{1 + \overline{\vec{\mathbf{u}}}^2 + f_0{}^2/2}. \tag{25}$$

Explicitly, the components of the four - vector of velocity are

$$\mathbf{u}^0(\Phi) = v_0 + \mathbf{u}^1(\Phi) \tag{26}$$

$$\mathbf{u}^1(\Phi) = \overline{\mathbf{u}^1} - (f_0/v_0)\{\overline{\mathbf{u}^2}\cos\alpha\,\cos\Phi - \overline{\mathbf{u}^3}\sin\alpha\,\sin\Phi\} + ({f_0}^2/4\,v_0)\,\cos(2\,\alpha)\,\cos(2\,\Phi) \tag{27}$$

$$\mathbf{u}^2(\Phi) = \overline{\mathbf{u}^2} - f_0\,\cos\alpha\,\cos\Phi \tag{28}$$

$$\mathbf{u}^3(\Phi) = \overline{\mathbf{u}^3} + f_0\,\sin\alpha\,\sin\Phi. \tag{29}$$

Particle motion in plane wave fields as described by the Lorentz equation obviously is periodic in velocity space, so that particle energy is subject to an upper limit. Equations (27) to (29) can easily be integrated

$$\mathbf{x}^1(\Phi) = \frac{1}{v_0}[\overline{\mathbf{u}^1}\,\Phi - (f_0/v_0)\{\overline{\mathbf{u}^2}\cos\alpha\,\sin\Phi + \overline{\mathbf{u}^3}\sin\alpha\,\cos\Phi\} + ({f_0}^2/8\,v_0)\,\cos(2\,\alpha)\,\sin(2\,\Phi)] + c^1 \tag{30}$$

$$\mathbf{x}^2(\Phi) = \frac{1}{v_0}[\overline{\mathbf{u}^2}\,\Phi - f_0\,\cos\alpha\,\sin\Phi] + c^2 \tag{31}$$

$$\mathbf{x}^3(\Phi) = \frac{1}{v_0}[\overline{\mathbf{u}^3}\,\Phi - f_0\,\sin\alpha\,\cos\Phi] + c^3 \tag{32}$$

with constants of integration c^1, c^2, and c^3.

In these solutions of the Lorentz equation, the velocity of particle drift $\overline{\vec{\mathbf{u}}}$ in the plane wave field under consideration, is related to the initial phase Φ_0 and the initial velocity $\vec{\mathbf{u}}(\Phi_0)$ as is obvious from equation (26) to (29):

$$\overline{\mathbf{u}^1} = \mathbf{u}^1(\Phi_0) + (f_0/v_0)\{\overline{\mathbf{u}^2}\cos\alpha\,\cos\Phi_0 - \overline{\mathbf{u}^3}\sin\alpha\,\sin\Phi_0\} - ({f_0}^2/4\,v_0)\,\cos(2\,\alpha)\,\cos(2\,\Phi_0) \tag{33}$$

where

$$\overline{\mathbf{u}^2} = \mathbf{u}^2(\Phi_0) + f_0\,\cos\alpha\,\cos\Phi_0 \tag{34}$$

$$\overline{\mathbf{u}^3} = \mathbf{u}^3(\Phi_0) - f_0\,\sin\alpha\,\sin\Phi_0 \tag{35}$$

with v_0 determined from (26)

$$v_0 = \mathbf{u}^0(\Phi_0) - \mathbf{u}^1(\Phi_0). \tag{36}$$

A case of special interest is the one with vanishing initial velocity $\vec{\mathbf{u}}(\Phi_0) = 0$, for which by (36)

$$v_0 = 1. \tag{37}$$

The mean values of the velocity components according to (33) to (35) in this case are

$$\overline{\mathbf{u}^1} = (f_0{}^2/4)\{2+\cos(2\,\alpha)\,\cos(2\,\Phi_0)\} \tag{38}$$
$$\overline{\mathbf{u}^2} = f_0\,\cos\alpha\,\cos\Phi_0 \tag{39}$$
$$\overline{\mathbf{u}^3} = -f_0\,\sin\alpha\,\sin\Phi_0 \tag{40}$$

They may be inserted into (27) to (29) to obtain the velocity components.

Another special case of interest is the one for vanishing drift velocity $\overline{\vec{\mathbf{u}}} = 0$. Equations (27) to (29) then reduce to

$$\mathbf{u}^1(\Phi) = (f_0{}^2/4\,v_0)\,\cos(2\,\alpha)\,\cos(2\,\Phi) \tag{41}$$
$$\mathbf{u}^2(\Phi) = -f_0\,\cos\alpha\,\cos\Phi \tag{42}$$
$$\mathbf{u}^3(\Phi) = f_0\,\sin\alpha\,\sin\Phi \tag{43}$$

with v_0 according to (25) determined by

$$v_0 = \sqrt{1+(f_0{}^2/2)}. \tag{44}$$

This, of course, is the case of closed orbits since it can be seen from equations (30) to (32) that all spatial coordinates are periodic functions of phase

$$\mathbf{x}^1(\Phi) = (f_0{}^2/8\,v_0{}^2)\,\cos(2\,\alpha)\,\sin(2\,\Phi)+c^1 \tag{45}$$
$$\mathbf{x}^2(\Phi) = -(f_0/v_0)\,\cos\alpha\,\sin\Phi+c^2 \tag{46}$$
$$\mathbf{x}^3(\Phi) = -(f_0/v_0)\,\sin\alpha\,\cos\Phi+c^3. \tag{47}$$

Closed particle orbits in linearly polarized waves are shown for illustration in figure 1.

2 Reflection of Charged Particles by Spherical Wave Fields

It is interesting to note that certain features of particle dynamics in spherical wave fields may be inferred from results obtained for particle dynamics in plane wave fields, although these features essentially arise from the spherical character of the wave fields. Motivation for studying these non - linear features primarily arises from the interest in rotating magnetized bodies as astrophysical particle accelerators. Discussion in the following two chapters correspondingly refer to the wave zone of a magnet rotating with its vector of angular velocity $\vec{\omega}$ perpendicular to its vector of magnetic dipole moment $\vec{\mu}$.

It is convenient to introduce, instead of the parameters ω and μ, the following two parameters, each with the dimension of a length, namely the light radius

$$r_L = c/\omega \tag{48}$$

used here as a unit of length which characterizes the state of rotation of the magnetic dipole moment and the typical radius

$$r_T = (e\,\mu/m\,c^2)^{1/2} \tag{49}$$

characterizing the electromagnetic properties of the system. The typical radius is defined either for electrons when m is the electron mass or for protons when m is the proton mass, respectively.

The local plane wave amplitude can then be written

$$f_0 = {r_T}^2/r_L\,r_0. \tag{50}$$

The plasma border r_P which I will consider first is defined as the minimum value of radial distance $r_P = min\ r$ of the particle from the rotating magnet. Another way of looking at this definition is to consider a magnet rotating within a sufficiently thin fully ionized plasma. If both the frequency of rotation ω and the magnetic dipole moment μ are of sufficiently high values the rotating magnet is expected to sweep away all charged particles from its immediate vicinity evacuating this region of space up to a radial distance $r = r_P$. Modifications due to effects of collective particle motion are expected for higher values of plasma density.

The estimate, which delivers r_P as a function of r_L and r_T is based on the suggestion that a particle orbit within the given spherical wave field at the point of reflection is osculated by a closed particle orbit in the corresponding plane wave field. The problem of finding r_P as a function of r_L and r_T thereby is reduced to finding a condition for an orbit to be closed in the corresponding local plane wave field, According to (44) therefore,

$$v_0 = \sqrt{1 + (1/2)({r_T}^2/r_L\,r_0)^2}. \tag{51}$$

As a consequence of (41), the point of reflection is characterized by

$$0 = ({f_0}^2/4\,v_0)\,\cos(2\,\alpha)\,\cos(2\,\Phi),$$

so that for all types of polarization, (except for circular polarization, $\alpha = \pm\pi/4$), $\Phi = \pi/4$. Introducing this value of phase for $r_0 = r_P$ into (42) and (43) yields

$$|\vec{\mathbf{u}}|^2 = \beta^2\,\gamma^2 = \gamma^2 - 1 = {f_0}^2/2$$

which with the help of (50) leads to the plasma border

$$r_P = (2^{-1/2})(\beta\,\gamma)^{-1}({r_T}^2/r_L). \tag{52}$$

For circular polarization, $\alpha = \pm\pi/4$, (42) and (43) immediately lead to the same result.

3 Decoupling of Charged Particles from Spherical Wave Fields

The dynamics of electrically charged particles initially at rest within the wave zone of a rotating magnet is subject to a competition between two quantities: One is the decay length of the spherical wave field, i.e. the interval of the radial coordinate on which the field amplitudes decay by the factor 1/2, which obviously is equal to the initial radial coordinate r_0. The other quantity, which enters into this competition, is just the distance

$$\delta x = 2\pi\,\overline{\mathbf{u}^1}\,r_L/v_0 \tag{53}$$

by which the particle according to (30) would be shifted during one period in radial direction through drift motion in the corresponding plane wave field.

For $r_0 \ll \delta x$ the test particle is expected to exhibit a largely monotonous development of energy since it essentially decouples from the spherical wave before entering into the decelerating regime.

Alternatively for $r_0 \gg \delta x$ the test particle is expected to exhibit oscillatory development of energy before approaching asymptotically its final value of energy since it has to pass through various accelerating and decelerating regimes before the wave amplitudes have decayed substantially.

The acceleration boundary by definition is located where these two spacial regimes merge, which is the case for

$$r_0 = r_B = \delta x = 2\pi\,\overline{\mathbf{u}^1}\,r_L/v_0. \tag{54}$$

Considering an initial position $x^1 = r_B$, $x^2 = x^3 = 0$, such that the electric field is of maximum strength, that is for $\Phi_0 = (\pi/2)$ and initial velocity $\vec{\mathbf{u}}(\Phi_0) = 0$, (38) reduces to

$$\overline{\mathbf{u}^1} = (f_0{}^2/4)\,\{1 + 2\,\sin^2\alpha\} \tag{55}$$

with the local wave amplitude as given by (50). Inserting this expression into the definition of the acceleration boundary as given by (54) finally delivers

$$r_B = (\pi/2)^{1/3}\,r_L\,(r_T/r_L)^{4/3}\,\{1 + 2\,\sin^2\alpha\}^{1/3}. \tag{56}$$

A schematic diagram illustrating these features of particle dynamics within the spherical wave field of a rotating magnet is shown in figure 2.

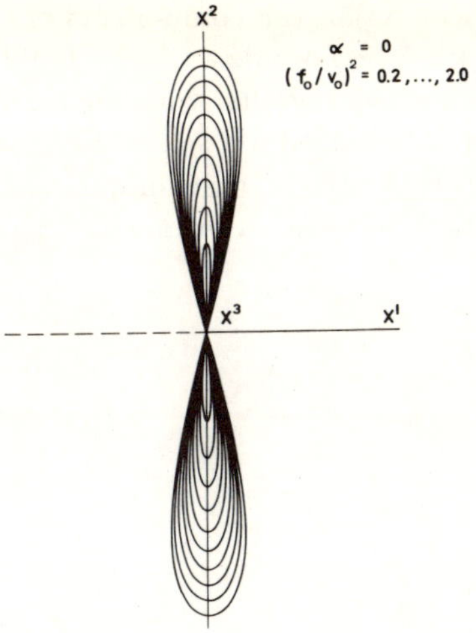

CLOSED ORBITS
IN PLANE ELECTROMAGNETIC WAVE
LINEAR POLARIZATION

x^1 : DIRECTION OF PROPAGATION
x^2 : DIRECTION OF ELECTRIC VECTOR

Fig. 1

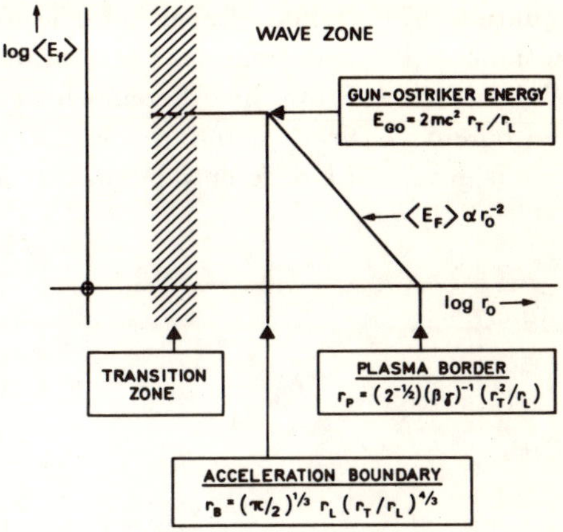

MEAN FINAL PROTON ENERGY
AS FUNCTION
OF INITIAL POSITION
IN WAVE ZONE
OF ROTATING MAGNET

Fig.2

4 Taking into Account Radiation Reaction

If effects of radiation reaction are taken into account, the appropriate equation of motion is the Lorentz - Dirac - Landau equation

$$\underline{\dot{\mathbf{u}}} = \underline{\underline{\mathbf{F}}}\,\underline{\mathbf{u}} + l_0\,\{\dot{\underline{\underline{\mathbf{F}}}}\,\underline{\mathbf{u}} + \underline{\underline{\mathbf{F}}}^2\,\underline{\mathbf{u}} + (\underline{\underline{\mathbf{F}}}\,\underline{\mathbf{u}}, \underline{\underline{\mathbf{F}}}\,\underline{\mathbf{u}})\,\underline{\mathbf{u}}\}, \tag{57}$$

where the point stands for differentiation with respect to the (dimensionless) eigentime $\mathbf{s}$ and

$$l_0 = 2\,r_e/3\,r_L \tag{58}$$

with the classicle particle radius

$$r_e = e^2/m\,c^2 \tag{59}$$

calculated either for the electron / positron when m is the electron mass or for the proton / antiproton when m is the proton mass. As before r_L is an arbitrary unit of length within the given frame of reference. Again the components of the field tensor are functions of the phase $\Phi = \mathbf{x}^0 - \mathbf{x}^1$. The Lorentz - Dirac - Landau equation (57) is a non - linear differential equation which together with the correct initial conditions determines the coordinate $\underline{\mathbf{x}}$ of the charged particle as functions of eigentime $\mathbf{s}$. From the difference between the zeroth and the first components of the Lorentz - Dirac - Landau equation the relation between eigentime and phase now is governed a differential equation for

$$v = d\Phi/d\mathbf{s}, \tag{60}$$

namely

$$\dot{v} = l_0 \, (\underline{\underline{\mathbf{F}}}\,\underline{\mathbf{u}}, \underline{\underline{\mathbf{F}}}\,\underline{\mathbf{u}}) \, v. \tag{61}$$

With

$$\underline{\underline{\mathbf{F}}}\,\underline{\mathbf{u}} = (\vec{f}\,\vec{\mathbf{u}}, \vec{f}\,\vec{\mathbf{u}}, v\, f^2, v\, f^3) \tag{62}$$

this differential equation can be written in the form

$$\dot{v} = -l_0 \, |\vec{f}|^2 \, v^3, \tag{63}$$

and be integrated once

$$\frac{1}{v} = \mathbf{s}' = \frac{1}{v_0} + l_0 \int_{\Phi_0}^{\Phi} |\vec{f}(\Phi')|^2 \, d\Phi' \tag{64}$$

and again

$$\mathbf{s} = \frac{1}{v_0}(\Phi - \Phi_0) + l_0 \int_{\Phi_0}^{\Phi} \int_{\Phi_0}^{\Phi''} |\vec{f}(\Phi')|^2 \, d\Phi' \, d\Phi'', \tag{65}$$

which is an explicit expression for the eigentime $\mathbf{s}$ as a function of the phase Φ. In order to substitute eigentime by phase as independant variable in the Lorentz - Dirac - Landau equation it is necessary to calculate the vector of velocity

$$\underline{\mathbf{u}} = \dot{\underline{\mathbf{x}}} = v\, \underline{\mathbf{x}}' \tag{66}$$

and the vector of acceleration

$$\begin{aligned}
\dot{\underline{\mathbf{u}}} &= \frac{d}{d\mathbf{s}}(v\, \underline{\mathbf{x}}') \\
&= \dot{v}\, \underline{\mathbf{x}}' + v^2\, \underline{\mathbf{x}}'' \\
&= v^2\, \underline{\mathbf{x}}'' + l_0 \, (\underline{\underline{\mathbf{F}}}\,\underline{\mathbf{u}}, \underline{\underline{\mathbf{F}}}\,\underline{\mathbf{u}}) \, v\, \underline{\mathbf{x}}' \\
&= v^2\, \underline{\mathbf{x}}'' + l_0 \, (\underline{\underline{\mathbf{F}}}\,\underline{\mathbf{u}}, \underline{\underline{\mathbf{F}}}\,\underline{\mathbf{u}}) \, \underline{\mathbf{u}},
\end{aligned} \tag{67}$$

where again the prime stands for differentiation with respect to the phase Φ.

As a result of this substitution the Lorentz - Dirac - Landau equation is obtained in the form

$$\underline{\mathbf{x}}'' = \mathbf{s}' \underline{\underline{\mathbf{F}}}\, \underline{\mathbf{x}}' + l_0 (\underline{\underline{\mathbf{F}}}' \underline{\mathbf{x}}' + \mathbf{s}' \underline{\underline{\mathbf{F}}}^2 \underline{\mathbf{x}}'). \tag{68}$$

which obviously has become a linear differential equation for the coordinates $\underline{\mathbf{x}}$ of the electrically charged masspoint in space time as functions of the phase Φ. Remarkably enough, the nonlinearity of the Lorentz - Dirac - Landau equation in the case of plane wave fields can completely be packed into the relation between phase and eigentime so that the remaining equations of motion become linear.

The latter can be integrated

$$\underline{\mathbf{x}}' = exp \int_{\Phi_0}^{\Phi} (\mathbf{s}' \underline{\underline{\mathbf{F}}} + l_0 \{ \underline{\underline{\mathbf{F}}}' + \mathbf{s}' \underline{\underline{\mathbf{F}}}^2 \})\, d\Phi'\, \underline{\mathbf{x}}_0', \tag{69}$$

so that $\underline{\mathbf{x}}'$ is obtained explicitely as a function of Φ.

Taking into account that all tensors occuring in the exponent of this expression are wave - like matices corresponding to the form of the field tensor (3) it is obvious that only few terms survive of the infinite number of terms occuring in the Taylor expansion of the matrix exponential function.

Introducing as abbreviations

$$\vec{a}(\Phi) := \int_{\Phi_0}^{\Phi} \vec{f}(\Phi')\, d\Phi', \tag{70}$$

$$M(\Phi) := \int_{\Phi_0}^{\Phi} |\vec{f}(\Phi')|^2\, d\Phi', \tag{71}$$

$$\vec{K}(\Phi) := \int_{\Phi_0}^{\Phi} M(\Phi')\, \vec{f}(\Phi')\, d\Phi', \tag{72}$$

$$\vec{b}(\Phi) := \int_{\Phi_0}^{\Phi} \mathbf{s}'(\Phi')\, \vec{f}(\Phi')\, d\Phi' + l_0(\vec{f}(\Phi) - \vec{f}(\Phi_0)) \tag{73}$$

and the corresponding wave - like matrix

$$\underline{\underline{\mathbf{B}}}(\Phi) = \begin{pmatrix} 0 & 0 & b^2 & b^3 \\ 0 & 0 & b^2 & b^3 \\ b^2 & -b^2 & 0 & 0 \\ b^3 & -b^3 & 0 & 0 \end{pmatrix} \tag{74}$$

and

$$D(\Phi) := \int_{\Phi_0}^{\Phi} \mathbf{s}'(\Phi')\, |\vec{f}(\Phi')|^2\, d\Phi' \tag{75}$$

the exponential function in equation (69) can be written in the form

$$\begin{aligned} exp\{\underline{\underline{\mathbf{B}}}(\Phi) + l_0\, D(\Phi)\, \underline{\underline{\mathbf{Q}}}\} &= \underline{\underline{\mathbf{I}}} + \underline{\underline{\mathbf{B}}} + l_0\, D\, \underline{\underline{\mathbf{Q}}} + 1/2\, \underline{\underline{\mathbf{B}}}^2 \\ &= \underline{\underline{\mathbf{I}}} + \underline{\underline{\mathbf{B}}} + \{(1/2)\, |\vec{b}|^2 + l_0\, D\}\, \underline{\underline{\mathbf{Q}}}. \end{aligned} \tag{76}$$

The velocity vector thereby is found in the form

$$\begin{aligned} \underline{\mathbf{u}} = v\, \underline{\mathbf{x}}' &= \frac{v}{v_0}\, (\underline{\underline{\mathbf{I}}} + \underline{\underline{\mathbf{B}}} + \{(1/2)\, |\vec{b}|^2 + l_0\, D\}\, \underline{\underline{\mathbf{Q}}})\, \underline{\mathbf{u}}_0 \\ &= (1 + l_0\, v_0\, M(\Phi))^{-1}\, [\underline{\underline{\mathbf{I}}} + \underline{\underline{\mathbf{B}}} + \{(1/2)\, |\vec{b}|^2 + l_0\, D\}\, \underline{\underline{\mathbf{Q}}}]\, \underline{\mathbf{u}}_0. \end{aligned} \tag{77}$$

5 Evaluation for Monochromatic Plane Wave Fields

For the electric vector of a monochromatic plane wave of arbitrary polarization as given by (15)

$$|\vec{f}|^2 = ({f_0}^2/2)\, \{1 - \cos(2\,\alpha)\, \cos(2\Phi)\}. \tag{78}$$

In the following discussion the initial value chosen for the phase Φ of the particle is

$$\Phi_0 = 0 \tag{79}$$

corresponding to the initial value of the electric vector

$$\vec{f_0} = \vec{f}(\Phi_0) = \vec{f}(0) = f_0 \cdot (0, 0, \sin\alpha)^T. \tag{80}$$

To calculate the transverse components of the velocity vector (77) it is necessary to evaluate from (70) and (71)

$$\begin{aligned} \vec{a}(\Phi) &= f_0 \cdot (0, \cos\alpha\, (1 - \cos\Phi), \sin\alpha\, \sin\Phi)^T, & (81) \\ M(\Phi) &= ({f_0}^2/2)\, \{\Phi - (1/2)\, \cos(2\,\alpha)\, \sin(2\,\Phi)\}. & (82) \end{aligned}$$

As a consequence of (64) the relation between phase and eigentime in this case is determined by

$$\mathbf{s}'(\Phi) = (1/v_0) + l_0\, ({f_0}^2/2)\, \{\Phi - (1/2)\, \cos(2\,\alpha)\, \sin(2\,\Phi)\}. \tag{83}$$

Furthermore from (75)

$$\begin{aligned} D(\Phi) &= ({f_0}^2/4\, v_0)\, \{2\,\Phi - \cos(2\,\alpha)\, \sin(2\,\Phi)\} \\ &\quad + l_0\, ({f_0}^4/32)\, \{2\,\Phi - \cos(2\,\alpha)\, \sin(2\,\Phi)\}^2, \end{aligned} \tag{84}$$

from (72)

$$\vec{K}(\Phi) = (f_0{}^3/2) \begin{pmatrix} 0 \\ \cos\alpha \{\sin\Phi - \Phi\cos\Phi - (1/3)\cos(2\alpha)\sin^3\Phi\} \\ \sin\alpha\{\cos\Phi - 1 + \Phi\sin\Phi + (1/3)\cos(2\alpha)(\cos^3\Phi - 1)\} \end{pmatrix} \quad (85)$$

and from (73)

$$\vec{b}(\Phi) = (1/v_0)\,\vec{a}(\Phi) + l_0\{\vec{K}(\Phi) + \vec{f}(\Phi) - \vec{f}(\Phi_0)\}. \quad (86)$$

With these quantities the transverse components of the velocity vector (77) are

$$\mathbf{u}^2(\Phi) = [1 + l_0\, v_0\, M(\Phi)]^{-1}\{\mathbf{u}_0{}^2 + v_0\, b^2(\Phi)\} \quad (87)$$
$$\mathbf{u}^3(\Phi) = [1 + l_0\, v_0\, M(\Phi)]^{-1}\{\mathbf{u}_0{}^3 + v_0\, b^3(\Phi)\}, \quad (88)$$

where $\mathbf{u}_0{}^2 = \mathbf{u}^2(\Phi_0)$ and $\mathbf{u}_0{}^3 = \mathbf{u}^3(\Phi_0)$ are the initial values of the transverse components of the velocity vector.

As can be seen particle motion as described by the Lorentz - Dirac - Landau equation is non - periodic in velocity space, since $M(\Phi)$, $\vec{K}(\Phi)$, and therefore $\vec{b}(\Phi)$ are non - periodic functions of phase. This behaviour is different from what is known of particle motion as described by the Lorentz equation.

6 Asymptotic Behaviour for Large Values of Phase

In what follows I will discuss the asymptotic behaviour of the components of the velocity vector for sufficiently large values of phase, namely

$$1 \ll (1/2)\, l_0\, v_0\, f_0{}^2\, \Phi \qquad \textit{and} \qquad 1 \ll \Phi. \quad (89)$$

Within this regime of large values of phase the relevant parameters defined above reduce to

$$\vec{K}(\Phi) \longrightarrow (f_0{}^3/2)\,\Phi\,(0, -\cos\Phi\cos\alpha, \sin\Phi\,\sin\alpha)^T, \quad (90)$$

and therefore

$$\vec{b}(\Phi) \longrightarrow l_0\,(f_0{}^3/2)\,(0, -\Phi\,\cos\Phi\,\cos\alpha, \Phi\,\sin\Phi\,\sin\alpha)^T, \quad (91)$$

and

$$|\vec{b}(\Phi)|^2 \longrightarrow l_0{}^2\,(f_0{}^6/4)\,\Phi^2\{\cos^2\alpha\,\cos^2\Phi + \sin^2\alpha\,\sin^2\Phi\}, \quad (92)$$
$$M(\Phi) \longrightarrow (f_0{}^2/2)\,\Phi, \quad (93)$$
$$D(\Phi) \longrightarrow l_0\,(f_0{}^4/8)\,\Phi^2, \quad (94)$$

and the transverse components of the velocity vector reduce to

$$\mathbf{u}^2(\Phi) \longrightarrow \frac{\mathbf{u}_0{}^2}{(1/2)\, l_0\, v_0\, f_0{}^2\, \Phi} - f_0 \cos\alpha \cos\Phi \longrightarrow -f_0 \cos\alpha \cos\Phi, \qquad (95)$$

$$\mathbf{u}^3(\Phi) \longrightarrow \frac{\mathbf{u}_0{}^3}{(1/2)\, l_0\, v_0\, f_0{}^2\, \Phi} + f_0 \sin\alpha \sin\Phi \longrightarrow f_0 \sin\alpha \sin\Phi. \qquad (96)$$

Obviously for $\mathbf{u}_0{}^2 = \mathbf{u}_0{}^3 = 0$ and / or in the limit $\Phi \longrightarrow \infty$ the transverse components of the velocity vector become periodic functions of phase identical to those for closed particle orbits as given by (42) and (43) and described by the Lorentz equation under neglection of radiation reaction.

The Lorentz factor and the longitudinal component of the velocity vector as described by (77) become

$$\gamma = \mathbf{u}^0 = [1 + l_0\, v_0\, M]^{-1}\, [\mathbf{u}_0{}^0 + (\vec{\mathbf{u}}_0, \vec{b}) + \{(1/2)\, |\vec{b}|^2 + l_0\, D\}\, v_0], \qquad (97)$$

$$\mathbf{u}^1 = [1 + l_0\, v_0\, M]^{-1}\, [\mathbf{u}_0{}^1 + (\vec{\mathbf{u}}_0, \vec{b}) + \{(1/2)\, |\vec{b}|^2 + l_0\, D\}\, v_0], \qquad (98)$$

so that for large values of phase the Lorentz factor reduces to

$$\gamma = \mathbf{u}^0(\Phi) \longrightarrow l_0\, (f_0{}^2/4)\, \Phi\, [1 + f_0{}^2\, \{\sin^2\alpha\, \sin^2\Phi + \cos^2\alpha\, \cos^2\Phi\}]. \qquad (99)$$

As can be seen the Lorentz factor asymptotically increases proportional to phase with superposition of a positive oscillatory term, the amplitude of which also increases proportional to phase. Somewhat unexpectedly radiation reaction thereby constitutes a mechanism for global acceleration which may become important in future generations of particle accelerators.

This phenomenon is illustrated by particle orbits in plane, monochromatic wave fields of linear polarization shown in figure 3 and figure 4 for differently chosen initial conditions. (The constant of radiation reaction l_0 in this case has arbitrarily been enhanced for better demonstration of global acceleration due to radiation reaction.)

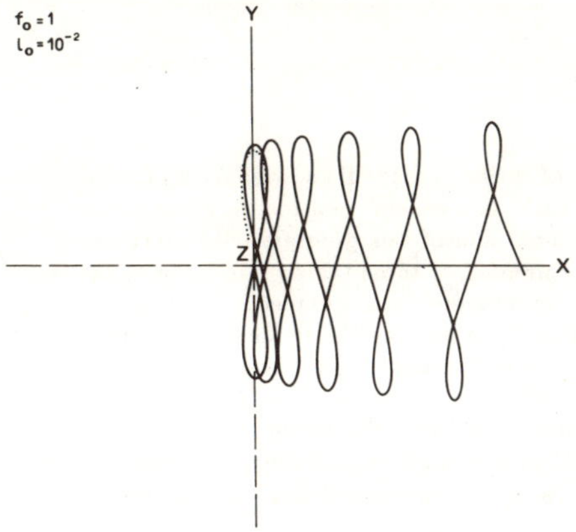

ELECTRON ORBITS
IN PLANE ELECTROMAGNETIC WAVE
OF LINEAR POLARIZATION
WITH RADIATION REACTION

BROKEN LINE : CORRESPONDING CLOSED ORBIT
WITHOUT RADIATION REACTION
(INITIAL PHASE CORRESPONDING TO ZERO FIELD STRENGTH)

Fig. 3

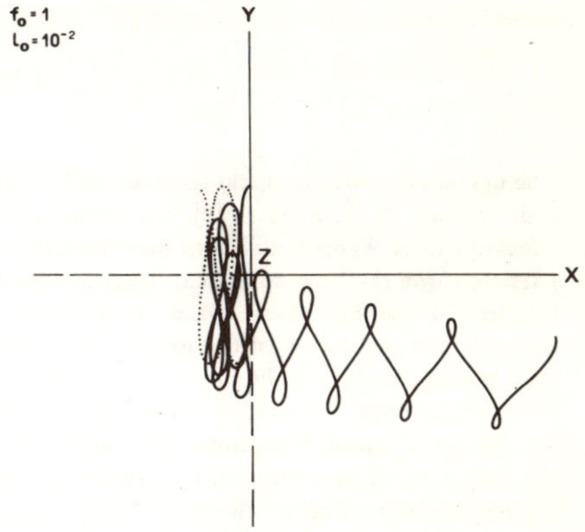

ELECTRON ORBITS
IN PLANE ELECTROMAGNETIC WAVE
OF LINEAR POLARIZATION
WITH RADIATION, REACTION

INITIAL VELOCITY : $U_o^1 = -0.2$
(INITIAL PHASE CORRESPONDING TO MAXIMUM FIELD STRENGTH)

Fig.4

SIXTRACK - A Single Particle Tracking Code

F. Schmidt, CERN, Geneva, Switzerland

Abstract

The new single particle tracking code SIXTRACK [1], an offspring of the well known RACETRACK code written by A. Wrulich [2], has been installed at CERN as a general tool to study non-linear effects in large hadron colliders like the LHC [3]. The main aim of this program is to carry one particle through those complicated structures over large number of turns taking into account the full six-dimensional phase space including synchrotron oscillations in a symplectic manner. It allows to the prediction of the long-term dynamic aperture by evaluating the Lyapunov exponent. Parameters of interest like non-linear detuning and smear are determined via a post-processing of the tracking data. An analysis of the first order resonances [4] can be done and correction schemes for several of those resonances can be calculated. Moreover there is the feature to calculate a one-turn map of such complicated structures as the LHC to very high order, using the differential algebra techniques of M. Berz [5]. This map allows a subsequent theoretical analysis like normal form procedures which are provided by E. Forest [6].

1 Introduction

For the new generation of large proton accelerators like the LHC superconducting magnets have to be used to achieve the high fields needed to bend protons in the energy regime of 10 TeV. These magnets have unavoidable high order multipole errors which limit the dynamic aperture and thereby the usable regime of transverse amplitudes where the protons are stable. SIXTRACK is one of the programs to tackle these single particle stability problems. The aim is to estimate the dynamic aperture sufficiently in a safe and fast way, by making use of today's most powerful computers with their vectorization facilities. After the description of the features of SIXTRACK, the next section shows how the tracking studies can be done most efficiently and finally a small tracking example is given.

2 Description of SIXTRACK

The Single Particle Tracking Code SIXTRACK is optimized to carry two particle [1] through an accelerator structure over a large number of turns.

The main features of SIXTRACK are:

1. Treatment of the full six-dimensional motion including synchrotron motion in a symplectic manner [7]. The energy can be ramped at the same time considering the relativistic change of the velocity [8].
2. Detection of the onset of chaotic motion and thereby the long-term dynamic aperture by evaluating the Lyapunov exponent.
3. Postprocessing procedure allowing
    - calculation of the Lyapunov exponent

[1]Two particles are needed for the detection of chaotic behavior.

- calculation of the average phase advance per turn
- FFT analysis
- resonance analysis
- calculation of the average, maximum and minimum values of the Courant Synder emittance and the invariants of linearly coupled motion
- calculation of smear
- plotting using the CERN packages HBOOK, HPLOT and HIGZ [9, 10, 11]

4. Calculation of first-order resonances and of correction schemes for the resonances [4].
5. Calculation of the one turn map using the differential algebra techniques of M.Berz [5].
6. A vectorized version is available, where the two particles, with a number of different amplitudes, the different momentum deviations and several seeds for the random distribution of multipole errors can be treated in parallel [12].
7. Operational improvements:
   - free format input
   - optimization of the calculation of multipole kicks
   - improved treatment of random errors

To save storage space and to facilitate program development the program is kept in Patchy [13] format. There are different versions of SIXTRACK for Apollo,Cray,IBM and the VAX.

Very large accelerator structures like the LHC with its 2500 multipoles demand large memory space (8 Mbytes). On top of that the differential algebra package needs another 12 Mb to allow the calculation of a one-turn map of the accelerator to 15th order. This was made manageable by a special treatment of large common blocks [14].

The IBM at CERN needs for one turn of two particles and for the full LHC lattice with all multipole errors in the dipoles and correction elements approximately 125 ms. Therefore 10^4 turns take 24 minutes of cpu time which is the upper limit that can be reached without going to dedicated emulators that can run 24 hours a day. There is, however, a vectorized version of the program, which allows a speedup by approximately a factor of 10 when 60 particles are treated in parallel [12]. The time needed to determine with the differential algebra option the one-turn map with four parameters (four-dimensional transverse motion) of the same LHC lattice is given in the table below as a function of the order to which the map is calculated.

Table 1: CPU time for One-turn Map Calculation

order	cpu time (seconds)
3	5
5	26
7	133
9	565

Increasing the order by two takes five times as much cpu time, which agrees with the increase of the number of differential algebra multiplications and additions [15].

3 Tracking Recipe

The following rather crude recommendations for getting answers about specific non-linear problems in the design phase of an accelerator reflect (partly) cumbersome experiences during such a design phase (HERA).

1. There is a tendency to produce large numbers of tracking studies for the sake of statistics. This is probably good for the final version of a lattice. In case of unexpected new effects which appear (!) it is hard to judge if the old data is still valid. It is therefore preferable to limit the studies to a small ($<< 1000$) but sufficient set of parameters, so that there is a chance to redo the analysis! It goes without saying that a careful judgment of what is sufficient is mandatory.
   - The variation of the amplitude in phase space is good enough. Don't vary the phase space angle! Tracking roughly 1000 turns will scan the phase space sufficiently, so that the angle information comes for free.
   - A worst but still realistic case should be studied. For instance tracking at one maximum relative momentum deviation.
   - An optimization of the tunes should be done first. This fixes the tunes to one pair of values. These optimized tunes might also diminish the effect of different random seeds.
   - If the distribution of machines with different seeds is narrow, study only one machine.
   - A thorough study for a low number of turns (10^4) should be done first, followed by a few well chosen cases at a large number of turns ($10^5 - 10^7$).
   - After a coarse variation of the amplitudes, a fine scan is necessary to find the dynamic aperture to the desired precision. The long-term runs should be done in this step of the analysis including a thorough study of the phase space. Complicated resonance structures and the dynamic aperture as a function of phase space angles can be found that way.
2. The tracking data are obtained over hours or even days of CPU-time and they hold a surprising amount of information, so that they all should be saved on tape. These data should be analyzed in any possible way and should be at hand at any time. Moreover during a longer period of tracking studies it often happens that new ways of analyzing the data emerge.
3. Criteria for early detection of instability should be searched, checked and used to reduce the tracking effort or even reach beyond the normal tracking scope. SIXTRACK offers the possibility to detect the onset of chaotic motion. It uses the evolution of the distance in phase space of two particles (example see next section). New direct methods are being investigated [16] to evaluate the Lyapunov exponent. Wether the predictive power can thereby be increased remains to be seen. The pragmatic approach followed here has undergone exhaustive testing:
   - The onset of chaotic motion can normally be detected after only 10^4 turns. In addition there are early indications that suggest chaotic behavior, so that a prolonged run can clarify the situation.
   - Even when the maximum distance in phase space is reached (opposite positions in phase space), the 2 types of motions can be easily distinguished. For regular particles the distance in phase space decreases to its original value, while in the chaotic case the distance in phase space remains at a statistical mean value.
   - The method needs the information of only 2 particles tracked in parallel, which can be done fast especially when the vectorization is used.

- Only one of the two particles has to be chaotic to find the chaotic behavior.
- The initial distance is dictated by rounding errors to stay above roughly 10^{-6} . Setting this distance to its minimum value allows to check for chaotic behavior just above the granularity of phase space imposed by these rounding errors.
- The effect of rounding errors is self-compensated, as long as the initial distance stays above its minimum value. The influence of the unavoidable presence of rounding errors has to be considered in any kind of method used to predict the long-term behavior.

4. Finally one should ask the theorists for complementary analytic tools and rules. These should be simple and take advantage of the strength of today's computer power. I would like to mention two examples that fulfill these requirements:

   - The D.A. package of M. Berz [5]
   - The Bengtsson rule for FFT's [17]

4 Tracking example

A simple tracking example is shown with its input file (4.1), its output file (4.2) and some corresponding plots in (4.3).

4.1 Input Example

For the description of the different input blocks see ref [1].

```
FREE FORMAT    TITLE: EXAMPLE
PRINTOUT OF INPUT PARAMETERS------------------------------------------
NEXT------------------------------------------------------------------
SINGLE ELEMENTS-------------------------------------------------------
B      0   0.0000000     0.00000      50.0000
QD2    2   0.0000000     0.009536     0.77000
QF2    2   0.0000000    -0.009536     0.77000
MU    11   1.            1.00000      0.00000
SEX    3   0.0500000     0.00000
NEXT
BLOCK DEFINITIONS
  1  1
B1    QD2 B QF2
B2    QF2 B QD2
NEXT
STRUCTURE INPUT
          MU  B1 SEX B2
NEXT
MULTIPOLE COEFFICIENTS---------------------------------------------
MU            10.0      3.5765
      0.         0.00000  0.         0.00000
      0.         0.00000  0.         0.00000
      0.405E-3   0.00000  0.         0.00000
      -.5E-5     0.00000  0.         0.00000
      -.56E-4    0.00000  0.         0.00000
      0.0        0.00000  0.         0.00000
      0.3E-5     0.00000  0.         0.00000
      0.         0.00000  0.         0.00000
      -.1E-5     0.00000  0.         0.00000
NEXT
TRACKING PARAMETERS---------------------------------------------------
 10000   0   2       11.0     11.5    0    1
   1         1         0   0
      0          0          1          1          1   50000
NEXT
INITIAL COORDINATES---------------------------------------------------
        2   0.        0.         1.
            0.
            0.
            0.
            0.
            0.
            0.
            0.
            0.000001
            0.
            0.
            0.
            0.
       450000.
       450000.
       450000.
NEXT
ITERATION-ACCURACY----------------------------------------------------
       50 1D-14 1D-15
    10 1D-10 1D-10
    10 1D-5 1D-6
    1D-8  1D-12 1D-10
NEXT
POSTPROCESSING--------------------------------------------------------
EXAMPLE
  1000     0 0  1     .08 .08
   0.     0.  1  1  20   .005    1 .10
  7878   -1   0     1 1 1 1
NEXT
ENDE==================================================================
```

4.2 Output Example

The preprocessing part is shown first.

```
----------------------------------------------------------------------------------------------------------------------------

      OOOOOOOOOOOOOOOOOOOOOO
      OO                  OO
      OO  PREPROCESSING   OO
      OO                  OO
      OOOOOOOOOOOOOOOOOOOOOO

----------------------------------------------------------------------------------------------------------------------------

VERSION 7.4/3.2 OF GKSGRAL STARTED
     ---- ENTRY CLORB ----/DPP= 0.00000 /CLOX/   0.00000   0.00000 /CLOZ/   0.00000   0.00000 /ITERAT.=  2/ ACCURACY= 0.000000D+00
     ---- ENTRY CLORB ----/DPP= 0.00000 /CLOX/   0.00000   0.00000 /CLOZ/   0.00000   0.00000 /ITERAT.=  2/ ACCURACY= 0.000000D+00
     ---- ENTRY ORBIT ----/NO MONITORS SPECIFIED
     ---- ENTRY CLORB ----/DPP= 0.00000 /CLOX/   0.00000   0.00000 /CLOZ/   0.00000   0.00000 /ITERAT.=  2/ ACCURACY= 0.000000D+00
     ---- ENTRY CLORB ----/DPP= 0.00000 /CLOX/   0.00000   0.00000 /CLOZ/   0.00000   0.00000 /ITERAT.=  2/ ACCURACY= 0.000000D+00

----------------------------------------------------------------------------------------------------------------------------

      TRACKING FOR CONSTANT MOMENTUM DEVIATION

           ------ NO ACCELERATION ------

            TUNE            CLO            CLOP            BETO            ALFO
       X    0.1222387       0.0000000      0.0000000       92.9575455      0.0000000
       Z    0.1222387       0.0000000      0.0000000      203.5812131      0.0000000

      REL. MOMENTUM DEVIATION=     0.00000
    ========================================
```

Followed by the initial coordinates and the final coordinates for a regular (right side) and chaotic (left side) case.

```
       ---- INITIAL COORD. OF TWIN-TRAJECTORIES

                11.5000000000000000000000000000000000            11.0000000000000000000000000000000000
                -0.0000000000000000004942488456867054            -0.0000000000000000004727597654394574
                17.0186209425976926000000000000000000            16.2786809016151821000000000000000000
                 0.0000000000000000001669894918209964             0.0000000000000000001597290791331270
                 0.0000000000000000000000000000000000             0.0000000000000000000000000000000000
                 0.0000000000000000000000000000000000             0.0000000000000000000000000000000000
                11.5000000000000000000000000000000000            11.0000000000000000000000000000000000
                 0.0000009999999999995057306000000000             0.0000009999999999995272240000000000
                17.0186209425976926000000000000000000            16.2786809016151821000000000000000000
                 0.0000000000000000001669894918209964             0.0000000000000000001597290791331270
                 0.0000000000000000000000000000000000             0.0000000000000000000000000000000000
                 0.0000000000000000000000000000000000             0.0000000000000000000000000000000000
            450000.0000000000000000000000000000000000        450000.0000000000000000000000000000000000
            449999.9999999999971000000000000000000000        449999.9999999999971000000000000000000000
            449999.9999999999971000000000000000000000        449999.9999999999971000000000000000000000

----------------------------------------------------------------------------------------------------------------------------

      OOOOOOOOOOOOOOOO
      OO            OO
      OO  TRACKING  OO
      OO            OO
      OOOOOOOOOOOOOOOO

----------------------------------------------------------------------------------------------------------------------------

      NUMBER OF REVOLUTION        10000

              -2.4876861557186029600000000000000000             6.4135052511566816900000000000000000
               0.0757560933282033777000000000000000            -0.0892929076372112190000000000000000
             -15.9695070845667739000000000000000000            -6.6013594190857034500000000000000000
               0.0812648120998927942000000000000000             0.0813569392448562634000000000000000
               0.0000000000000000000000000000000000             0.0000000000000000000000000000000000
               0.0000000000000000000000000000000000             0.0000000000000000000000000000000000
              10.7169921168215441000000000000000000             6.4077425454377838800000000000000000
              -0.0796775728660257115000000000000000            -0.0893061436073242232000000000000000
               5.9738417719087710800000000000000000            -6.6086811751542611200000000000000000
              -0.0754077087252212203000000000000000             0.0813539663958259968000000000000000
               0.0000000000000000000000000000000000             0.0000000000000000000000000000000000
               0.0000000000000000000000000000000000             0.0000000000000000000000000000000000
          450000.0000000000000000000000000000000000         450000.0000000000000000000000000000000000
          449999.9999999999971000000000000000000000         449999.9999999999971000000000000000000000
          449999.9999999999971000000000000000000000         449999.9999999999971000000000000000000000
```

Finally part of the postprocessing for the two particles are shown (chaotic on the left and regular on the right respectively) and a summary of the postprocessing is given.

```
oooooooooooooooooooooo
oo                  oo
oo  POSTPROCESSING  oo
oo                  oo
oooooooooooooooooooooo
```

ANALYSING THE INCREASE OF THE DISTANCE IN PHASE-SPACE

TURNS	DISTANCE	SLOPE	RESIDUAL
2000	0.2253779764D-03	0.4554389715	0.0000000000
3000	0.1081182799D-02	1.3750724792	0.3952417374
4000	0.6399160375D-02	3.0346260071	2.1428222656
5000	0.4981834333D-02	2.2196264267	2.3270816803
6000	0.8028940085D-01	4.1395473480	10.5333480835
7000	0.3407768847D+00	6.2189645767	25.0942840576
8000	0.4788764947D+00	6.7490110397	22.5940704346
9000	0.4507363285D+00	6.2723016739	21.2880401611
10000	0.6438836450D+00	5.8029022217	21.1567993164

TURNS	DISTANCE	SLOPE	RESIDUAL
2000	0.9092427661D-04	0.3869932294	0.0000000000
3000	0.2666317995D-03	0.9347699285	0.1416964531
4000	0.3394974411D-03	0.9024477601	0.0693726540
5000	0.5020486718D-03	1.0815429687	0.0974925160
6000	0.5503330639D-03	0.9851537347	0.0972504616
7000	0.7196859601D-03	1.0547361374	0.0967649221
8000	0.7402482154D-03	0.9820190072	0.1136143804
9000	0.9506629146D-03	1.0501804352	0.1352081299
10000	0.9737567472D-03	1.0120038986	0.1287498474

AVERAGED PHASE-ADVANCE

```
X-PHASE :     0.1175695955    +/_    0.0003315228
Z-PHASE :     0.1196627812    +/_    0.0001993022

START-QX :   0.1222386779   CHANGE IN X : -.4669082376D-02
START-QZ :   0.1222386779   CHANGE IN Z : -.2575896685D-02
```

```
X-PHASE :     0.1189145300    +/_    0.0001260177
Z-PHASE :     0.1211689637    +/_    0.0001953892

START-QX :   0.1222386779   CHANGE IN X : -.3324147915D-02
START-QZ :   0.1222386779   CHANGE IN Z : -.1069714140D-02
```

THE AVERAGED PHASE-ADVANCES ARE CLOSER THEN 0.5000D-02 TO THE FOLLOWING RESONANCES UP TO 20 ORDER

NX * QX	+ NZ * QZ	- P	= DELTA
1	-1	0.0	-.2093D-02
2	-2	0.0	-.4186D-02
14	3	2.0	0.4963D-02
15	2	2.0	0.2869D-02
16	1	2.0	0.7763D-03
17	0	2.0	-.1317D-02
18	-1	2.0	-.3410D-02

NX * QX	+ NZ * QZ	- P	= DELTA
1	-1	0.0	-.2254D-02
2	-2	0.0	-.4509D-02

SUMMARY OF THE POSTPROCESSING

```
-----------------------------------------------------------------------------------------------------------------------------
|        |               |              |                |          |NORMALIZED| SLOPE    |                  |           |                      |
| TURN   |    LINEAR     |    BETA-     |   AMPLITUDES   | MOMENTUM |PHASESPACE| OF THE   |    NONLINEAR     |  NEAREST  |       SMEAR OF       |
| NUMBER |    TUNES      | FUNCTIONS    |                | DEVIATION| DISTANCE |DISTANCE  |    DETUNING      | RESONANCE |   THE EMITTANCES     |
-----------------------------------------------------------------------------------------------------------------------------
|        |               |     [M]      |      [MM]      |          |          |          |                  |     |ORD.|  [%]    |     [%]    |
-----------------------------------------------------------------------------------------------------------------------------
|   10000|X    0.12224|X    92.9575|X   11.000000|0.0000D+00|0.9738D-03|   1.0120|X -.33241D-02|X   1|  2 |X  26.423|X+Z   5.664|
|        |Z    0.12224|Z   203.5812|Z   16.278681|          |          |         |+/- 0.126D-03|Z  -1|    |Z  34.450|            |
|        |            |            |             |          |          |         |Z -.10697D-02|     |    |         |            |
|        |            |            |             |          |          |         |+/- 0.195D-03|     |    |         |            |
-----------------------------------------------------------------------------------------------------------------------------
|   10000|X    0.12224|X    92.9575|X   11.500000|0.0000D+00|0.6439D+00|   5.8029|X -.46691D-02|X  16| 17 |X  45.228|X+Z   6.683|
|        |Z    0.12224|Z   203.5812|Z   17.018621|          |          |         |+/- 0.332D-03|Z   1|    |Z  33.745|            |
|        |            |            |             |          |          |         |Z -.25759D-02|     |    |         |            |
|        |            |            |             |          |          |         |+/- 0.199D-03|     |    |         |            |
-----------------------------------------------------------------------------------------------------------------------------
```

4.3 Plot Example

In figure 1 a typical example of the evolution of the distance in phase space is shown of a regular and chaotic particle. Figure 2 and figure 3 show the corresponding horizontal phase space and the physical phase space projections respectively.

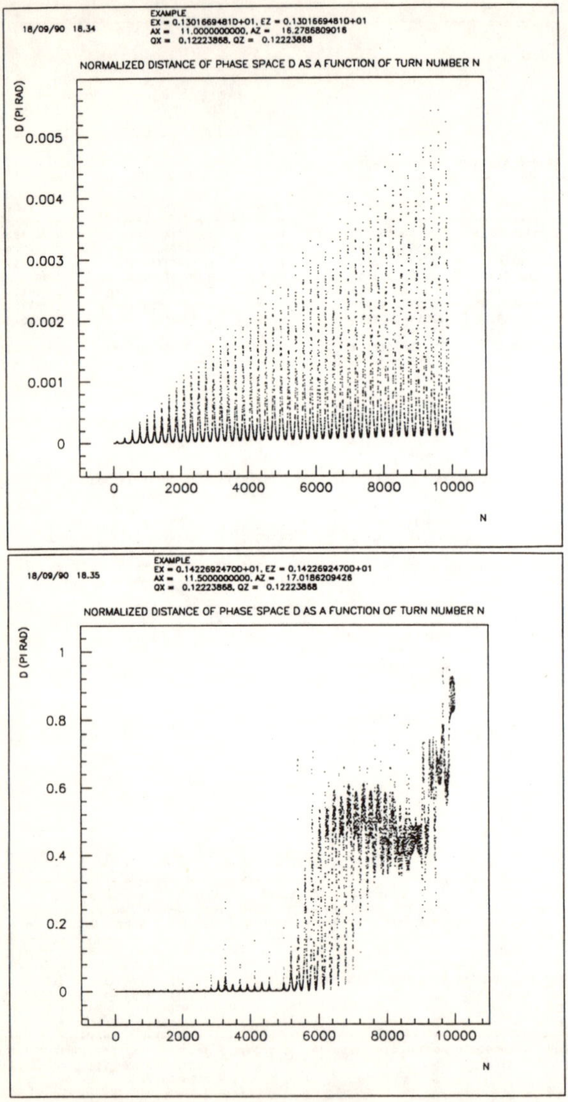

Figure 1: Evolution of the distance of phase space for regular (upper part) and chaotic (lower part) motion

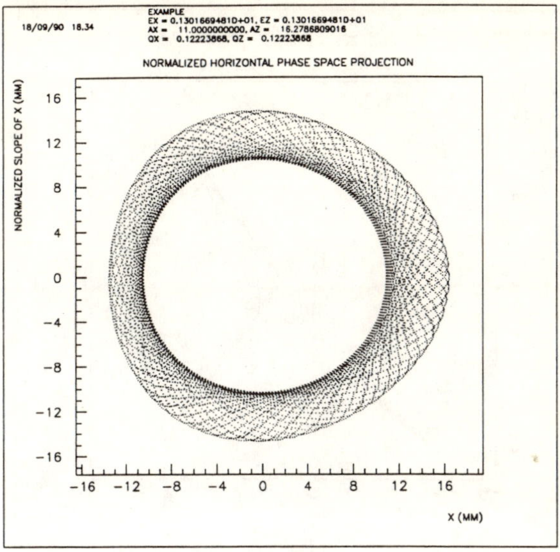

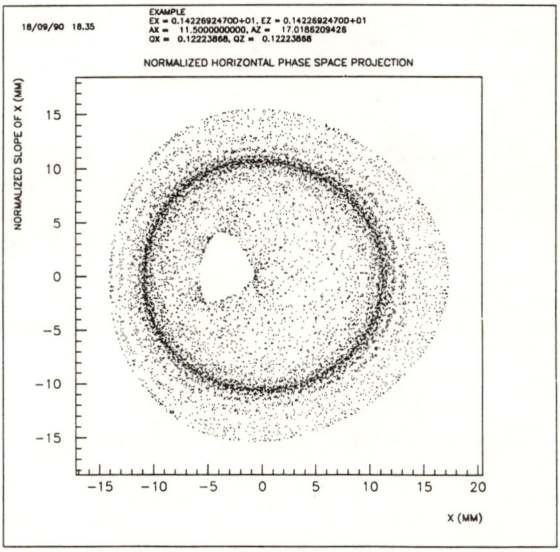

Figure 2: Horizontal phase space projections for the regular (upper part) and the chaotic (lower part) cases

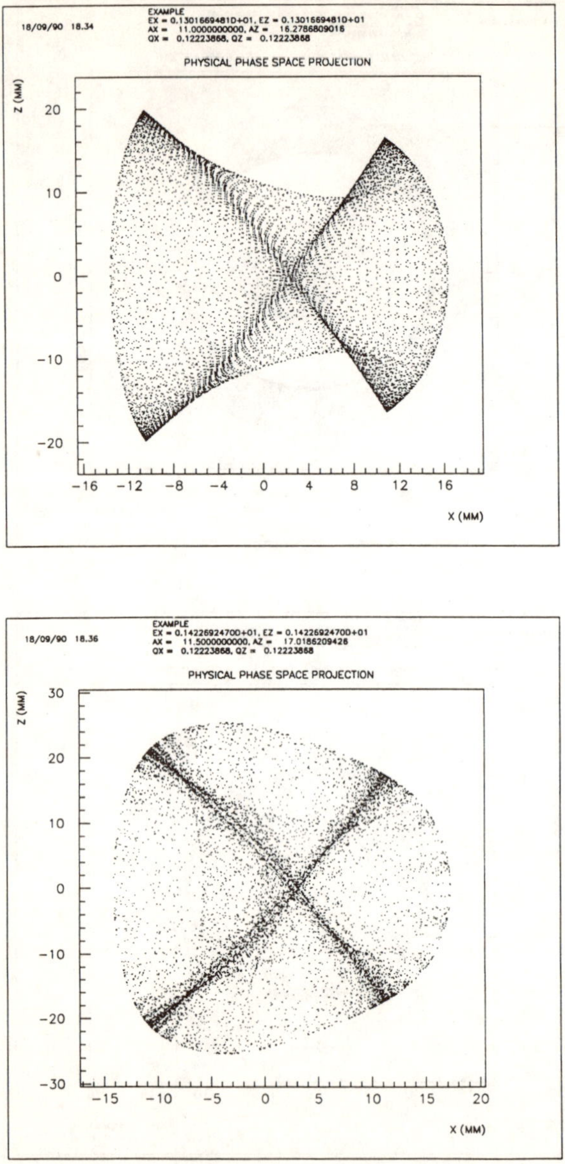

Figure 3: Physical phase space projections for the regular (upper part) and the chaotic (lower part) cases

References

[1] F. Schmidt, *SIXTRACK, Version 1, Single Particle Tracking Code treating Transverse Motion with Synchrotron Oscillations in a Symplectic Manner*, CERN/SL/90-11 (AP)

[2] A. Wrulich, *RACETRACK, A Computer Code for the Simulation of Nonlinear Motion in Accelerators*, DESY 84-026 (1984)

[3] G. Brianti, K. Hubner *The large Hadron Collider in the LEP Tunnel*, CERN 87-05 (1987)

[4] G. Guignard, *A General Treatment of Resonances in Accelerators*, CERN 78-11 (1978)

[5] M. Berz, *Differential Algebra Description of Beam Dynamics to very high Orders*, Particle Accelerators, 1989, Vol. 24, pp. 109-124

[6] M. Berz, E. Forest, J. Irwin, *Normal Form Methods for Complicated Periodic Systems: A Complete Solution using Differential Algebra and Lie Operators*, Particle Accelerators, 1989, Vol. 24, pp. 91-107

[7] G. Ripken, *Non-Linear Canonical Equations of Coupled Synchro-Betatron Motion and their Solution within the Framework of a Non-Linear 6-Dimensional (Symplectic) Tracking Program for Ultra-Relativistic Protons*, DESY 85-084 (1985)

[8] D.P. Barber, G. Ripken, F. Schmidt, *A Non-Linear Canonical Formalism for the Coupled Synchro-Betatron Motion of Protons with Arbitrary Energy*, DESY 87-036 (1987)

[9] R. Brun, D. Lienart, *HBOOK User Guide*, CERN Program Library Y250 (1987)

[10] R. Brun, N.C. Somon, *HPLOT User Guide*, CERN Program Library Y251 (1988)

[11] R. Bock, R. Brun, O. Couet, N.C. Somon, C.E. Vandoni, P. Zanarini, *HIGZ User Guide*, CERN Program Library Q120

[12] F. Schmidt, M. Vaenttinen, *Vectorization of SIXTRACK*, to be published

[13] H. J. Klein, J. Zoll, *PATCHY, Reference Manual, Version 4.09*, CERN Program Library (1983)

[14] H.Renshall, private communication

[15] J. Irwin, private communication

[16] F. Zimmermann, private communication

[17] J. Bengtsson, *Non-Linear Transverse Dynamics for Storage Rings*, CERN 88-04

LONG-TIME TRACKING WITH MAD VERSION 8

F. CHRISTOPH ISELIN
SL/AP Group, CERN, CH-1211 Geneva 23, Switzerland

ABSTRACT

This paper presents the algorithms available in MAD for tracking. It discusses their advantages and drawbacks for long-term tracking and gives benchmarks results for tracking.

1 Tracking Algorithms in MAD

The MAD program[1] contains three different tracking algorithms. This chapter is an attempt to evaluate their merits and their proper use.

1.1 TRANSPORT Maps

A TRANSPORT map consists of a 6×6-matrix R representing the linear effects, and a $6 \times 6 \times 6$-tensor T representing the quadratic effects on the orbit. The effect of a TRANSPORT map on an orbit vector Z can be written as

$$Z_i^2 = \sum_{k=1}^{6} R_{ik} Z_k^1 + \sum_{k=1}^{6} \sum_{l=1}^{6} T_{ikl} Z_k^1 Z_l^1, \qquad i = 1, \ldots, 6. \tag{1}$$

In general TRANSPORT maps are not symplectic, and it violates Liouville's theorem. This method is therefore not suited for long-term tracking.

1.2 Lie transformations

Given a differentiable function f, we define the Lie operator $:f:$ acting on a function g by the Poisson bracket

$$:f\!: g = [f, g] = \sum_{k=1}^{3} \left(\frac{\partial f}{\partial q_k} \frac{\partial g}{\partial p_k} - \frac{\partial f}{\partial p_k} \frac{\partial g}{\partial q_k} \right), \tag{2}$$

the iterated Lie operator as

$$:f\!:^i g = :f\!: (:f\!:^{i-1} g) = \underbrace{:f\!: :f\!: :f\!: \ldots :f\!:}_{i \text{ applications of } :f:} \; g, \tag{3}$$

and the Lie transformation of a function g as

$$e^{:f:}g = \sum_{i=0}^{\infty} \frac{:f:^i}{i!} g. \tag{4}$$

A Lie transformation is usually written in factored form as

$$Z^2 = e^{:f_2:} e^{:f_3:} e^{:f_4:} \dots Z^1. \tag{5}$$

where each f_i is a homogeneous polynomial in Z of degree i.

MAD computes Lie transformations for single elements up to the f_4 terms and is able to concatenate maps carrying up to the f_6 terms.

Lie-algebraic tracking is done in a symplectic way, but the computation is slowed down by a factor 5 to 100 with respect to TRANSPORT maps, depending on the order of the map and the kind of elements involved. In its actual form it is too slow for long-term tracking. However, by concatenation a considerable speed-up is possible, and the speed becomes comparable to TRANSPORT tracking.

1.3 Thin Lens Maps

The fastest tracking method is the method of thin lenses. Here all magnetic elements are replaced by thin lenses separated by drift spaces. Thus all transformations are symplectic, but at the cost of an approximation to the lattice. Thanks to symplecticity and speed the method is well suited for long-term tracking.

2 Tracking Output from MAD

MAD allows selection amongst absolute phase space coordinates, normalized phase space coordinates, linear invariants, phases, and turn number. It plots any one versus any other of these values for the same particle and turn. Plots may be grouped by turn or by particle, or all turns may be plotted for all particles on the same plot. Printed output after each turn, or after selected elements is available on request.

3 Time Critical Sections of the Program Code

In the first place all program code has been written such as to take advantage of the computer hardware, i. e. of parallel and vector facilities. This still left a few bottlenecks which could be identified and removed in MAD.

3.1 Building Transfer Maps for Dipoles

Transfer map for dipoles consist of several parts:

- Fringe field at entrance
- Body of the magnet
- Fringe field at exit
- Roll angle (optional)

Each part has to be evaluated separately, and combined with the others. There is a significant saving if the transfer maps are accumulated once and for all before beginning to track. There is however no big loss if the maps for other elements are recomputed on the fly during tracking.

3.2 Building Lumps

Concatenation of "lumps", that is of maps for sequences of elements can take a lot of time, even when done before tracking. It pays off to pre-compute smaller parts which are used many times in different places.

3.3 Application of Maps

It has been found that in tracking often 1% of the code uses 99% of the time. Many computer systems contain tools to identify such bottlenecks. The CFT77 compiler for example offers a "flowtrace" option which lists the time used by all subroutines in the program, together with the number of calls. In our experience it pays to make full use of such tools. In MAD it was possible to gain considerably in execution speed by rewriting just a few routines:

- The routine to apply a general TRANSPORT map could be vectorized, thus saving a factor of 4 in its execution time, or a factor of 3 in overall time.
- In tracking by Lie-algebra, many homogeneous polynomials must be transformed by the transformation

$$f(\zeta) = f(MZ) \tag{6}$$

  i. e. their argument vector is multiplied by a matrix M. This routine accounted for up to 50% of the CPU time. It could be improved by a factor varying from 2 to 100 depending on the order of the polynomial.

- The routine for thin multipoles could be vectorized, and unnecessary subroutine calls eliminated, resulting in an overall gain of a factor 4 for thin-lens lattice tracking. The speed for such lattices is now slower than the SSCTRACK program by only a factor of two; but the full generality of available elements could be kept; it is still possible to mix thin and thick elements in the same lattice.

4 Benchmarks

In order to compare the speed for various tracking methods, some benchmarks were set up, all consisting of a single cell, tracked for 5 or 50 turns:

- *REAL1 lattice:* A pure thin multipole lattice. The cell quadrupoles and sextupoles are represented by four thin multipoles with quadrupole and sextupole components, and the long dipoles are modelled including their field errors by six thin multipoles with dipole through dodecapole components. The cell contains a RF cavity.
- *REAL2 lattice:* Each thin multipole representing a dipoles is replaced by a sandwich containing two dipoles and a thin multipole to model the field errors. Otherwise REAL2 is like REAL1.
- *REAL3 lattice:* Like REAL2, but the four thin multipoles representing the quadrupoles and sextupoles are replaced by four long quadrupoles and two long sextupoles.
- *REAL4 lattice:* This lattice is like REAL3, but the thin multipoles for the field errors are omitted.
- *REAL5 lattice:* Like REAL1, but with four long quadrupoles and two long sextupoles.
- *LUMP3 lattice:* Like REAL3, but each cell is handled as a LUMP. This causes ignoring multipoles higher than octupole. The numeric results resemble the results for the REAL4 lattice, but they are obtained at significantly higher speed.

The results are tabulated in Table 1 with all CPU times scaled to 50 turns.

From the benchmarks the CPU time needed to track one particle through one element have been estimated and listed in Table 2. It is seen that the `LIE3` method is on average 7 to 15 times slower than the `TRANSPORT` method, and the `LIE4` method takes another factor of about 1.5 to 2.

Some improvements are still possible:

Table 1: CPU times in seconds for 50 turns through the various benchmark lattices

version	method	REAL1	REAL2	REAL3	REAL4	REAL5	LUMP3
Cray	TRANS.	.634	2.07	2.29	2.09	.926	.163
IBM S	TRANS.	4.06	14.2	14.8	12.3	4.22	.900
IBM V	TRANS.	5.05	9.83	9.77	6.59	4.70	.460
Cray	LIE3	3.28	44.2	61.9	61.9	18.5	.147
IBM S	LIE3	7.09	55.1	75.3	72.4	24.2	2.43
IBM V	LIE3	9.39	74.8	104.	100.	33.6	3.02
Cray	LIE4	4.67	61.6	87.9	87.2	46.0	6.31
IBM S	LIE4	10.9	109.	157.	156.	50.7	4.42
IBM V	LIE4	12.1	115.	161.	157.	51.0	4.47

Table 2: CPU times in microseconds per element and per particle

version	method	thin lens	dipole	quadrupole sextupole	RF cavity	lump
Cray	TRANSPORT	2.1	7.5	7.3	18.6	10.2
IBM S	TRANSPORT	28.	53.	43.	20.	56.
IBM V	TRANSPORT	39.	25.	23.	28.	30.
Cray	LIE3	2.1	214.	215.	184.	100.
IBM S	LIE3	28.	250.	251.	163.	152.
IBM V	LIE3	39.	341.	347.	196.	189.
Cray	LIE4	2.1	297.	303.	271.	200.
IBM S	LIE4	28.	511.	542.	401.	276.
IBM V	LIE4	39.	536.	545.	366.	279.

- Use the (symplectic) TRANSPORT method for RF cavities, even when tracking with one of the LIE methods.

- For the LIE methods the vectorized IBM version is slower than the scalar one. This is due to vectorization of short loops which would run faster in scalar mode. Suppression of vectorization for those loops will make the vectorized program faster than the scalar version.

- Introduction of new representations of Lie-algebraic maps, like factorization into kick maps, could save considerable factors.

5 Conclusions

When used properly, MAD can be an adequate tool for long-term tracking. Care must be taken to preserve symplecticity. Furthermore, Lie-algebraic methods should not be used without reasonable use of concatenation. Finally there is always room for improvements.

References

1. H. Grote and F. C. Iselin. *MAD Version 8, User's Reference Manual*, CERN/SL/90-13 (AP).

SYMPLECTIC TRACKING THROUGH CIRCULAR ACCELERATORS WITH HIGH ORDER MAPS

MARTIN BERZ
Department of Physics and Astronomy
and National Superconducting Cyclotron Laboratory
Michigan State University
East Lansing, Mi 48824

and

Exploratory Studies Group
Lawrence Berkeley Laboratory
Berkeley, CA 94720
(previous address)

ABSTRACT

It is discussed how high order transfer maps generated using differential algebraic methods can be used for symplectic tracking. Contrary to the usual tracking, the map approach makes it possible to study the specific properties of the system with as few approximations as desired without prohibitive extra effort. For example, the full Hamiltonian can be used, and the elements can be treated with finite length and with their fringe fields. Furthermore, tracking through maps is usually significantly faster than element by element tracking.

Different schemes for fast symplectic tracking suited for different degrees of nonlinearity of the original map are presented. The fact that different approaches sometimes produce different long term results shows that symplectification is not the cure of all evil and should be used cautiously. It also suggests to use a symplectification scheme which changes the original map by the least amount possible.

1. Introduction

The recently developed differential algebraic methods [1,2,3] allow the determination of the Taylor series representation of transfer maps

$$\vec{z}_f = \mathcal{M}(\vec{z}_i) \tag{1}$$

describing the evolution of the phase space vector $\vec{z}$ of general optical systems to arbitrary order.

The maps can be used to extract many quantities of interest in accelerator simulations like tune shifts, chromaticities and smear [4,1]. Within the same context it is also possible to study the dependence of these quantities on parameters. The parameter dependences in many cases allow a very direct way to optimize the system.

The practical usefulness of these differential algebraic methods is intimately related to a powerful and general software environment. For this purpose, an object oriented structured language was created which allows a direct use of the differential algebra operations. For the sake of portability, the compiler was written in FORTRAN 77. Using the environment, the design and simulation code COSY INFINITY was created [5,6]. Due to the openness of this approach, it is easy to use for standard studies yet allows utmost flexibility for more advanced problems.

While our philosophy is to try to solve as many problems as possible by merely studying the map of the system, sometimes it is still desirable to be able to use the crude but robust tool of tracking. For this purpose, in order to limit unphysical and difficult to estimate growth or shrinkage, it is important that the approximated map preserve phase space volumes, which can be achieved by preserving the symplectic structure of the underlying system in the approximative map. It is well known that a map is called symplectic if its Jacobian M satisfies the symplectic condition [7]:

$$M \cdot J \cdot M^t = J, \text{ or alternatively } M \cdot J = (M \cdot J)^t \tag{2}$$

where J has the form

$$J = \begin{pmatrix} 0 & 0 & 0 & -1 & 0 & 0 \\ 0 & 0 & 0 & 0 & -1 & 0 \\ 0 & 0 & 0 & 0 & 0 & -1 \\ 1 & 0 & 0 & 0 & 0 & 0 \\ 0 & 1 & 0 & 0 & 0 & 0 \\ 0 & 0 & 1 & 0 & 0 & 0 \end{pmatrix} \tag{3}$$

In the case of regular tracking, the preservation of phase space volume is usually achieved by using a symplectic integrator. The well-known leapfrog algorithm turns out to be a symplectic integrator of second order, and it has been possible to develop kick based integrators with a similar flavor of order four [8] and six [9].

It is also well known that truncated Taylor series maps do only preserve symplecticity up to the order of the map. Thus, unless the orders are so high that for all particles of interest, the approximation is accurate to machine precision (which in many cases can be achieved), there will be some loss of symplecticity that can after many turns lead to unphysical shrinkage or growth of phase space.

There are different ways to symplectify transfer matrices. It has recently been shown [10,11] that it is possible to factor the transfer map into a sequence of kick maps followed by rotations such that the composition of these maps equals the given map to the pre-specified order. Since both kick maps and rotations are symplectic and their action on particles can be directly and easily evaluated, this constitutes a symplectic tracking algorithm.

In this paper we will present various different ways to perform symplectic tracking using Taylor maps. They all go back to the generating function representation of transfer maps, which incidentally has a long history in optics as the Eikonal.

A symplectic map can be described using the so-called generating functions [7] in mixed coordinates:

$$\begin{array}{ll} F_1 & (\vec{q}_i, \vec{q}_f) \\ F_2 & (\vec{q}_i, \vec{p}_f) \\ F_3 & (\vec{p}_i, \vec{q}_f) \\ F_4 & (\vec{p}_i, \vec{p}_f) \end{array} \tag{4}$$

which contain the information about the map as the solution of the implicit equations:

$$\begin{aligned} (\vec{p}_i, \vec{p}_f) &= J \cdot \vec{\nabla} F_1 \\ (\vec{p}_i, \vec{q}_f) &= J \cdot \vec{\nabla} F_2 \\ (\vec{q}_i, \vec{p}_f) &= J \cdot \vec{\nabla} F_3 \\ (\vec{q}_i, \vec{q}_f) &= J \cdot \vec{\nabla} F_4 \end{aligned} \tag{5}$$

if these exist. Furthermore, the map represented by any generating function, be it the right one or not, is always symplectic. An approximative generating function can be used to perform symplectic tracking in the following way: Use the underlying approximative map to compute first values of the final coordinates $(\vec{q}_f, \vec{p}_f)$. Depending on the accuracy of the map, the quadruple $(\vec{q}_i, \vec{p}_i, \vec{q}_f, \vec{p}_f)$ is already very close to a solution of the implicit equations 5. It is used as a starting point of a numerical solution of the implicit equations, and $\vec{q}_f$ and/or $\vec{p}_f$ are varied to determine an exact solution. This can be done by Newton's method, and usually one iterative step is enough to obtain machine accuracy.

To conclude this introduction, we present some notation. On the space of all infinitely often differentiable functions from R^v into R^m, we introduce a relation $=_n$ as follows: two functions are said to be equivalent if their derivatives agree to order n. Quite clearly this is an equivalence relation, and we denote its classes by $[f]_n$. Via

$$\begin{aligned} [f]_n + [g]_n &:= [f+g]_n, \\ c \cdot [f]_n &:= [c \cdot f]_n, \\ [f]_n \cdot [g]_n &:= [f \cdot g]_n, \end{aligned} \tag{6}$$

which are all well defined, the structure becomes a finite dimensional algebra, denoted by ${}_nD_v^m$. We note that the class into which the function falls uniquely describes its aberrations or nonlinearities through order n. Via $\partial_i[f]_n := [\partial f]_{n-1}$, the algebra even becomes a differential algebra. Standard functions like exp and sin can formally be introduced via $\exp[f]_n := [\exp(f)]_n$ etc, even though for practical purposes, other ways are needed; for details refer to [1]. For our purposes, the key advantage of the differential algebra is that it allows us to compute derivative classes of very complicated functions, even transfer maps described by numerical integrators, from simpler ones.

Now let f be origin preserving, i.e. $f(\vec{0}) = \vec{0}$. Then it is easy to show that this is equivalent to the fact that $[f]_n$ is nilpotent, in particular, $([f]_n)^{n+1} = 0$. One can also show that the differential algebra can be ordered, and that all nilpotent elements are infinitely small; it is even possible to do very interesting Calculus on certain algebraic closures of the differential algebra discussed here [12]; a very short overview over this is also given in [13].

2. Inverses of Transfer Maps

In order to determine generating functions for the maps presented here, it is necessary to invert transfer maps in their DA representation; i.e., given $[A]_n$ in ${}_nD_v^n$, find $[B]_n$ in ${}_nD_v^n$ such that $[A \circ B]_n = [E]_n$. Suppose first the map $[A]_n$ is nilpotent. We begin by splitting the map $[A]_n \in {}_nD_v^n$ into its linear and nonlinear nilpotent parts:

$$[A]_n = [A_1]_n + [A_2]_n. \tag{7}$$

Furthermore, we write the sought for inverse in ${}_nD_v^n$ as $[M]_n$.

$$[A^{-1}]_n = [M]_n \tag{8}$$

Composing the functions, we obtain

$$\begin{aligned} ([A_1] + [A_2]_n) \circ [M]_n &= [E]_n \Rightarrow \\ [A_1] \circ [M]_n &= [E]_n - [A_2]_n \circ [M]_n \Rightarrow \\ [M]_n &= [A_1^{-1}] \circ ([E]_n - [A_2]_n \circ [M]_{n-1}). \end{aligned} \tag{9}$$

In the last step use has been made of the fact that knowing $[M]_{n-1}$ gives $A_{2n} \circ [M]_n$ in ${}_nD_v^n$. The necessary computation of A_1^{-1} is a linear matrix inversion. Eq. 9 can now be used in a recursive manner to compute the M_i order by order.

In case the map $[A]_n$ is not nilpotent, one first treats the nilpotent part alone. Secondly, one uses the map generating program to find the values $(\vec{q}_i, \vec{p}_i)$ that map into the origin $(\vec{0},\vec{0})$, which can be done using Newton's method. The resulting $(\vec{q_i},\vec{p_i})$ are then obviously the constant part of the inverse map.

3. The Taylor Expansion of the Generating Functions

Suppose we are given a transfer map $\mathcal{M}$. In this section we will describe how the class of the respective generating functions can be determined uniquely from the class of $\mathcal{M}$. We here show the algorithm for the computation of an F_2-type generator. It is immediately apparent how other generators can be determined.

We denote with $\mathcal{M}_1$ the part of the transfer map describing the final positions, and with $\mathcal{M}_2$ the part describing the final momenta. Thus, we have $\mathcal{M} = (\mathcal{M}_1, \mathcal{M}_2)$. We do the same with the identity map: $\mathcal{E} = (\mathcal{E}_1, \mathcal{E}_2)$. In order to obtain "mixed" relations $(\vec{q_f},\vec{p_i}) = \mathcal{F}(\vec{q_i},\vec{p_f})$, we start by setting $\mathcal{N} = (\mathcal{E}_1, \mathcal{M}_2)$. Then,

$$(\vec{q_i},\vec{p_f}) = \mathcal{N}(\vec{q_i},\vec{p_i}). \tag{10}$$

In order for the sought for generating function to exist, it is necessary that the map $\mathcal{N}$ is invertible, which in the DA picture requires the linear matrix of $\mathcal{N}$ to be invertible. Note that while symplectic matrices are always invertible, this does not have to be the case for the linear matrix of $\mathcal{N}$; in fact, in many important cases in optics, certain types of the generating functions do not exist. It is however guaranteed that at least one always exists. In case $\mathcal{N}$ is invertible, we obtain

$$(\vec{q_i},\vec{p_i}) = \mathcal{N}^{-1}(\vec{q_i},\vec{p_f}). \tag{11}$$

Composing the map $(\mathcal{M}_1, \mathcal{E}_2)$ and the map $\mathcal{N}^{-1}$, we finally obtain the desired "mixed" relations:

$$(\vec{q_f},\vec{p_i}) = \left((\mathcal{M}_1, \mathcal{E}_2) \circ \mathcal{N}^{-1}\right)(\vec{q_i},\vec{p_f}) = \mathcal{F}(\vec{q_i},\vec{p_f}). \tag{12}$$

The generating function F_2 can be obtained as the potential of $\mathcal{F}$. If $\mathcal{M}$ is symplectic, this exists up to a constant (which is only relevant in section 5); if $\mathcal{M}$ is not symplectic, the integration of $\mathcal{F}$ over an arbitrary path yields a generating function that represents a symplectic transfer map "near" the original one.

This algorithm can be used for symplectification of transfer maps that are not symplectic, for example because of the limited accuracy of the integrator used for the computation of the map. Finally note that computing the symplectified $\mathcal{M}$ from $\mathcal{F}$ is the same algorithm as computing $\mathcal{F}$ from $\mathcal{M}$.

Altogether, the whole process of obtaining the gradient of the generating function can be performed to arbitrary order using only composition and inversion of differential algebraic transfer maps. The determination of the generating function itself is only an integration. It is also worth mentioning that as long as the inverse of the linear part of $\mathcal{N}$ exists, there is a generating function. In particular, this implies that in the differential algebraic view there is never more than one solution of the implicit equations 5.

As it turns out, the ease of computing a generating function with Differential Algebra is one of the strong points of the power series representation of the map. In the Lie representation, the computation of the generating function can not be done in a straightforward pattern and gets increasingly cumbersome with high orders. We also note here that it is possible to solve for the generating function directly via the Hamilton-Jacobi equation, without previously calculating a map. This has been demonstrated in [14].

For all practical applications it is very important how "well behaved" the nonlinear generating function is; in particular, it is important that its nonlinearities are not too large. While the nonlinearities of the transfer map are a more or less direct measure of the nonlinearity of the system, this is not the case for generating functions. By inspecting the algorithm for the computation of the generating function, it becomes apparent that the new results are concatenated over and over with the inverse of the linear generating function. While symplectic matrices are always restricted to unity determinant, this is not the case for the linear matrix of $\mathcal{N}$. If its determinant is larger than one, after a few orders of the iteration process, very large high order terms may be produced.

4. Retroactive Symplectic Extension

Using the Taylor series expansion terms of the generating function as described in the last section, it is rather straightforward to compute a transfer map that agrees to the given transfer map to order n, yet is symplectic to higher orders than the old transfer map. Even if the old transfer map violates symplecticity noticeably because of truncation errors, it is possible to have the extended map satisfy the symplectic condition to higher and higher orders. Depending on the case, it is often possible to obtain machine accuracy symplecticity for the phase space regions of interest.

To this end, one first computes any suitable one of the four generating functions to the same order as the original map, following the algorithm discussed in the last section. While the result is not the proper generating function for the true map, it has the same Taylor expansion as the proper one, and agrees to it better and better the higher the order. One now approximates the real generating function by its Taylor series, and computes the map that is generated from it using the above algorithm. Up to oder n, the old map is reproduced; but there is no reason to stop the computation there. Just continuing produces higher and higher orders extending the original transfer map, and the kth order extension is such that the map is now symplectic through order k.

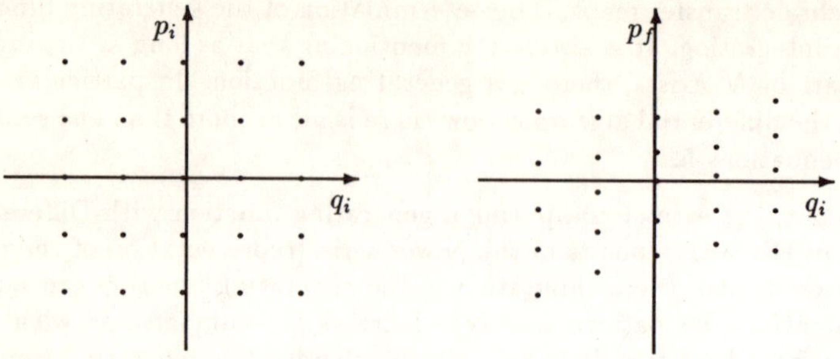

Figure 1: The distribution of the expansion points of the transfer maps of the system (left) and the expansion points of the local generating functions (right), where only the q_i spacing is equidistant.

In the end we obtain a map of higher order than the one we started out with and that is symplectic to any requested order. If the missing higher order terms are sufficiently small, what we have produced is an explicit symplectic integrator. This algorithm is particularly useful for systems in which the computation of the map is rather expensive to high orders, but whose inherent nonlinearity is not too high. In particular, this could be the case for machines consisting of many different elements, or compact machines with very complicated fields requiring detailed integrations.

5. The Superposition of Local Generating Functions

The symplectic extension technique outlined in the last section is fruitful for the study of problems that are not too nonlinear. In certain cases, however, the technique may be impractical because too high orders would be required. In this case the technique discussed in this section may prove useful.

When using generating functions for symplectic tracking, it is of course not mandatory that they actually have the same Taylor expansion as the true generating function. Indeed, in the case of nonlinear problems in which the function $\mathcal{N}$ is not well behaved, it may be advantageous to produce generating functions that are smoother overall. This can be achieved by a superposition of local generating functions.

To this end, a representative ensemble of nodes in phase space is chosen, preferably in a regular way as in Figure 1. For each of these nodes, a transfer map is computed to a certain order. Then, for each of the nodes the respective generating function is computed. Each of these generating functions is uniquely determined except for its constant value c_i.

A total generating function can now be determined by a smooth interpolation of the local polynomial type generating functions in the non-equidistant mesh. This has the form

$$F(\vec{q}_i, \vec{p}_f) = \sum_{j=1}^{n} F_j(\vec{q}_i, \vec{p}_f) \cdot w_j(\vec{q}_i, \vec{p}_f), \tag{13}$$

where the w_i are smooth weighting factor functions that ensure that the influence of F_i only extends to the respective next nodes and not far beyond. For example, they can be Gaussians centered at $(\vec{q}_i, \vec{p}_f)$, with widths determined by the distances to the next nodes and a height chosen accordingly.

While in the case of a single generating function, the unknown constant term was irrelevant, here it is significant since it is multiplied by the position dependent weighting function and thus shows up in the implicit solution $(\vec{q}_f, \vec{p}_f)$. So it is necessary to choose the c_i in a self-consistent way.

One solution to this problem is to demand that at each node, the predictions of all the neighboring nodes are as close together as possible (which is somewhat similar to the finite element potential solving problem). This yields a least squares problem for the c_i which can be solved using conventional techniques. Naturally, the higher the orders of the individual F_i, the better their prediction at the neighboring nodes will be, and the smaller the resulting sum of squares will be.

Altogether, one obtains a generating function that is not Taylor series like, and one can cover large and rather nonlinear areas of phase space. Thus, at least in principle, the technique is suitable for the symplectic tracking through any Hamiltonian system.

6. References

1. M. Berz. Arbitrary order description of arbitrary particle optical systems. *Nuclear Instruments and Methods*, Proc. Charged Particle Optics III, 1990.

2. M. Berz. Differential Algebraic description of beam dynamics to very high orders. *Particle Accelerators* **24**(1989)109

3. M. Berz. Differential Algebraic treatment of beam dynamics to very high orders including applications to spacecharge. *AIP Conference Proceedings* **177**(1988)275

4. E. Forest, M. Berz, and J. Irwin. Normal form methods for complicated periodic systems: A complete solution using Differential Algebra and Lie operators. *Particle Accelerators* **24**(1989)91

5. M. Berz. COSY INFINITY, an arbitrary order general purpose optics code. *AIP Conference Proceedings*, in print, 1990.

6. M. Berz. Computational aspects of design and simulation: COSY INFINITY. *Nuclear Instruments and Methods*, Proc. Charged Particle Optics III, 1990.

7. H. Goldstein. *Classical Mechanics.* Addison-Wesley, Reading, Massachusetts, 1980.

8. Ron Ruth. private communication.

9. Etienne Forest. private communication.

10. A. Dragt. In *Nonlinear Problems in Future Accelerators*, these proceedings.

11. J. Irwin. In *Nonlinear Problems in Future Accelerators*, these proceedings.

12. M. Berz. Analysis auf einer nichtarchimedischen Erweiterung der Reellen Zahlen. Report (in german), Universität Gießen, 1988.

13. E. Forest and M. Berz, Canonical Integration and Analysis of Periodic Maps using Non-Standard Analysis and Lie Methods, *Proceedings Second Workshop on Lie Methods in Optics*, 1988, in print

14. G. D. Pusch. *Differential Algebraic Methods for Obtaining Approximate Numerical Solutions to the Hamilton-Jacobi Equation.* PhD thesis, Virginia Polytechnic Institute and State University, Blacksburg, Virginia, 1990.

A TYPICAL "Ztrack" LONG-TERM TRACKING RESULT FOR THE SSC APERTURE STUDY

Yiton Yan and George Bourianoff
Superconducting Super Collider Laboratory
Dallas, Texas 75237
U. S. A.

and

Lindsay Schachinger
Lawrence Berkeley Laboratory
Berkeley, California, 94720
U. S. A.

ABSTRACT

A large amount of supercomputer CPU time has been used for tracking particles in the Superconducting Super Collider (SSC) using a vectorized, multi-tasked post-Teapot tracking program called "Ztrack." Typically, hundreds of particles with appropriate initial displacements (relative to the closed orbits) are tracked for a hundred thousand turns. One then simply makes a survival plot (turn at which particle is lost versus initial displacements) to determine the dynamic aperture. Occasionally, particles are tracked to a million turns for very selective cases. These numerical studies aid in determining the best aperture for the SSC.

Introduction

Critical to the successful performance of the SSC is injection when the proton beams coast for about 40 minutes. A key issue is the aperture, or the cross-sectional area in which the proton motion is stable. If this cross section area is large (relative to the size of the proton beam emerging from the injector), the proton beams will have longer lifetimes. Clearly it is desirable to have as large an aperture as possible, given cost constraints. The aperture depends on the choice of FODO cell characteristics, as well as on the field uniformity in the magnets. Magnets with larger coil diameter have a larger uniform field region, but are more expensive. In this letter, we show a typical long-term tracking result from our recent studies of the dependence of dynamic aperture on magnet coil diameter.

As long-term tracking for the SSC was started in 1988, all the tracking codes available to us were only optimized for scalar computation. Tracking a hundred particles up to ten-thousand turns in the SSC had been a very difficult task. Therefore, "Ztrack"[1] was developed to take advantage of the computational speed of supercomputers (half a year later, another tracking code "SSCTRK"[2] was also developed at the SSC Laboratory as a parallel tool to the program Ztrack for long-term tracking of the SSC). "Ztrack" is an element-by-element post-Teapot[3] tracking program. It is vectorized and multi-tasked. The user needs to prepare two input files for "Ztrack": (a) a machine file "Zfile" provided by the systematic program "Teapot," which describes the non-linear lattice to be studied; and (b) a name list command file "Zcmd" which tells how the tracking is to be performed, such as how many particles, how particle coordinates are initialized, how particles are handled once they get lost, etc. Usually hundreds of particles with well distributed initial displacement amplitudes are tracked, simultaneously, element-by-element for a hundred-thousand turns or more. A survival plot (turn at which particles are lost versus initial displacements) is then obtained for dynamic aperture study.

Numerical Description

In the following, we give a brief description of the steps which were used to create the machines which were subsequently tracked using Ztrack.

Step 1. Preparation of the Linear Lattice

The machine files are prepared with the program Teapot. A standard "MAD"[4] input that describes a thin-lens representation of the SSC injection linear lattice[5] is first prepared. Teapot then reads in the input and converts all thick elements to thin ones. Note that all the correctors are included in the lattice.

Step 2. Addition and Correction of Random and Systematic Multipole Errors

Once the linear lattice is defined, random and systematic multipole errors are added to the dipoles in the lattice. Here as an example, two sets of random and systematic errors representing the 4 cm and the 5 cm coil-diameter dipoles are presented in Table 1. In this study, there are no corrections other than the use of chromaticity sextupoles and tune quadrupoles to correct the systematic b_2 and the working tunes (Note that also the normal and skew b_1 and a_1 are set to zero with the assumption of decoupling).

Table 1. Random, Systematic, and Orbit Errors and Their Corrections

Case	4 cm	5 cm
Half Cell Length	90 m	90 m
Injection Energy	2 TeV	2 TeV
Dipole Coil Diameter	4 cm	5 cm
Random Errors	$\sigma_{b_2} = 0.4$ $\sigma_{b_3} = 0.3$ $\sigma_{b_4} = 0.7$ $\sigma_{a_2} = 0.6$ $\sigma_{a_3} = 0.7$ $\sigma_{a_4} = 0.2$	$\sigma_{b_2} = 0.272$ $\sigma_{b_3} = 0.18$ $\sigma_{b_3} = 0.35$ $\sigma_{a_2} = 0.41$ $\sigma_{a_3} = 0.41$ $\sigma_{a_4} = 0.1$
Systematic Errors	$b_2 = -3.0$ $b_3 = 0$ $b_4 = 0.2$ $b_6 = -0.05$	$b_2 = -1.9$ $b_3 = 0$ $b_4 = 0.09$ $b_6 = -0.017$
Misalignment Errors	$\sigma_{a_0} = 0.000834$ $\sigma_{b_0} = 0.0012$ $\sigma_\theta = 0.0005$ rad (dipole) $^*\sigma_x = \sigma_y = 0.001$ m	$\sigma_{a_0} = 0.000834$ $\sigma_{b_0} = 0.0012$ $\sigma_\theta = 0.0005$ rad (dipole) $^*\sigma_x = \sigma_y = 0.001$ m
Corrections	Use only chromaticity sextupoles to correct the systematic b_2. σ_{b_2} reduced to 0.4	Use only chromaticity sextupoles to correct the systematic b_2. σ_{b_2} reduced to 0.272
Tunes	(0.285, 0.265)	(0.285, 0.265)
RMS orbit distortion	1 mm	1 mm

*For dipoles, quadrupoles, and beam monitors.

Step 3. Addition of Misalignments and Correction of Orbit

After the random and systematic multipole errors are added to the lattice and the chromaticities and the tunes are adjusted to the desired values, the next step is the addition of random errors due to misalignments (see Table 1). These errors include: (1) random steering errors (σ_{a_0} and σ_{b_0}); (2) rotations (σ_θ); and (3) displacements (σ_x and σ_y). After the addition of errors due to misalignments, the orbit is distorted and corrected to an rms orbit deviation of 1 mm in both the horizontal and vertical planes with respect to the reference orbit. In some other cases not shown here, decoupling is simultaneously performed between the horizontal and vertical planes. The above numerical steps are all performed on a SUN workstation.

Step 4. Retuning

Once the orbit is corrected to a desired value of rms orbit deviation, a machine file is written and transfered to a Cray computer at DOE National Energy Research Supercomputer Center (NERSC) where Teapot reads in the machine file and readjusts the chromaticities and the working tunes. After the chromaticities and the working tunes are well adjusted, a machine file "Zfile" that is suitable for the program Ztrack is written.

Step 5. Tracking

Ztrack was used to track particles. Based on the command input file "Zcmd," Ztrack adds an rf cavity if synchrotron oscillations are required. A TOPDRAWER[5] file is written for the survival plot at the end of the tracking. For protection from loss of tracking information due to unexpected computer problems, for retrieving information during the tracking, and for continuing for longer turn tracking after the requested turns of tracking has been performed, a restart file is written every "ntsave" turns and at the end of tracking.

A Typical Long-Term Tracking Result

While referring interested readers to the "SCDR"[6] for more long-term tracking results, presented here is a typical result obtained from "Ztrack" long-term tracking for the SSC aperture study (magnet and orbit errors are shown in Table 1).

Figure 1 shows survival plots for 100 00 turn tracking from Ztrack. We studied two 2-TeV injection lattices, one with 4-cm coil-diameter dipoles and the other with 5-cm coil-diameter dipoles. The two lattices have the same random seed. The only difference between the two lattices is in the multipole content due to the different size of magnet aperture. The rf cavity system was turned on to maintain bunching. All particles tracked are initiated with a same synchrotron oscillation amplitude of 3 σ_{rms} (σ_{rms}: rms half bunch length) but with an appropriate distribution on initial horizontal (x-axis) and vertical (y-axis) displacements ($y/\sqrt{\beta_y} = x/\sqrt{\beta_x}$ and $P_x = P_y = 0$ at injection, which is at the middle of a horizontal focusing quadrupole).

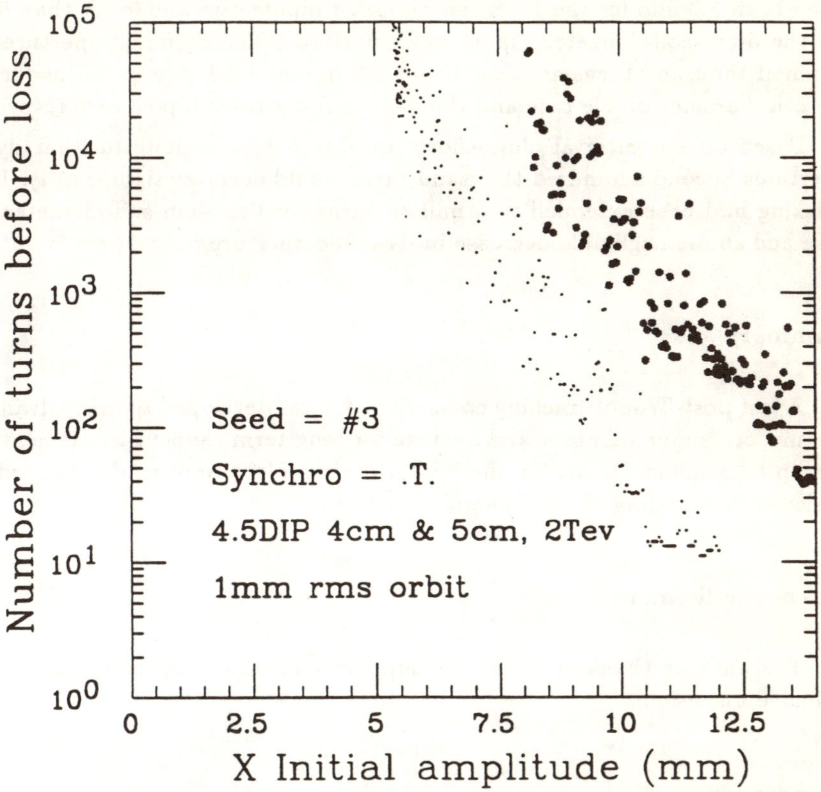

Figure 1. Hundred-thousand-turn survival plots for two 2-TeV SSC injection lattices, comparing the dynamic aperture for two different magnet apertures, one with 4-cm coil-diameter dipoles (shown in '.' symbol) and the other with 5-cm coil-diameter dipoles (shown in '*' symbol). With the 4-cm coil-diameter dipoles, no particles with initial x displacement amplitude of less than 5.3 mm were lost. With the 5-cm coil-diameter dipoles, no particles with initial x displacement amplitude of less than 8.1 mm were lost. By increasing the magnet aperture, the dynamic aperture for 100 000 turns enlarges from about 5.3 mm to about 8.1 mm in radius, which increases the machine's linearity. This plot shows only the protons that were lost before 100 000 turns are reached.

While many of the particles with higher displacement amplitudes were lost in the earlier turns, all of the particles with initial x displacement amplitudes lower than 5.3 mm for the 4-cm coil-diameter dipole case and lower than 8.1 mm for the 5-cm coil-diameter dipole case survived. The dynamic apertures for a hundred-thousand turns are thus about 5.3 mm and 8.1 mm in radius for the 4-cm coil-diameter dipole case and the 5-cm coil-diameter dipole case respectively.

Based on the survival plots shown in Figure 1, it is doubtful that dynamic apertures beyond a hundred-thousand turns would decrease significantly. Indeed, tracking had been extended to a million turns for the 4-cm coil-diameter dipole case and shows negligible decrease in dynamic aperture.[1]

Summary

A fast post-Teapot tracking code "Ztrack" was developed to take advantage of the modern supercomputer architecture for long-term element-by-element tracking up to a million turns for the SSC. Survival plots were made that led to an easier understanding of the dynamic aperture.

Acknowledgement

The authors thank A. Chao for support and encouragement and for many valuable discusions.

References

1. Y. Yan, "Supercomputing for the Superconducting Super Collider," SSCL-239, September 1989; also to be pulished in Energy Sciences Supercomputing by DOE National Energy Research Supercomputing Center (NERSC); L. Schachinger and Y. Yan, "Recent SSC Dynamic Aperture Measurements from Simulations", SSC-N-664, September 1989;
2. D. Ritson, "SSCTRK: a thin-element-lumping tracking program for the SSC", private communication.
3. L. Schachinger and R. Talman, "Teapot: A Thin-Element Accelerator Program for Optics and Tracking," Particle Accelerators, Vol. 22, 1987.
4. D.C. Carey and F.C. Iselin, "A Standard Input Language for Particle Beam and Accelerator Computer Programs," Proc. of the 1984 Summer

Study on the Design and Utilization of the Superconducting Supercollider, Snowmass, CO; June 1984.

5. R. B. Chaffee, "TopDrawer," SLAC Computation Group Note CGTM No. 178 (1980).

6. "SCDR: Site-Specific Conceptual Design of the Superconducting Super Collider," SSC-SR-1051, June 1990.

SSCTRK: A PARTICLE TRACKING CODE FOR THE SSC*

D. RITSON
Stanford Linear Accelerator Center,
Stanford University, Stanford, CA 94309, USA
and SSCL, Dallas, TX 75237–3946, USA

1. Introduction

Critical to the SSC performance is an understanding of the factors controlling the dynamic aperture of the main storage ring. The most critical problems are expected to arise at injection time prior to ramp-up to full energy. At injection the beam will coast for up to one hour. This long time is required to fill sequentially the two rings. It is known that the effective aperture of a proton storage will progressively degrade with storage time.

The SSC storage time at injection is $\sim 1/2$ hr or $\sim 7 \times 10^6$ turns around the machine. The SSC aperture will be 4–5 cm, approximately half that of the Tevatron. The multipole field errors, both random and systematic are thus substantially greater than those of the Tevatron. The tune of the SSC will be about five times that of the Tevatron and the sensitivity to systematic tune shift will then be worse by a factor of 5. Therefore it cannot be argued that success of the Tevatron guarantees success for the SSC.

While many indirect methods are available to evaluate dynamic aperture there appears at this time to be no reliable substitute to tracking particles through realistic machine lattices for a number of turns determined by the storage times. Machine lattices are generated by "Monte Carlo" techniques from the expected rms fabrication and survey errors. Any given generated machine can potentially be a lucky or unlucky fluctuation from the average. Therefore simulation to serve as a predictor of future performance must be done for an ensemble of generated machines. Further,

* Work supported by Department of Energy contract DE–AC03–SF00515.

several amplitudes and momenta are necessary to predict machine performance. Thus to make Monte Carlo type simulations for the SSC requires very considerable computer resources. Hitherto, it has been assumed that this was not feasible, and alternative indirect methods have been proposed or tried to answer the problem.

We reexamined the feasibility of using direct computation. Previous codes have represented lattices by a succession of thin elements separated by bend-drifts. With "kick drift" configurations, tracking time is linear in the multipole order included, and the code is symplectic. Modern vector processors simultaneously handle a large number of cases in parallel. Combining the efficiencies of kick drift tracking with vector processing, in fact, makes realistic Monte Carlo simulation entirely feasible.

SSCTRK uses the above features. It is structured to have a very friendly interface, a very wide latitude of choice for cases to be run in parallel, and, by using pure FORTRAN 77, to interchangeably run on a wide variety of computers. We describe below the program structure operational checks and results achieved.

2. Description of Program

The code splits naturally into two parts. A Monte Carlo generator to provide one (or more) machine lattices and tracking code to examine the long term stability of particles launched into the generated machine lattice(s).

The Monte Carlo generator sets up a nominal thin lens FODO lattice chromatically corrected by sextupoles to provide the desired numbers of elements, tune values and radius. Files of orbit errors and multipole errors are then generated from specified rms and systematic field and orbit errors. These are combined together (including multipoles caused by orbit error feed down) into effective multipoles centered around the equilibrium orbit. The program then makes user specified corrections, employing multipole families, to provide a final file that should correspond to that to be attained operationally during the commissioning of the accelerator.

The generator is fast, and it provides lattices with the expenditure of seconds or less of cpu time. The generator can be used to sequentially generate specified lattices or cases and to accumulate them into an array indexed by the case number that is to be used for parallel (vector) processing.

Kick drift tracking then tracks the particles over the required number of turns through the set of lattices provided by the generator. The tracking simulation can include synchronous motion, ramping in energy, and power supply ripple for the focusing quadrupoles relative to the bend magnets.

At each turn, quantities such as averages and rms deviations of the Courant and Snyder invariants are accumulated. After n turns (a specified number), these quantities are dumped into an output minifile. The minifile data is then put through a postprocessor to provide graphics or other outputs. The final orbit data are also dumped to permit resumption of running if so desired.

3. Physics Inputs

We represent the lattice by a succession of thin elements. This naturally provides sympleticity and, in view of the fact that the focusing quadrupoles only occupy 5% of the arcs of the lattice, it is a reasonable approximation. At present the IRs are represented as unit transformations. This is a reasonable approximation under injection conditions, but it will require modification to simulate the low beta interaction optics and future code will include IP effects. The RF acceleration is modeled by a single "thin" RF cavity located in a dispersion-free location. The drift-bend transfer matrices are approximated with a standard six-by-six linear transfer matrix acting on the vertical, the horizontal coordinates, and the longitudinal coordinate and momentum.

A SSC half-cell contains a quadrupole, a chromatic sextupole corrector and five bend magnets. All of these components may be incorrectly placed or contain systematic or random multipole errors.

To economize cpu time, we use the Simpson Rule procedures of Neuffer and of Forest and Peterson to concatenate lattice errors into the ends and cell centers. Essentially, this procedure redistributes the errors in the five dipoles in a half-cell into three positions. The procedure takes a series of multipoles of order n s_{in} at points t_i, where t_i is the displacement from cell center and $2t_0$ is the half-cell length, and to replace them with three values S_{1n}, S_{2n}, S_{3n} lumped at the beginnings, midpoints and ends of the half-cells.

The S_{in}, $i = 1, 2, 3$ are determined for each half-cell by the three conditions

$$S_{1n} + S_{2n} + S_{3n} = \sum s_{in} \tag{1}$$

$$-t_0 . S_{1n} + t_0 . S_{3n} = \sum t_i . s_{in} \tag{2}$$

$$t_0^2 . S_{1n} + t_0^2 . S_{3n} = \sum t_i^2 . s_{in} \quad . \tag{3}$$

This procedure takes account of the variations of beta and eta values over the cells and cancels the deviations of the these parameters to second order for random errors and to third order for systematic errors. The efficacy of these procedures for random errors is detailed in the *1988 Rome Particle Accelerator Conference*, by Peterson and Forest, page 827. The results are good to 5–10%. We have compared this to a noncorrelated concatenation that assigns random errors to the beginning, center and end of the cells in the ratio of $1 : \sqrt{2} : 1$. Both schemes give results equivalent on the 5% level. Etienne Forest analytically finds both schemes to be nearly equivalent. For systematic errors in the SSC, we have compared analytic first-order predictions of tune shifts in the distributed case with errors lumped according to the Neuffer procedure for both octupole and decupole systematics, and we have found agreement to a few percent. We therefore assign an upper bound to the error on the dynamic aperture of $< 10\%$ smaller than that associated with predicting tolerances to be achieved in practice.

4. Code Checks

As is standard, the code is double precision. Detailed below are checks or procedures used to ensure integrity of the code.

4.1 Overall Organization

A potential source of problems is the unavoidable complexity of code written to model a real machine with all its underlying warts and blemishes. To minimize the complexity SSCTRK was designed as a "linear" code. The main program steps sequentially through subroutine calls that take the specifications and translate theme into a design lattice structure. The program assembles this structure into a "real" object flawed by systematic and random errors. It applies corrections to this object and finally checks this object by tracking to find the maximum usable aperture. Each of these steps is done using main subroutine calls. The subroutines are simple, perform general tasks and contain an absolute minimum of internal branches or special purpose assumptions. Any operation or piece of code occurs only once in the program. The only subsidiary subroutines or functions called by these main subroutines are mathematical functions or subroutines such as a random number generator.

Tricks or shortcuts in coding, equivalence statements, and backward "goto" statements are deliberately avoided, and at all stages the code is designed to be explicit and to conform fully to FORTRAN 77 protocol. To further ensure full transportability, no external system specific routines are invoked.

The complete SSCTRK code (excluding comments) has less than one thousand lines of FORTRAN. This amount of code is easy to debug, edit and transfer between machines.

Identical running on IBM, CRAY, SX2 and SUN computers is already a strong check on the generality of the code.

4.2 Standard Code Checks

We have made use of many of the standard techniques currently available to ensure error free coding. These include:

(1) The "IMPLICIT NONE" convention. This convention forces the programmer to explicitly assign integer, real etc. types to all variables. The compiler automatically flags any variable that is not explicitly typed (spelling errors for instance) or whose type conflicts with a previously typed variable of the same name.

(2) Load options to check for improper initialization of variables. These options per-

mit loading of memory banks with zeros or alternately with "overflow" numbers. Incorrectly initialized variables are flagged by the program running differently with different load options or abending with error messages.

(3) Quadruple precision compilation. The code can be compiled with a quadruple precision option and compared to the standard double precision results to ensure that round off precision is indeed sufficient.

(4) The debug option. All sections of the code were stepped and looped through under the "debug" option to ensure that operations were indeed proceeding as expected.

4.3 Redundancy checks

Central to the integrity of the code is the central tracking routine. Kick code can be written very simply in complex notation. As the tracking routine dominates the cpu usage, there is a high premium on optimizing its performance. In practice, therefore, this routine utilizes real quantities and is arranged to minimize the multiplication and additions required for evaluation. For operation with the CRAY 2, where the FORTRAN compiler was less than optimal, an assembly coded routine was used (this, in fact, gained a factor of 2 in speed). These substantially more complicated and opaque routines were checked against the results from a slower but transparent tracking routine implemented in complex notation, and they gave identical results within the double precision roundoff errors.

The fact that the code runs "identically" on a variety of different computers has already been alluded to.

Twiss parameters and tunes are calculated directly from the lattice characteristics. They are independently evaluated from the orbits obtained in tracking. Agreement of these Twiss parameters and tunes obtained from orbit data provide an important redundancy check on the code.

Chromatic tune shifts resulting from systematic errors are obtainable analytically to first order, and it has been checked that the code produces correct results for these terms.

Finally, intensive useage by a number of people over an extended period is an excellent tool for uncovering potential flaws in the underlying structures.

4.4 Reverse tracking

An interesting code check is provided by taking a particle through a number of turns, reversing its direction and returning it through the same number of turns to ensure that it returns to the identical starting amplitude. This test did indeed return the particle to the identical starting pointing within the precision expected from the previously determined roundoff errors.

5. Long-Term Running

A standard worry in long-term simulation is whether artificial damping or beam blowup is caused by nonphysical artifacts of the code.

In principle the code is fully symplectic. There are, however, a number of approximations that have been used.

Fully accurate tracking involves a "finite momentum" correction. This correction is of the order of v_s/v where v_s is the longitudinal component of velocity, and v, the magnitude of the velocity. This is a very small correction indeed but it results in a loss by a factor of 2–3 in tracking speed. Most codes do not include this correction, and we have also omitted it for most running. We have, however, checked with a million turn thirty-two case run that inclusion of this factor did not effect the dynamic aperture determination.

We have made long runs with linear lattices that show no growth or change in the Courant Snyder invariants. This demonstrates that roundoff errors that cause departures from symplecticity are sufficiently small in double precision simulations as to not cause problems. It also provides a check that coding artifacts do not appear to be producing spurious problems.

6. Computation Requirements

6.1 Memory Usage

Considerable care was taken to minimize the memory requirements of the program. Typically, memory requirements are dominated by the array used to describe the lumped multipole content around the machine plus, of course, a standard overhead of a few megabytes. An initial parameter statement sets the size of this multipole array.

For a lattice of eight hundred quadrupoles there will be, including midcell locations, sixteen hundred sets of multipoles. If multipoles up to sixth order are included for each lumped element the required storage is twelve REAL*8 numbers per location or a total of $\sim 154,000$ bytes per case. The program therefore is relatively modest in its demands for memory, even when used in a vector mode.

6.2 Speed

SSCTRK has made extensive SSC simulation studies. To economize super computer use we have assembly coded the actual short tracking routine. We give below in Table 1 typical cpu tracking times required for a model machine with eight hundred half-cells, lumping multipole errors into the center and ends of the half-cells, and multipole order up to six included.

Current technologies are rapidly evolving, and it can be expected that the above performance figures will improve by substantial factors. Usually most of the required information can be obtained running between 32 to 64 cases for a 10^5 turns each

Table 1: Benchmark computer speeds for SSCTRK.

Scalar processors	
SUN SPARC workstations	15 particle turns/sec
IBM 3080	80 particle turns/sec
IBM 6000 workstation	100 particle turns/sec
Vector processors	
CRAY 2	1,000 particle turns/sec
CRAY YMP	2,000 particle turns/sec
SX2	3,000 particle turns/sec

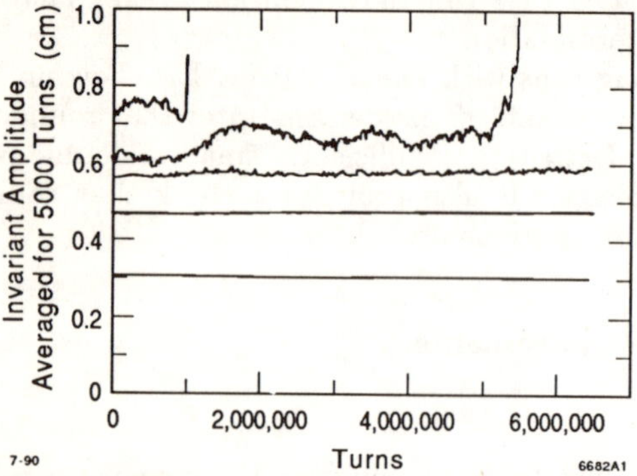

Figure 1: The evolution of the root of Courant and Schnyder invariant for individual particles averaged for 5,000 turns plotted against turn number out to six million turns.

requiring ~ 1 hr of supercomputer time. At complete evaluation of evaluations of dynamic aperture using 32 parallel cases and 5×10^6 turns per case can be run in about one day of cpu time.

7. Typical Results

Figure 1 shows the behavior of tracks launched at different amplitudes. Plotted is the root of the Courant and Schnyder invariant averaged over 5,000 turns versus turn number. At small amplitudes well within the dynamic aperture the amplitude is unchanging. At larger amplitudes the motion begins to show chaotic irregularities and beyond the dynamic aperture the particle motion is chaotic and loss occurs after 150,000 turns. An informative display of results is provided by a "loss-plot." Figure 2 shows the history of a group of particles at four closely spaced amplitudes injected into an ensemble of eight Monte Carlo generated lattices for 4 cm aperture machines. The particles at amplitudes well above the dynamic aperture. After relatively few turns the particles are lost from the machine aperture. If a particle is lost it is restarted at a smaller initial amplitude. This process continues until it

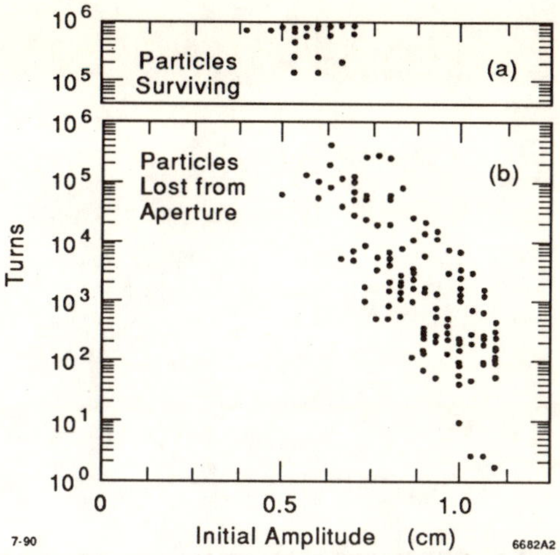

Figure 2: Particles initially launched at 4 amplitudes close to 1.1 cms into eight Monte-Carlo generated lattice structures. After a particle leaves the aperture it is restarted at 1 mm in. Plotted is the logarithm of the number of turns prior to loss versus aperture. (a) Surviving particles and (b) lost particles.

is injected at an amplitude just within the long-term dynamic amplitude. Plotted in Fig. 2(a) is the logarithm of the number of turns made by the surviving particles versus aperture; and in Fig. 2(b), the logarithm of the number of turns before loss versus aperture. The envelope of the curve in Fig. 2(b) effectively delineates the edge of the dynamic aperture as a function of turns. Typically, an increase in the number of turns required, by a factor of 30, will decrease the dynamic aperture by 1 mm.

8. Conclusions and Future Directions

On the basis of the above SSCTRK appears to provide a sound reliable basis to predict the SSC performance. This has been achieved via straightforward computer tracking. Computer capabilities are increasing at a remarkable speed and over the next few years one can predict with confidence that tracking simulations for long storage times will become increasingly simple.

For the future, detailed IP configurations and operational simulation capability will be added to the code.

MATHEMATICAL ASPECTS AND PHYSICAL APPLICATIONS OF THE TUNESHIFT CORRECTION

E. Todesco

Dipartimento di Fisica della Università di Bologna, Italy
Istituto Nazionale di Fisica Nucleare, Sezione di Bologna.

Abstract

We review the problem of the tuneshift correction of symplectic maps and its applications to accelerator physics. The normal form techniques used to compute the correcting terms are outlined both in two and four dimensional phase space. A theorem which gives an upper bound to the growth of the coefficients of the correcting map is proved. Two ways of correcting the tuneshift are discussed: a compensation obtained by adding multipolar terms and a procedure based on the composition with a normal form. Both cases are analysed numerically, showing for the quadratic map a factorial divergence in agreement with the proved theorem. Applications of the tuneshift correction to real machines are discussed, referring in particularly to LHC.

1. Introduction

The aim of this paper is to discuss some mathematical properties and physical applications of the tuneshift correction of a polynomial symplectic map $\mathcal{M}$. Such a model can be considered the perturbation of a isochronous linear map; when one adds a non-linear part the frequency becomes amplitude-dependent: the shift in the frequency due to the non-linear terms is called tuneshift.

It is well known that the tuneshift has a destabilizing effect on the dynamics, as it can drive the frequency on resonant values; for this reason in order to increase the stability domain of a map $\mathcal{M}$ in principle it is necessary to reduce its tuneshift as much as possible.

The physical applications connected with the tuneshift corrections are very relevant: one of the most important fields is accelerator physics, as a magnetic lattice for hadrons has its natural model in a polynomial symplectic map [1]. The magnetic elements used in the machines built in the past decades were linear up to a very good approximation; therefore the linear theory for maps (i.e., the Courant–Snyder formalism [2]) was enough to understand the behaviour of an accelerator. Nowadays the introduction of superconducting magnets has made accelerators very non-linear machines, requiring good schemes for the tuneshift correction and both numerical and analytical tools to understand the dynamics of the beam. The possibility of compensating the tuneshift by using non-linear multipoles was successfully tested at the Super Proton Synchrontron (SPS, CERN) during this year [3]: a very strong reduction of the tuneshift caused by extraction sextupoles was obtained using octupoles; an increase in the stability domain was observed.

The tuneshift correction is carried out by adding to $\mathcal{M}$ a correcting term $\mathcal{M}^{(c)}$ such that the tuneshift of $\mathcal{M} + \mathcal{M}^{(c)}$ is minimized. In order to compute the coefficients of $\mathcal{M}^{(c)}$ which best correct the tune one can follow two strategies: a numerical minimization using the tracking [4] or an analytical approach [5],[6],[7]. The tracking method attacks the tuneshift in a certain number of points scattered in the phase space, finding a local minimum with numerical tools. The first analytical approach to the tuneshift correction was based on classical perturbative theory for hamiltonian flows [5]; it had the drawback of being limited to the analysis of the low-order effects. The introduction of perturbation theory for maps, based on normal forms, has given the possibility of automatically computing every perturbative order. This approach is based on the expansion of the tuneshift in a power series in the non-linear invariants; the correction procedure attacks the coefficients of this expansion order by order, computing analytically the global minimum [6]. This method allows an exact compensation of a single order or a minimization if an insufficient numbers of parameters is available in the map $\mathcal{M}^{(c)}$; it can be carried out up to every order. Both the tracking and the normal form methods were applied to the study of the Large Hadron Collider (LHC, CERN) [4],[6]; it was checked that the numerical solutions were in good agreement with the analytical ones. In this paper we will deal with the normal form approach, showing how to compute explicitly correction schemes for both simple and realistic models of a magnetic lattice.

From a mathematical point of view the problem of the correction of the tuneshift poses some very interesting questions; the main point is the convergence of the correction procedure. In fact if the map corrected at all orders is convergent in an open set, then its normal form is linear; in this case a theorem proved by Rüssmann [8] and later revisited by Gallavotti [9], ensures the convergence of the normal form series, and therefore the integrability, in an open set around the origin. On the other hand, the divergence of the correction implies that there is an optimal order up to which to carry out the minimization.

A still open question is the connection between tuneshift compensation and dynamic aperture, i.e. the boundary of stability of the motion; numerical results seem to suggest that the tuneshift correction leads to a wider dynamic aperture. Nevertheless this result is strongly dependent on the linear frequency of the map, and the increase is not dramatic. From the point of view of accelerator physics, the problem of the connection between tuneshift minimization and dynamic aperture is rather academic. In fact hadron accelerators should work in a domain which is quasi-linear, and phase space zones where the motion is strongly perturbed, even if bounded, will never be chosen as safe operational points. The physical quantity which should be taken into account is the domain of quasi-linearity, which is called linear aperture. The definition of this quantity is based upon the following bounds on the tuneshift $\delta\nu$ and on the deformation of the orbits σ (called smear):

$$\begin{aligned} |\delta\nu| &< 0.005 \\ \sigma &< 10\% \end{aligned} \tag{1.1}$$

these bounds have been obtained from experiments. Both tracking [4] and normal form simulations [6] have shown that the tuneshift correction leads to a strong increase in the linear aperture, as expected.

Finally, it must be pointed out another effect of the tuneshift compensation: the destabilization of low order resonances. It is well known that a given resonance p/q is stable if there are some detuning terms of order less than q. For this reason it is clear that when we compensate exactly a tuneshift order, setting its coefficients to zero, the corresponding resonances become unstable. The analysis of the interpolating hamiltonian of the resonant normal form [10] shows that the correction of the first tuneshift order makes resonances 4 and 5 unstable, of the second tuneshift order makes resonances 6 and 7 unstable, and so on. In fact a map with a linear normal form is unstable for each rational value of the linear frequency. Whilst the destabilization of low order resonances causes no problems in accelerator physics, as the linear tune of a machine is far from these rational values, it can become dangerous for high order resonances, which will anyway fall close to the working point. In order to solve this problem it is advisable not to correct the tuneshift completely beyond the first two orders; moreover, a complete correction of the higher orders would be not only dangerous for the stability of the beam but also not feasible from a practical point of view, requiring a correction scheme with too many parameters to handle.

The scheme of the paper is the following: in Section 2 we briefly review the tuneshift definition and the analytical estimate using the normal form theory. In Section 3 we discuss the two dimensional case, showing a procedure to correct the tuneshift at every order using multipolar kicks; we prove a theorem which gives an upper bound to the growth of the coefficients of the correcting map and we present some numerical results which confirms the factorial divergence proved by the theorem. Section 4 reviews the four dimensional case, showing how to correct the tuneshift using multipolar elements; some results obtained for a realistic model of the LHC are shortly outlined. In Section 5 we analyse the possibility of correcting the tuneshift of the map $\mathcal{M}$ by composing it with a normal form; even if this procedure has no practical applications, it can be relevant form a mathematical point of view. We study the convergence of this scheme, finding out different behaviours according to the structure of $\mathcal{M}$.

2. Tuneshift definition and non-resonant normal forms

In this Section we briefly introduce the tuneshift definition for a polynomial symplectic maps in a $2d$ dimensional phase space:

$$\mathbf{z}' = \mathcal{M}(z, z^*) = e^{i\omega}\left(\mathbf{z} + \sum_{k\geq 2} \mathcal{M}_k(\mathbf{z}, \mathbf{z}^*)\right) \qquad \mathbf{z}, \mathbf{z}^* \in \mathbf{C}^d, \tag{2.1}$$

where $\omega \in \mathbf{R}^d$ and $\mathcal{M}_k$ are homogeneous polynomials of order k.

The amplitude-dependent tune is defined as the average phase advance of an orbit starting from a given point $\mathbf{z}_0$:

$$\nu(\mathbf{z}_0) = \lim_{N\to+\infty} \frac{1}{N}\sum_{i=1}^{N} \alpha_i \qquad \nu \in \mathbf{R}^d, \tag{2.2}$$

where α_i is the phase shift between two successive iterates of the map $\mathbf{z}_0^{(i)}$ and $\mathbf{z}_0^{(i-1)}$. The tuneshift is the difference between the amplitude-dependent tune and the linear tune $\frac{\omega}{2\pi}$:

$$\delta\nu(\mathbf{z}_0) = \nu(\mathbf{z}_0) - \frac{\omega}{2\pi} \qquad \delta\nu \in \mathbf{R}^d. \tag{2.3}$$

According to this definition (we will not discuss the condition of existence of the limit (2.2)) one can easily compute the amplitude-dependent tuneshift with a numerical method, iterating the map $\mathcal{M}$.

Indeed, it is possible to give an analytic estimate of the tune using perturbation theory. The basic idea is to perform a change of coordinates using a symplectic function ϕ from the map space $(\mathbf{z}, \mathbf{z}^*)$ to the normal form space (ς, ς^*) where $\mathcal{M}$ is transformed to an amplitude dependent rotation $\mathcal{N}$:

$$\varsigma' = \mathcal{N}(\varsigma, \varsigma^*) = \exp(i\Omega(\rho_1, ..., \rho_d))\varsigma \qquad \varsigma \in \mathbf{C}^d. \tag{2.4}$$

ρ_i are the amplitudes in the normal space, i.e. the non-linear invariants:

$$\rho_i = \varsigma_i \varsigma_i^* \qquad i = 1, ..., d. \tag{2.5}$$

The map $\mathcal{N}$ is called non resonant normal form; Ω is the amplitude dependent tune. Both $\mathcal{N}$ and ϕ are polynomials in the variables ς, ς^* truncated at a given order. A theorem [11] guarantees that the normal form can be computed at every order provided that the linear frequencies ω are not resonant. According to this approach the tuneshift is splitted in different contributions that are homogeneous polynomials in the non-linear invariants ρ_i.

3. Multipolar correctors for the two dimensional case

In this Section we analyze the correction problem in the case of a two dimensional phase space. In order to correct a generic non-linear symplectic map $\mathcal{M}$ (2.1) it is enough to use a counterterm whose structure is a sum of odd orders kicks

$$z' = \mathcal{M}^{(c)}(z, z^*) = e^{i\omega}\left(z + i\sum_{k\geq 1} \alpha_{2k+1}(z + z^*)^{2k+1}\right) \qquad z \in \mathbf{C}, \tag{3.1}$$

as in this case one has one tuneshift coefficient every two map orders to compensate. From a physical point of view this model corrresponds to correct the tuneshift of an accelerator represented by a two dimensional map $\mathcal{M}$ by inserting in a given point of the lattice an element which is composed by octupoles, dodecapoles, 16th poles and so on. This model, even if is very simple, can be referred to real machines: in fact in the SPS experiment of 1990 [3] octupoles were succesfully used to correct the tuneshift caused by sextupoles in conditions which simulated a two dimensional phase space (flat beam, no coupling).

The normal form tecniques allow to compute the values of the multipolar correctors α_{2k+1} which compensate the tuneshift. In order to discuss the equations that solve this problem, we first introduce the projector over the subspace of homogeneous polynomials of order n, denoted by $[\cdot]_n$:

$$[\mathcal{M}^{(c)}]_n = \mathcal{M}_n^{(c)} \tag{3.2}$$

and the projector Π_1 on the normal form subspace, defined as the set of maps which commutes with an irrational rotation:

$$\Pi_1 \mathcal{N} = \mathcal{N} \iff \mathcal{N} \circ e^{i\omega} = e^{i\omega} \circ \mathcal{N}. \tag{3.3}$$

In order to compute the coefficients of $\mathcal{M}^{(c)}$ one has to solve the functional equation of the normal forms [1]

$$[(\mathcal{M} + \mathcal{M}^{(c)}) \circ \phi - \phi \circ \mathcal{N}]_n = 0 \tag{3.4}$$

imposing the linearity of $\mathcal{N}$, i.e.

$$\mathcal{N}(\varsigma, \varsigma^*) = e^{i\omega}\varsigma. \tag{3.5}$$

Using the notations (3.2), (3.3) we write the equations which determine $\mathcal{M}^{(c)}$ and ϕ at order n; the first one reads:

$$\Pi_1 \left[\sum_{k=2}^{n} \mathcal{M}_k \circ \phi + \sum_{k=3}^{n-1} \mathcal{M}_k^{(c)} \circ \phi \right]_n + \Pi_1(\mathcal{M}_n^{(c)}) = 0 \qquad n \quad odd. \tag{3.6}$$

Having computed $\Pi_1 \mathcal{M}_n^{(c)}$ and the coefficient α_n, one solves the second equation for the conjugation function ϕ:

$$\Delta(1-\Pi_1)\phi_n = (1-\Pi_1) \left[\sum_{k=2}^{n} \mathcal{M}_k \circ \phi + \sum_{k=3}^{n} \mathcal{M}_k^{(c)} \circ \phi \right]_n \qquad n \in \mathbf{N}; \tag{3.7}$$

$\Pi_n \phi_n$ is partially determined by imposing the conditions of symplecticity on the function ϕ itself.

We now study the convergence of the correcting map; it is convenient to use the norm of the max:

$$\|\phi_n\| = \max_{j_1+\ldots+j_{2d}=n} |\phi_{j_1 \ldots j_{2d}}|. \tag{3.8}$$

Using the method of majorant series, it is possible to give a rigorous upper bound to $\|\mathcal{M}_n^{(c)}\|$; we proved a very general theorem which holds for every type of correcting map and in a phase space of arbitrary dimension:

Theorem Let $\mathcal{M}$, $\mathcal{M}^{(c)}$ be polynomial symplectic maps in $2d$ dimensional phase space; let $\mathcal{M}$ be a generic non-linear map and $\mathcal{M}^{(c)}$ chosen so that the normal form of $\mathcal{M} + \mathcal{M}^{(c)}$ is linear. Then one has

$$\|\mathcal{M}_n^{(c)}\| \leq A \frac{\epsilon_n}{\eta_n} B^n (n-2)!, \tag{3.9}$$

where $\frac{\epsilon_n}{\eta_n}$ is the contribution given by the small divisors and A, B suitable constants. We only give a sketch of the proof, referring to [12] for a more detailed exposition. The idea of the proof is to separate the contribution ϵ_n of the small divisors from the term χ_n given by the functional equations (3.6) and (3.7), according to the techniques developed for the proof of the Nekhoroshev theorem for maps [13],[14]. One assumes

$$\begin{aligned} \|\mathcal{M}_n^{(c)}\| &\leq \frac{\epsilon_n}{2\eta_n} \chi_n \\ \|\phi\| &\leq \epsilon_n \chi_n. \end{aligned} \tag{3.10}$$

The ϵ_n is defined acccording to the recurrence:

$$\epsilon_n = 2\eta_n \max_{k=2,\ldots,n-1} \epsilon_k \max_{j_1+\ldots+j_k=n} \epsilon_{j_1} \ldots \epsilon_{j_k}, \tag{3.11}$$

where η_n is the inverse of the smallest divisor at order n:

$$\eta_n = \max_{k=0,\ldots,n} |e^{i\omega(2k-n-1)} - 1|^{-1}. \tag{3.12}$$

Substituting the Ansatz (3.10) for the orders less than n in Eqs. (3.6) and (3.7) one finds the recurrence which defines χ_n in order to satisfy (3.10) at order n:

$$\chi_n = \sum_{k=2}^{n} \sum_{j_1+\ldots+j_k=n} \chi_{j_1}\cdots\chi_{j_k} + \sum_{k=3}^{n-1} \frac{\chi_k}{2\eta_k} \sum_{j_1+\ldots+j_k=n} \chi_{j_1}\cdots\chi_{j_k}. \tag{3.13}$$

The growth of the χ_n can be estimated through a straightforward but rather long inductive proof; one finds

$$\begin{aligned} \chi_n &\leq A(n-2)\chi_{n-1} \\ A &\geq \frac{3\eta_2}{\eta_2 - 1}. \end{aligned} \tag{3.14}$$

This estimate completes the proof of the theorem.

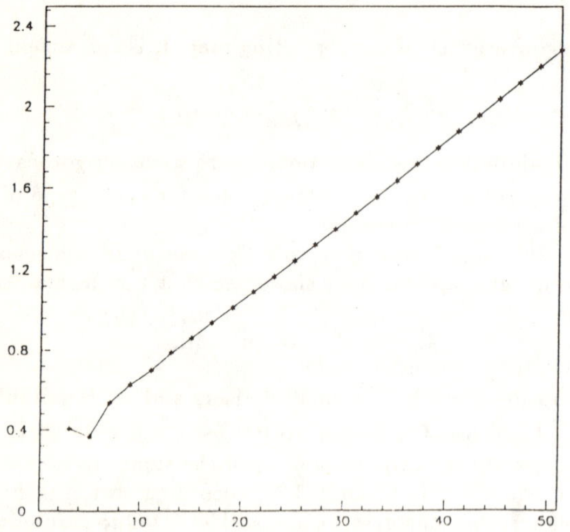

Fig. 1 - Linear growth of $(\alpha_{2n+1})^{\frac{1}{2n+1}}$ for the Henon's map with golden mean frequency

This theorem does not ensure the divergence of the correction scheme, providing only an upper bound to the growth of the $\|\mathcal{M}_n^{(c)}\|$. A numerical check was performed in the two dimensional case by using the correction procedure developed at the beginning of this Section applied to the Henon's map (i.e., a linear term plus a quadratic kick) with the linear frequency set to the golden mean value. The series of the α_{2n+1} [see Eq. (3.1)] was found to be divergent, as it is shown in Fig. 1 where we plotted the quantity $(\alpha_{2n+1})^{\frac{1}{2n+1}}$; the observed linear growth of this quantity is consistent with a factorial growth of α_{2n+1}:

$$(\alpha_{2n+1})^{\frac{1}{2n+1}} \propto (2n-1)^{\frac{2n-1}{2n+1}} \propto n. \tag{3.15}$$

This result is in agreement with the theorem, as $\|\mathcal{M}^{(c)}_{2n+1}\|$ and α_{2n+1} behave in the same way due to the formula:

$$\|\mathcal{M}^{(c)}_{2n+1}\| = \frac{(2n+1)!}{(n+1)!n!}\alpha_{2n+1} \longrightarrow_{n\to+\infty} 2^{2n+1}\alpha_{2n+1}. \tag{3.16}$$

The perfect linear dependence observed in Fig. 1 shows that the small divisors are not giving any contribution, whilst all the divergence is caused by the growth due to the non-linear functional equation which was estimated using the serie χ_n.

4. Multipolar correctors for the four dimensional case and applications to LHC

In the case of a four dimensional phase space the tuneshift structure leads to a more complicated situation: a complete compensation of each tuneshift order is not as simple as in the two dimensional case, where only one parameter for each tuneshift order is needed, and a more complex structure of the correcting map $\mathcal{M}^{(c)}$ is needed.

The tuneshift in the normal form space reads:

$$\begin{aligned} \delta_x &= \sum_{i\geq 1} \Omega^x_i(\rho_1,\rho_2) \\ \delta_y &= \sum_{i\geq 1} \Omega^y_i(\rho_1,\rho_2). \end{aligned} \tag{4.1}$$

where Ω^x_i, Ω^y_i are homogeneous polynomials of order i. The simplecticity of the normal form gives the following bounds on the derivatives of Ω:

$$\frac{\partial\Omega^x}{\partial\rho_2} = \frac{\partial\Omega^y}{\partial\rho_1} \qquad \Longleftrightarrow \qquad \Omega^x = \frac{\partial G}{\partial\rho_1}, \Omega^y = \frac{\partial G}{\partial\rho_2}, \tag{4.2}$$

where G is a polynomial in the non-linear invariants:

$$G(\rho_1,\rho_2) = \sum_{i\geq 1} G_i(\rho_1,\rho_2) = \sum_{i\geq 1}\sum_{k=0}^{i} g_{i,k}\rho_1^k\rho_2^{i-k}. \tag{4.3}$$

According to what discussed above, each tuneshift order i is characterized by $i+2$ real coefficients $g_{i+1,0},\dots,g_{i+1,i+1}$. In order to correct completely a tuneshift order i we need at least $i+2$ free parameters in the map $\mathcal{M}^{(c)}$; if we use a sum of odd multipolar kicks like in the two dimensional case (octupoles, dodecapoles, ...) we have only one free parameter to correct each tuneshift order. One can have more free parameters by using also even kicks (sextupoles, decapoles, ...) and different set of correctors inserted in different points of the lattice: this is the procedure used for planned large hadron colliders [4],[5]. We now denote by $\mathbf{K}_i$ the vector containing the free gradients of the corrector elements used to compensate the order i of the tuneshift. If the number of the components $\mathbf{K}_i$ is greater or equal to $i+2$ then one can solve the system:

$$\begin{cases} g_{i+1,i+1}(\mathbf{K}_i) = 0 \\ \dots \\ g_{i+1,0}(\mathbf{K}_i) = 0. \end{cases} \tag{4.4}$$

In this way one fixes the $\mathbf{K}_i$ in order to make the tuneshift of order i equal to zero. When one uses an order-by-order procedure the system (4.4) is linear or of low order in the gradients and can be solved analytically.

If the number of the components of $\mathbf{K}_i$ is less than $i+2$ one has to build a scalar function χ_i of the coefficients $g_{i+1,0}, \dots, g_{i+1,i+1}$ and then minimize it. A possible choice is to average the tuneshift over the sum R of the invariants:

$$\chi_i(g_{i+1,i+1}, \dots, g_{i+1,0}) = \frac{1}{R^i}\left(\int_0^R t_{2i}(\rho_1, R-\rho_1; h_{i+1}^{i+1}, \dots, h_0^{i+1})d\rho_1\right)^{\frac{1}{2}}, \tag{4.5}$$

where $t_{2i}(\rho_1, \rho_2; h_0^{i+1}, \dots, h_{i+1}^{i+1})$ is the sum of the squares of the tuneshift of order i divided by two. Also in this case the order by order procedure gives for the minimization of χ_i a set of equations that are linear or of low order in the gradients which allow an analytic computation of the solution.

The outlined method was used to compute correction schemes for the LHC [6]; a realistic model of the machine was considered, including insertions, chromaticity correction and cell asymmetry. The correction was carried out over the first two tuneshift orders using three families of correctors; different schemes were studied, considering the possibility of compensating completely only the first order or both the first and the second one. In both cases the gain in the linear aperture was more than a factor two.

5. Normal form correctors

In Sections 3 and 4 we used a serie of multipolar kicks to correct the map $\mathcal{M}$; it must be pointed out that in this way one adds not only the terms that are needed to minimize the tuneshift, but also other non-linearities. In fact, if we examine the case of a two dimensional phase space, a kick of order $2n+1$ is the sum of $2n+2$ monomials:

$$i\alpha_{2n+1}(z+z^*)^{2n+1}; \tag{5.1}$$

among these terms only $z^{n+1}z^{*n}$ affects the tuneshift of order n. One remarks that the other monomials of the multipolar kick usually have a stabilizing effect, as they correct the corresponding terms of the errors; nevertheless, it can be interesting from a mathematical point of view to analyse the possibility of correcting the tuneshift only by using terms which are in the form:

$$i\alpha_{2n+1}z^{n+1}z^{*n}, \tag{5.2}$$

i.e. monomials which belong to the normal form subspace as defined in (3.3). One cannot simply add these terms to $\mathcal{M}$, as the symplecticity condition is not fulfilled: a possible way is to compose $\mathcal{M}$ with a correcting map $\mathcal{N}^{(c)}$ which is in normal form. In this case one minimizes the tuneshift of the map

$$\mathcal{M} \circ \mathcal{N}^{(c)}; \tag{5.3}$$

the free parameters of $\mathcal{N}^{(c)}$ are enough to compensate exactly the tuneshift at each order. We write the explicit formulas in the two dimensional case, where we define the truncated normal form:

$$\mathcal{N}^{(c)}_{2n+1} = \exp\left(i \sum_{i=1}^{n} \Omega_{2i}(zz^*)^i\right) z. \tag{5.4}$$

The coefficient Ω_{2n} is computed by imposing:

$$\Pi_1[\mathcal{M} \circ \mathcal{N}^{(c)}_{2n+1}]_{2n+1} = 0 \tag{5.5}$$

whose solution is:

$$\Omega_{2n} = -ie^{-i\omega}\Pi_i[\mathcal{M} \circ \mathcal{N}^{(c)}_{2n-1}]_{2n+1}. \tag{5.6}$$

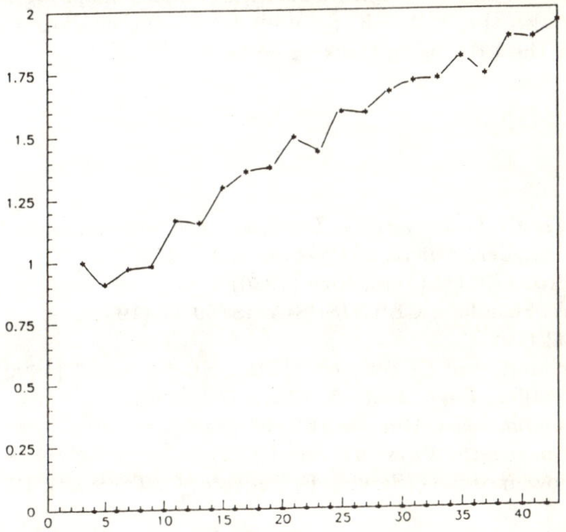

Fig. 2 - Growth of $(\|\mathcal{M}_{2n+1}\|)^{\frac{1}{2n+1}}$ for the cubic map with golden mean frequency

The growth of the coefficients of $\mathcal{N}^{(c)}$ has a factorial upper bound according to the theorem proved in section 3, which holds for every type of correcting map. A numerical analysis of the Henon's map in two dimensional phase space was carried out; the observed behaviour is a factorial divergence exactly like in the example discussed in section 3. This can be rather surprising as we are correcting the map $\mathcal{M}$ by adding the smallest possible amount of non-linearity. We also analysed a purely cubic map: the mechanism of divergence in this case is different, as expected, as it is shown in Fig. 2. The growth of $(\|\mathcal{M}_{2n+1}\|)^{\frac{1}{2n+1}}$ is no longer linear, and exhibits some dips in correspondance of the rational approximations of the frequency given by the truncated continuous fraction; this means that, contrary to the previous cases, the small divisors play an important role. However, the numerical analysis is not sufficient to decide weather the procedure is convergent in this last case, and a deeper analysis is required.

Acknowledgements

We want to thank Prof. Turchetti and Dr. Scandale for many interesting and useful discussions. Dr. Bazzani made an important contribution by introducing us to the techniques needed for the proof of the Nekhoroshev thoerem; discussions with Dr. Marmi on the correction problem and its connections with complex maps were very interesting and stimulating. We also thank Dr. F. Schmidt for having checked the validity of the correction schemes for the LHC using tracking codes.

References

[1] A. Bazzani, P. Mazzanti, G. Servizi, G. Turchetti, *Il Nuovo Cimento* **102B**, 51 (1988).
[2] E. Courant and H. Snyder, *Annals of Physics.* **3**, 1 (1958).
[3] D. Brandt *et al.*, *Proc. EPAC Conf.*, Nice (1990).
[4] F. Galluccio and W. Scandale, CERN/SPS-AMS/89-51 (1990).
[5] D. Neuffer, SSC 132 (1987).
[6] W. Scandale, F. Schmidt and E. Todesco, CERN/SL-AP/90-07 (1990).
[7] E. Forest and D. Neuffer, *Phys. Lett.* **A 135**, 197 (1989).
[8] H. Rüssmann, *Mathematische Annalen* **169**, 55 (1967).
[9] G. Gallavotti, *Comm. Math. Phys.* **87**, 365 (1982).
[10] A. Bazzani, M. Giovannozzi, G. Servizi, E. Todesco, G. Turchetti, CERN/ SPS-AMS/ 89-24 (1989).
[11] A. Bazzani, *Celestial Mechanics* **42**, 107 (1987).
[12] A. Bazzani, G. Turchetti, E. Todesco, in preparation.
[13] A. Bazzani, S. Marmi and G. Turchetti, *Celestial Mechanics*, to appear (1990).
[14] G. Turchetti, *Number Theory and Physics* **47**, 223 (1990).

DYNAMIC APERTURE OF THE FUTURE TEVATRON COLLIDER

VLADIMIR VIŠNJIĆ
Fermi National Accelerator Laboratory
Batavia, IL 60510, USA

ABSTRACT

I study the dynamic aperture of four Tevatron Collider lattices by using computer simulation (single particle tracking) method with the particular emphasis on predicting the dynamic aperture for the next run which will have two interaction regions (B0 and D0) and two low beta straight sections. The results of tracking calculations indicate that in the first phase of the ongoing upgrade ($\beta_* \approx 50$cm at both B0 and D0 interaction regions), dynamic aperture of Tevatron should increase by about factor of two, however for the second phase ($\beta_* \approx 25$cm) it is predicted to decrease by 40% compared to the present one.

1. INTRODUCTION

In 1991 the Tevatron Collider is scheduled to go into operation with two interaction regions – B0 and D0 – at the same time providing higher luminosity than in the past. In order to achieve this goal, new low beta quadrupoles are going to be installed in both D0 and B0 straight sections. This should push the Tevatron performance to a new limit (β^* down to 25cm, β_{max} up to about 1500m) and the question is whether the machine can reasonably operate at such conditions. This question is usually phrased in terms of the dynamic aperture of the machine.

In this talk I will describe the dynamic aperture of four different Tevatron lattices which I shall call, for want of better names, by the year they were (or are intended to be) used. They are based on two physically different configurations and the lattice is then defined by specifying the configuration and the currents (*i. e.* the magnetic fields). The situation is summarized in the following table:

Name	ν_x	ν_y	Interaction Regions	β_x^*	β_y^*	$\mathcal{L}$ [$cm^{-2}sec^{-1}$]
1987	19.43	19.41	B0	65 cm	73 cm	1.1×10^{29}
1988/89	19.43	19.41	B0	51 cm	58 cm	2.06×10^{30}
1991	20.58	20.59	B0 and D0	42 cm	40 cm	5×10^{30}
Post '91	20.58	20.59	B0 and D0	25 cm	25 cm	$> 10^{31}$

Table 1: Some characteristics of the Tevatron lattices studied in this paper.

The talk is organized as follows. In Section 2, I describe the Tevatron model used in the numerical tracking and the method used to evaluate the aperture and the acceptance. The results (Section 3) indicate that the phase one future low beta lattice should have about twice the acceptance of the 1988/89 lattice, however attempting to reach the designed value for β_* of 25cm might lead to problems. The predicted value of the acceptance is about 40% smaller than that of the present Tevatron lattice. The main reason for this is the large value of β_{max} which effectively creates a bottleneck at low beta quadrupoles. Section 4 contains a brief discussion of short term and long term tracking in comparison with an experiment done at Tevatron with the original low beta lattice. In Section 5, I compare the tracking results with the experimental measurements of the dynamic aperture done at Tevatron in 1985. Section 5 contains the computation of the driving terms of low level resonances for the new low beta quadrupoles.

2. TRACKING

The tracking is done using the code TEVLAT[1] on the Fermilab Amdahl. The Tevatron is modelled as follows. First there are the linear elements of the lattice–the dipoles and the quadrupoles. Their strength is determined from the energy and the tune at which we want to do the tracking. Then the chromaticity sextupoles are turned on in order to compensate for the chromaticity of the linear lattice. Finally, the magnet errors in the form of higher multipole coefficient are read from the file containing the actual measurement data for each element. The new low beta quadrupoles have not yet been manufactured and in the simulation they are assigned the same multipole coefficients as the quadrupoles they are going to replace. In other words, we are using the same table of multipole coefficients for all four lattices of Table 1. TEVLAT treats the linear and the nonlinear elements differently— in the linear elements the actual trajectory is computed, while the nonlinear elements act as *kicks*, *i. e.* the instantaneous changes of x' and y'. The kicks due to the chromaticity sextupoles and the higher multipole moments of the dipoles and quadrupoles occur in the middle point of each element. If we wish to consider the linear lattice only, as we occasionally do in this paper, we shall set the sextupole strength and the higher multipole coefficients equal to zero. For most part of this study tracking is done in the four dimensional phase space. An exception is the tracking 10^5 turns in Section 4 where the synchrotron oscillations have been included.

In order to determine the dynamic aperture by tracking, we basically have to launch a particle at the initial coordinates (x_0, x'_0, y_0, y'_0) in the four dimensional phase space, let it propagate and define it as lost if it crosses a specified boundary in the xy plane. However, it would be both very expensive and impractical to treat the problem in its generality, *i. e.* to examine the full phase space at a large number of turns at many points. Rather, we scan a particular subspace of the phase space (*e. g.* the xy plane) and track for a relatively small number of turns (here 256). We must then address the question of how relevant the results are for the real machine.

Let us look for a moment at the question of tracking for a low number of turns. The plot in Fig. 1*a* shows the number of turns the particle survives as a function of the initial amplitude along the line $y_0 = 0$, $x'_0 = y'_0 = 0$. The particle launched at $x_0 = 5.26$mm survives 450 turns and at $x_0 = 4.8$mm or less more than 10^5 turns. If we are interested in the beam stability on the time scale of the order of 10 hours, we ought to plot these data as in Fig. 1*b*. These figure clearly shows the risk we take deriving conclusions from computer simulations. A simple straight line extrapolation translates into the reduction of about 40% in the aperture as we go from a few thousand to a few billion turns.

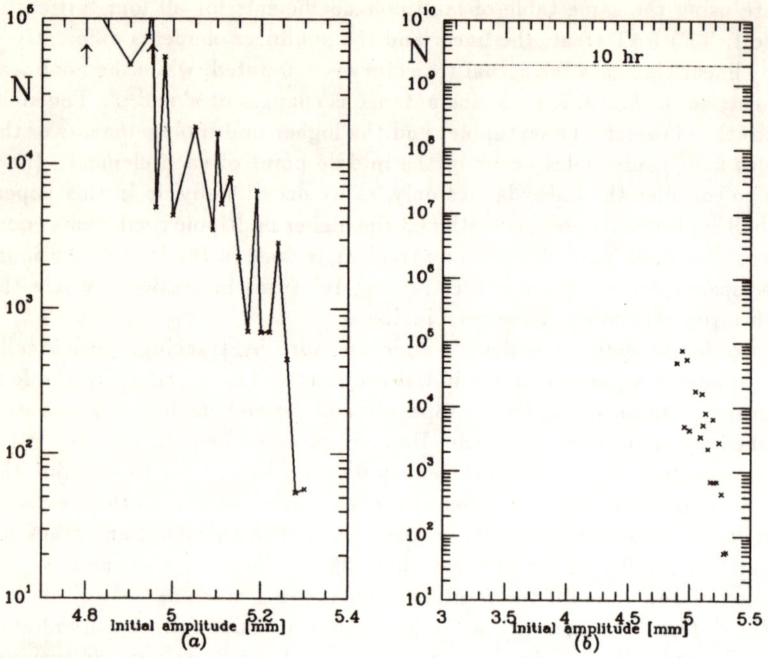

FIGURE 1 (*a*)The number of turns the particle survives as function of x_0 ($y_0 = 0$ and $x'_0 = y'_0 = 0$). The vertical arrows indicate that the particle with these amplitudes remained stable for 120 000 turns. It took about 7 hrs CPU time on the Amdahl to obtain the results shown here.(*b*) The same results plotted on the scale relevant for the Collider operation.

3. THE RESULTS

The results of the numerical simulation are shown in Figs. 2. What is plotted here is the largest initial amplitude for which the particle survives 256 turns. Since the particle is launched at a point s_0 where $\alpha \neq 0$, the amplitude is obtained as $x_0\sqrt{1+\alpha^2(s_0)}$ and normalized as usual by $\sqrt{\beta(s_0)}$. In other words, we consider the quantities

$$\zeta_x = x(s)\sqrt{\gamma_x(s)} \quad \text{and} \quad \zeta_y = y(s)\sqrt{\gamma_y(s)} \tag{3.1}$$

which are supposed to be independent of s and thus invariant for a given machine. In this formula

$$\gamma_x(s) = \frac{1+\alpha_x(s)^2}{\beta_x(s)} \tag{3.2}$$

and analog for γ_y. The dynamic aperture thus defined has the dimension of square root of length, but has the property of being linear in the maximal amplitude the machine can take. Since the natural unit for the amplitude is millimeters and for β meters, we see that for the aperture ζ it is square root microns. In order to obtain the maximal amplitude in millimeters at a given point of the accelerator from ζ, Fig. 2, one has to multiply ζ expressed in $(\text{microns})^{1/2}$ with $\sqrt{\frac{\beta[\text{meters}]}{(1+\alpha^2)}}$ at that point. The physical aperture (the beam pipe) is taken to be square with side 70 mm and is represented on the plots by a dotted line rectangle with dimensions $70\text{mm}/\sqrt{\beta_{xmax}} \times 70\text{mm}/\sqrt{\beta_{ymax}}$.

For each lattice, the tracking is done with and without the nonlinear elements, in particular the higher multipoles in the dipoles and quadrupoles (magnet errors). The linear lattice contains only the ideal linear elements, *i. e.* dipoles and quadrupoles and its dynamic aperture should coincide with the physical one. The full lattice contains in addition the chromaticity sextupoles and higher multipoles of the dipoles and quadrupoles.

In Figure 2, the tracking results for the lattices of Table I are shown. The open and solid squares represent the linear and the full nonlinear lattice, respectively.

Note that the aperture of the 1987 lattice, Fig. 2, is almost identical with the physical aperture. The apertures of the 1988/89 and 1991 lattices are still dominated by their physical apertures, and only for the Post'91 lattice do we see the true dynamic aperture inside the physical one. Comparing the *horizontal* dynamic apertures of the four lattices normalized to the 1988/89 one we obtain the ratio

$$1987 : 1988/89 : 1991 : \text{Post'91} = 1.15 : 1 : 1.46 : 0.82. \tag{3.3}$$

While the ζ's are convenient when we want to compare the dynamic aperture with the physical one, the *acceptance* is a more convenient quantity when we wish to compare it with the beam emittance. If we wish to express the results in terms of the acceptance we have only to take the square of the dynamic aperture multiplied by π:

$$A_x = \pi\zeta_x^2 \quad \text{and} \quad A_y = \pi\zeta_y^2. \tag{3.4}$$

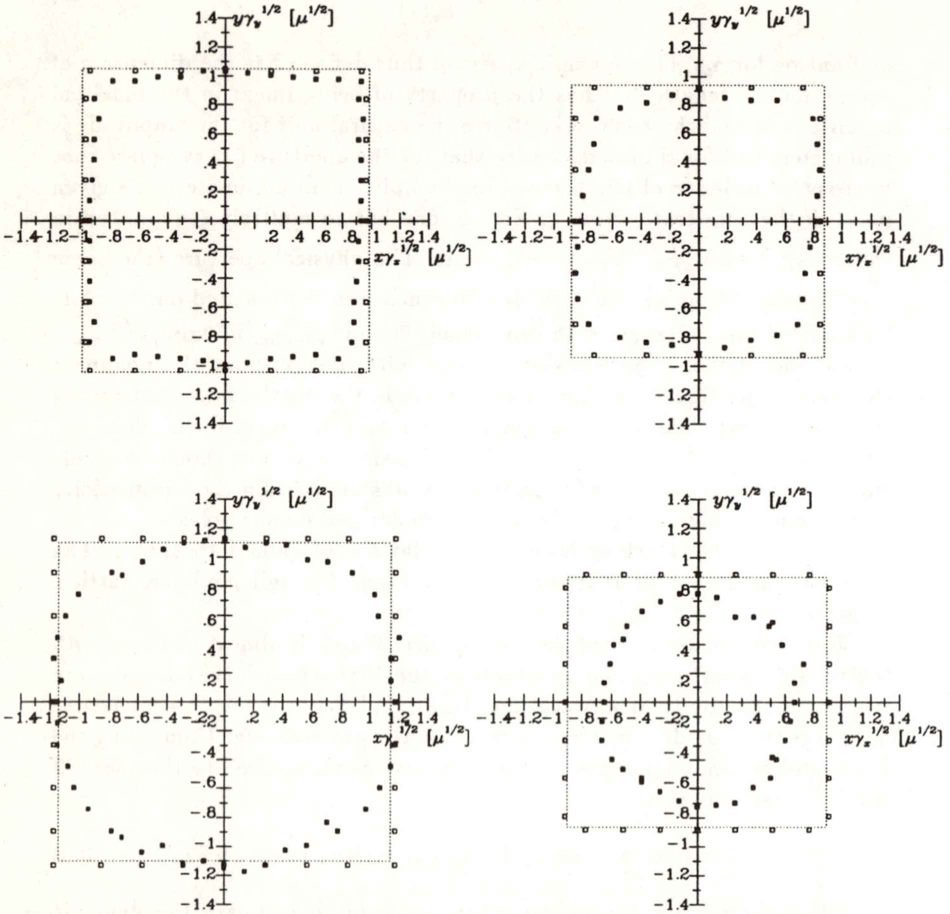

FIGURE 2 The dynamic aperture of the lattices from Table I. The open and solid squares denote the linear and the full lattice, respectively. The dotted rectangle is the physical aperture with the dimensions $70\text{mm}/\sqrt{\beta_{xmax}} \times 70\text{mm}/\sqrt{\beta_{ymax}}$.

In terms of the acceptance, the ratios of Eq. (3.3) read

$$1987 : 1988/89 : 1991 : \text{Post'91} = 1.32 : 1 : 2.13 : 0.67. \quad (3.5)$$

The graphs A_x *vs.* A_y are shown in Fig. 3.

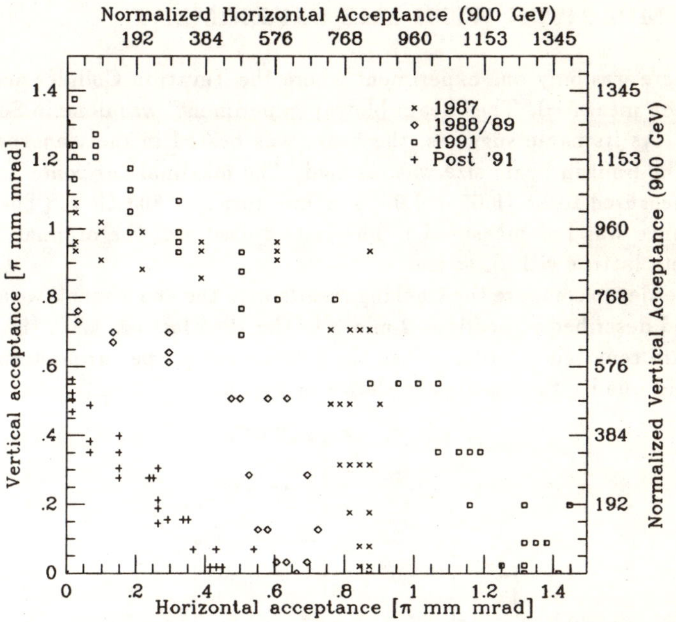

FIGURE 3 Vertical *vs.* horizontal acceptance for the four Tevatron lattices studied here.

In practice the *normalized* emittance which does not change with the beam energy ($\epsilon_N = \beta\gamma\epsilon$) is used more often than the emittance. The acceptance A of Eq. (3.4) is of course a characteristic of the given machine and has nothing to do with the beam energy. However, if we want to use the normalized emittance when talking about the beam, we should use the *normalized acceptance* to describe the machine,

$$A_N = \beta\gamma A. \quad (3.6)$$

The normalized acceptance can be thought of as the largest single particle normalized emittance for which the particle remains stable. Note that the normalized acceptance, unlike the normalized emittance, is not independent of the beam energy. To obtain the values for the normalized acceptance one has to multiply the value from Fig. 3 with $\beta\gamma$ of the beam. The values of the normalized emittances for the beam energy of 900 GeV are given in Fig. 3.

4. COMPARISON WITH THE EXPERIMENT

There was only one experiment where the Tevatron Collider acceptance has been measured. The "beam blowup experiment" was done in September 1985[2]. As its name suggests, the beam was heated in the transverse plane until the limit in beam size was reached. The maximal *horizontal* emittance was measured to be $(0.65 \pm 0.05) \times \pi$ mm mrad at 800 GeV. (The vertical emittance was not measured.) The lattice used was the original Tevatron low beta lattice with $\beta_* \approx 1$m.

In order to compare the tracking results with the experiment, we apply the method described in Sections 2 and 3 to the 1985 lattice. With the physical aperture removed in order to see the true dynamic aperture determined by the field quality the results are shown in Fig. 4.

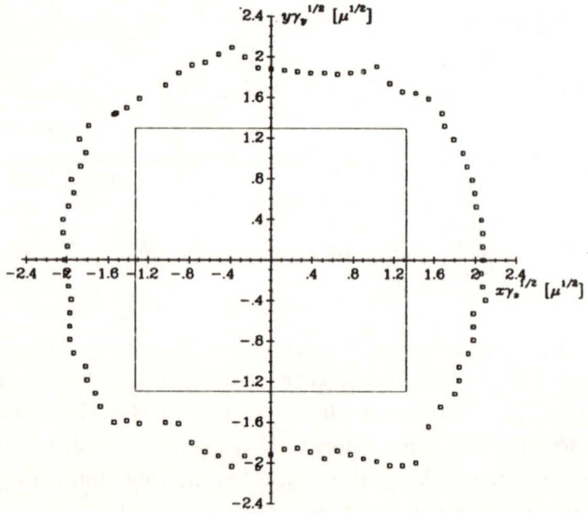

FIGURE 4 The dynamic aperture of the 1985 Tevatron lattice. The tracking was done with the physical aperture in the low beta straight section (the dotted line) removed. Note that the scale is different from that in the previous figures.

Before we compare these results with the experimental data, we should remark that they do not correspond to the same time scale—the experiment was done over a few hours or $O(10^8)$ turns, the tracking over 256 turns. We can improve our tracking results to some extent by tracking again for 100 000 turns along the positive x_0 axis and include the synchrotron oscillations. The results are similar to those of Fig. 1 and are shown in Fig. 5. Finally, the results of the tracking and the experimental values are summarized in Fig. 6.

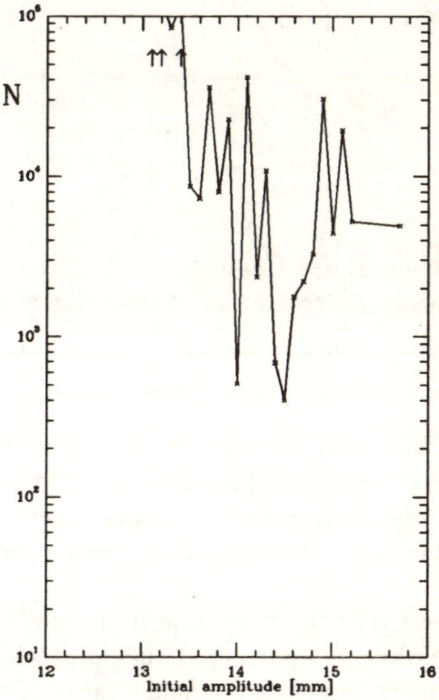

FIGURE 5 The number of turns the particle survives as function of x_0 (y_0 = 0 and $x'_0 = y'_0 = 0$). The vertical arrows indicate that the particles with these initial amplitudes remained stable for more than 100 000 turns. The synchrotron oscillations have been included in the computation.

The tracking results seem to overestimate the acceptance of the machine by approximately a factor of two as compared with the measurement. However, we should stress at this point that more than one piece of experimental data is needed before we can take this conclusion seriously. If we nevertheless reexamine the predictions of Fig. 3 in view of this uncertainty, we conclude that the normalized acceptance at 900 GeV of the 1991 lattice could go down to 700 π mm mrad and that of the Post'91 lattice down to 200 π mm mrad.

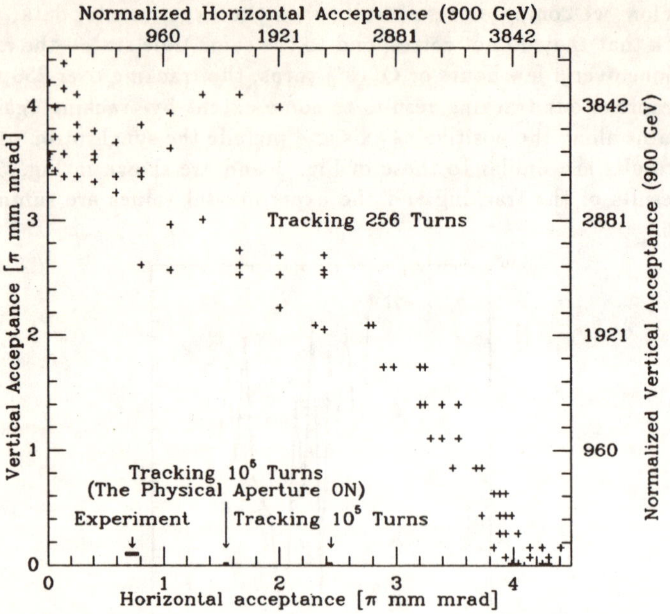

FIGURE 6 Vertical *vs.* horizontal acceptance for the original Tevatron low beta lattice. The open squares denote the results of the tracking of 256 turns with the physical aperture removed, the solid ones the results of the tracking of 10^5 turns with the physical aperture both removed and in place. Also shown are the results of a measurement.

The tracking results assert that the most important single factor limiting the dynamic aperture is the physical aperture at the point where $\beta = \beta_{max}$ (*i. e.* in low beta quadrupoles), rather than the field quality of those quadrupoles. This prediction can be tested experimentally by doing a beam blowup experiment accompanied by beam loss measurements in the vicinity of the points where $\beta = \beta_{max}$ (there are two such points at each B0 and D0).

5. COMPUTATION OF THE DRIVING TERMS

An estimate of what effect higher multipoles of the magnets can have on the beam can be obtained by computing the driving terms for the resonances. In this Section, the driving terms are computed for the low beta quadrupoles only, the rest of the machine is linear. The resonance width is given by[3]

$$\begin{aligned} G &= \left(\frac{-i}{\sqrt{2}}\right)^{m_x+m_y} \frac{1}{m_x+m_y} \binom{m_x+m_y}{m_x} \frac{B_0 R}{|B\rho|} \\ &\times \int \frac{d\theta}{2\pi} \left\{ \begin{array}{l} (-1)^{m_y/2} b_{m_x+m_y-1} \\ (-1)^{(m_y+1)/2} a_{m_x+m_y-1} \end{array} \right\} \beta_x^{m_x/2}(\theta) \beta_y^{m_y/2}(\theta) \exp i(\mathbf{m}\cdot\boldsymbol{\psi} - \Delta\theta). \end{aligned} \tag{5.1}$$

Here $\Delta = \mathbf{m}\cdot\nu - n$ is the distance from the resonance in tune space and the b (a) coefficients are chosen for m_y even (normal multipoles) or odd (skew multipoles), respectively. The relevant quantities are

$$\Delta_{crit} = G I^{\frac{m_x+m_y}{2}-1} \tag{5.2}$$

and the ratio of the resonance width and the distance from it,

$$\frac{\Delta_{crit}}{\Delta}, \tag{5.3}$$

where I is the amplitude, equal to $\epsilon_N/2\pi\beta\gamma$ where ϵ_N is the normalized emittance. If we replace the functions in the integrand by their averages*) the integral can be evaluated as the sum

$$\begin{aligned} G &= \left(\frac{-i}{\sqrt{2}}\right)^{m_x+m_y} \frac{1}{2\pi} \frac{1}{m_x+m_y} \binom{m_x+m_y}{m_x} \frac{1}{|B\rho|} \\ &\times \sum_{i=1}^{36} l_i \mathrm{grad}(i) \left\{ \begin{array}{l} (-1)^{m_y/2} b_{m_x+m_y-1}/b_1 \\ (-1)^{(m_y+1)/2} a_{m_x+m_y-1}/b_1 \end{array} \right\} \beta_x^{m_x/2}(i) \beta_y^{m_y/2}(i) \exp i(\mathbf{m}\cdot\boldsymbol{\psi} - \frac{\Delta s_i}{R}). \end{aligned} \tag{5.4}$$

---

*) Since we are dealing with positive definite functions, we approximate the average of the product by the product of the averages.

The summation is over the 18 low beta quads at each B0 and D0, $l_i = R\theta_i$ is the length of the i–th quad, s_i its position and grad$(i) = B_0 b_1$ its gradient. In deriving Eq.(5.4) from Eq.(5.1) we have multiplied and divided the expression on the right hand side by b_1. Then $B_0 b_1(i) \equiv$ grad(i) = the quadrupole field of magnet i and the harmonic coefficients $a_1, \ldots, a_8$ and $b_2, \ldots, b_{10}$ are normalized to b_1. The measured values of the harmonic coefficients vary in sign and in magnitude and in reality there will be cancellations among different magnets. This implies that the computed value of G may be larger or smaller than the real one depending on the ordering of the magnets in the ring. For this reason, in this calculation they are assigned the *largest absolute values* of the coefficients measured for each of the nine existing low beta quads at B0. In other words, we compute the largest value for G for the given multipoles. Another possible source of cancellations is the fact that the real and imaginary parts in Eq.(5.4) can be positive as well as negative and cancellations are likely to occur. The worst case scenario is that the actual signs of $a_1, \ldots, a_8$ and $b_2, \ldots, b_{10}$ exactly match the signs of the real and the imaginary parts and the contributions of all the magnets add up. Again, to make sure that the results of the calculation are on the safe side, in this computation we replace the real and the imaginary parts in Eq.(5.4) with their absolute values. Since we are interested in the absolute value of the resonance width, the overall phase is discarded and we compute separately the real and the imaginary part of G:

$$
\begin{aligned}
\mathrm{Re}G &= \left(\frac{1}{\sqrt{2}}\right)^{m_x+m_y} \frac{1}{2\pi} \frac{1}{m_x+m_y} \binom{m_x+m_y}{m_x} \frac{1}{|B\rho|} \\
&\times \sum_{i=1}^{36} l_i \mathrm{grad}(i) \left\{ \begin{array}{l} (-1)^{m_y/2} b_{m_x+m_y-1}/b_1 \\ (-1)^{(m_y+1)/2} a_{m_x+m_y-1}/b_1 \end{array} \right\} \beta_x^{m_x/2}(i) \beta_y^{m_y/2}(i) |\cos(\mathbf{m}\cdot\psi - \frac{\Delta s_i}{R})|.
\end{aligned} \tag{5.5a}
$$

$$
\begin{aligned}
\mathrm{Im}G &= \left(\frac{1}{\sqrt{2}}\right)^{m_x+m_y} \frac{1}{2\pi} \frac{1}{m_x+m_y} \binom{m_x+m_y}{m_x} \frac{1}{|B\rho|} \\
&\times \sum_{i=1}^{36} l_i \mathrm{grad}(i) \left\{ \begin{array}{l} (-1)^{m_y/2} b_{m_x+m_y-1}/b_1 \\ (-1)^{(m_y+1)/2} a_{m_x+m_y-1}/b_1 \end{array} \right\} \beta_x^{m_x/2}(i) \beta_y^{m_y/2}(i) |\sin(\mathbf{m}\cdot\psi - \frac{\Delta s_i}{R})|.
\end{aligned} \tag{5.5b}
$$

The computation is done for both normal and skew multipoles (from sextupole to 22-pole) at the beam energies of 150 GeV and 1 TeV. The results are shown in Table 2 for those resonances having the width 10^{-4} or larger.

m_x	m_y	$m_x + m_y - 1$	Δ	Δ_{cr}	Δ_{cr}/Δ
E=150 GeV					
NORMAL MULTIPOLES					
3	0	2	0.260	9.290×10^{-4}	3.573×10^{-3}
1	2	2	0.240	3.739×10^{-4}	1.558×10^{-3}
2	2	3	0.340	7.450×10^{-5}	2.191×10^{-4}
3	2	4	0.080	1.350×10^{-5}	1.688×10^{-4}
SKEW MULTIPOLES					
1	1	1	0.170	1.558×10^{-3}	9.167×10^{-3}
2	1	2	0.250	2.385×10^{-3}	9.543×10^{-3}
3	1	3	0.330	3.117×10^{-4}	9.447×10^{-4}
4	1	4	0.090	1.478×10^{-5}	1.642×10^{-4}
0	3	2	0.230	8.221×10^{-4}	3.574×10^{-3}
2	3	4	0.070	8.851×10^{-6}	1.264×10^{-4}
E = 1TeV					
NORMAL MULTIPOLES					
3	0	2	0.260	1.400×10^{-4}	5.388×10^{-4}
1	2	2	0.240	5.638×10^{-5}	2.349×10^{-4}
SKEW MULTIPOLES					
1	1	1	0.170	2.349×10^{-4}	1.382×10^{-3}
2	1	2	0.250	3.597×10^{-4}	1.439×10^{-3}
3	1	3	0.330	4.701×10^{-5}	1.424×10^{-4}
0	3	2	0.230	1.239×10^{-4}	5.390×10^{-4}

Table 2: The resonance widths for the most prominent resonances, *i. e.* those having $\Delta_{cr}/\Delta > 10^{-5}$.

REFERENCES

[1] Code written by A. Russel and modified by N. Gelfand, unpublished.
[2] Experiment done by D. Finley and R. Johnson, unpublished.
[3] See, *e. g.*, L. Michelotti, *Physics of Particle Accelerators I*, AIP Conference Proceedings 153 (1987), page 236.

NONLINEAR DYNAMICS WITH SEXTUPOLES IN LOW-EMITTANCE LIGHT SOURCE STORAGE RINGS

R. NAGAOKA*, K. YOSHIDA AND M. HARA
RIKEN-JAERI Synchrotron Radiation Facility Design Team
2-28-8 Honkomagome, Bunkyo-ku, Tokyo, 113 Japan

ABSTRACT

The third generation light source storage rings, which store the low-emittance electron beam, become highly nonlinear due to the strong sextupole fields intentionally introduced to correct the large chromatic aberration. These sextupoles severely limit the transverse stability of a single particle motion. It becomes therefore of crucial importance for storage ring designs to analyze the nonlinear system to obtain a large dynamic aperture. A familiar optimization scheme is reviewed from which a computer code is developed to generalize the procedure. Numerical studies with particle tracking stress the importance of understanding the higher-order effects of sextupoles that dominate the large amplitude motions. Two approaches are discussed in this context. One is to interprete the phenomena as a combined effect of lowest-order resonances, which is verified by a fictious tracking. Another is the perturbative method. Employing the theory developed by Poincare and von Zeipel, the formulation is made up to 4th-order perturbation to analyze the horizontal motion. Examples are shown with a practical low-emittance ring in which the theory reproduces satisfactorily the phase space trajectories observed in the tracking.

1. Introduction

A major goal of the third generation synchrotron light sources is to produce a highly brilliant synchrotron radiation from a ring which stores the low-emittance electron beam and which is provided with a number of dispersion free straight sections for insertion devices. Generally, the low-emittance is achieved by a strong horizontal focusing at every dipoles and by increasing the number of dipoles in the ring. These two aspects, however, bring about a large negative consequence in the chromaticity correction: The larger chromaticities due to the stronger quadrupoles, and the smaller dispersions due to the smaller bending angle inevitably call for the strong sextupoles to compensate the chromaticity. A serious drawback of this is the drastic reduction of the dynamic aperture. Although the origin of nonlinearity is explicitly known, it becomes therefore of great importance in the low-emittance storage ring design to analyze the nonlinear particle motion and to find a cure to the dynamic aperture.

A familar correction scheme is to re-optimize the sextupole fields by adding some families of sextupoles in the dispersion free sections. Since the storage rings usually consist of a number of identical cells, an effective approach is to make a harmonic analysis in the unit cell. One then attempts to compensate particular harmonics that drive the lowest-order resonances which make the motion unbounded. This has been shown to be quite successful in the case of a Chasman Green (CG) lattice.[1,2] One

* Present address: Sincrotrone Trieste, Padriciano 99, 34012 Trieste, Italy

of our aims is to develop a computer code to enable a systematic treatment of such optimization.

There are, however, many cases where the nonlinear motion is dominated by higher-order effects for which the former simple analysis loses its validity. Namely, one needs to analyze the higher-order effects in order to obtain a prescription of the optimization. Two approaches are made to understand the phenomena observed in the particle tracking. One is to consider higher-order effects as a combined effect of lowest-order resonances. The underlying idea is that if one could identify the harmonics that contribute to the higher-order effect, one might find a way to optimize the field. A fictitious tracking in harmonic expanded fields is made to verify this picture. Another is the perturbative method. The theory of Poincare and von Zeipel is employed to describe the horizontal motion.[3] We stress that a systematic formulation is made up to 4th-order perturbation. Comparison of phase space structure will be made between the theory and the tracking in the application to a practical light source ring.

2. The Optimization Scheme

The Hamiltonian describing the betatron oscillations in a ring with sextupoles can be expressed in terms of actions I_u and angles ϕ_u (u = x and y) as

$$H = \nu_x I_x + \nu_y I_y + (2I_x)^{3/2} \sum_{j=1,\,3} \sum_{m=-\infty}^{+\infty} \gamma_j A_{jm} \cos(j\phi_x - m\theta)$$

$$- 3(2I_x)^{1/2}\, 2I_y \sum_{m=-\infty}^{+\infty} [2B_{1m} \cos(\phi_x - m\theta) + B_{+m} \cos(\phi_+ - m\theta) + B_{-m} \cos(\phi_- - m\theta)], \quad (1)$$

with

ν_u : unperturbed betatron tune,

$\phi_\pm = \phi_x \pm 2\phi_y$,

$\gamma_j = 3 \ (j = 1)$,

$= 1 \ (j = 3)$,

$\theta = 2\pi s/C$ (C: cell length).

We assume here that the ring is composed of a super periodic cell, with $\theta = 0$ position satisfying the reflective symmetry of the structure. The first two terms in Eq. 1 express the linear part, and the remainings are the sextupole potential expanded into harmonics in which sextupoles are treated as thin lens. Coefficients of harmonics such as A_{jm} (j = 1, 3; $-\infty < m < +\infty$) depend on sextupole strengths, betatron functions and phases at the sextupoles.

From the harmonic expanded form of the potential, one finds that the particle can be driven by resonances such as $j\nu_x$ = integer and $\nu_x \pm 2\nu_y$ = integer. While, it is well known that when the betaron motion is close to a particular resonance, a good description can be made with a time-independent Hamiltonian in a rotating frame which keeps only the corresponding driving term.[4] Applying this so called "single resonance approximation" to the present system, one can argue, through fixed points analysis, the maximum stable amplitudes limited each of the resonances.

We have developed a computer code, "CATS" to perform the sextupole optimization, fully utilizing the single resonance ayalysis.[5] The procedure is as follows; Starting from the optics calculation, the code corrects the chromaticity, and performs the tracking to determine the dynamic aperture. The code then computes the boundaries of stable motion brought about by each of the harmonics which are compared with the dynamic aperture on the graphics. Also provided are the second-order tune shifts with amplitude and their harmonic decompositions. One then defines, in terms of these quantities, a function to be least-square fitted. The strength of new families of sextupoles in the disperion free section are determined by the fitting procedure. The whole process is to be iterated, by trial and errors, until a satisfactory solution is found.

We show an example with a Chasman Green lattice (Fig.1). In this case the

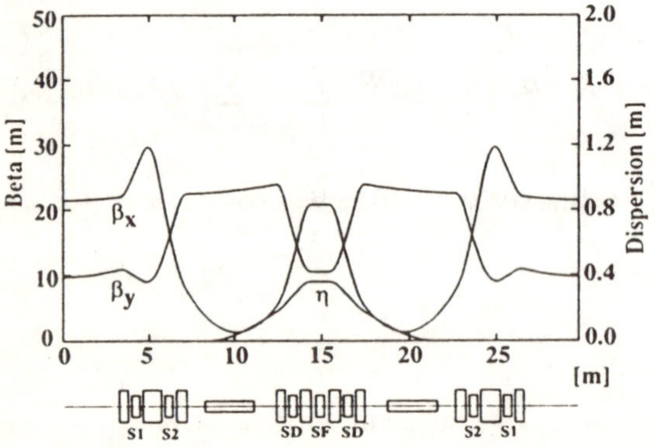

Fig. 1. Optics functions over a unit cell of CG lattice.

double horizontal focusing at dipoles to lower the emittance brings the horizontal tune across a cell close to unity. Having the chromaticity correcting sextupoles, the particle motion is strongly limited by the single resonance $\nu_x = 1$, which can be clearly seen in Fig. 2. The prescription of optimization to be given may be to reduce the magnitude of harmonics A_{11} and A_{33} by additional sextupoles, which also helps in reducing the tune shift with amplitude.[1,2] The resultant improvement of dynamic aperture is shown in Fig. 3.

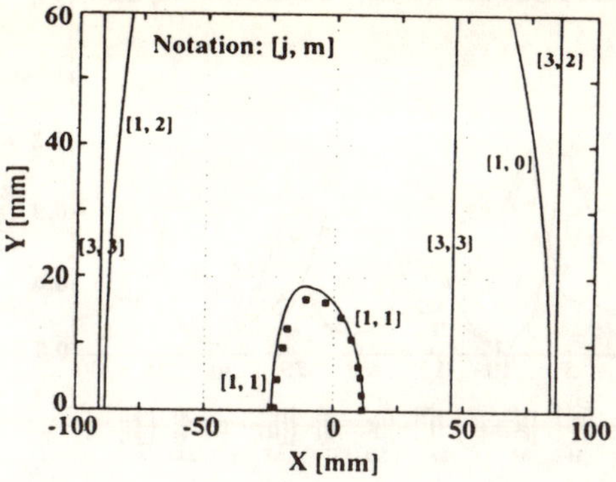

Fig. 2. Dynamic aperture of CG lattice (dark squares) in comparison with the boundaries of stable motion predicted in the lowest-order single resonance approximation (solid lines). Coupling resonances are not considered.

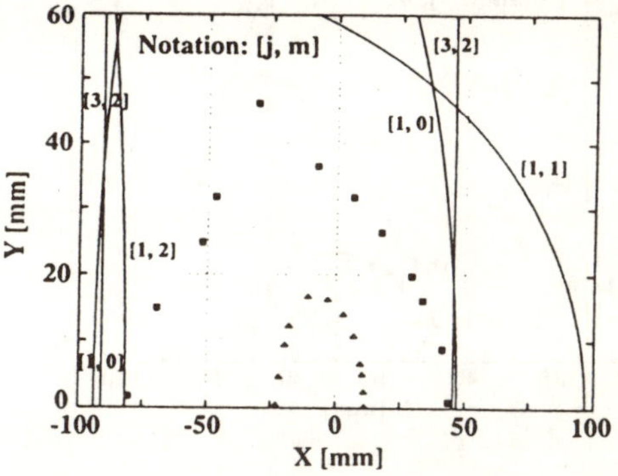

Fig. 3. Same as Fig. 2, obtained after the optimization. For comparison, dynamic aperture before the optimization are also plotted (triangles).

3. Analysis of Higher-Order Effects

3.1 *Particle Tracking in Harmonic Expanded Sextupole Field*

We now consider a Triple Bend Achromat (TBA) lattice to present another example of the analysis in the low-emittance ring (Fig. 4). Figure 5 illustrates the dynamic aperture of the lattice with chromaticity correcting sextupoles, in comparison with those predicted in the single resonance approximation. Although the boundary of [j, m] = [1, 1] component somewhat simulates the dynamic aperture, the discrepancy is

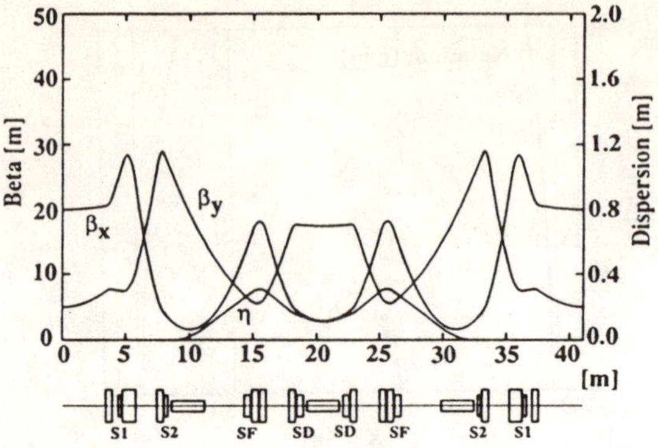

Fig. 4. Optics functions over a unit cell of TBA lattice.

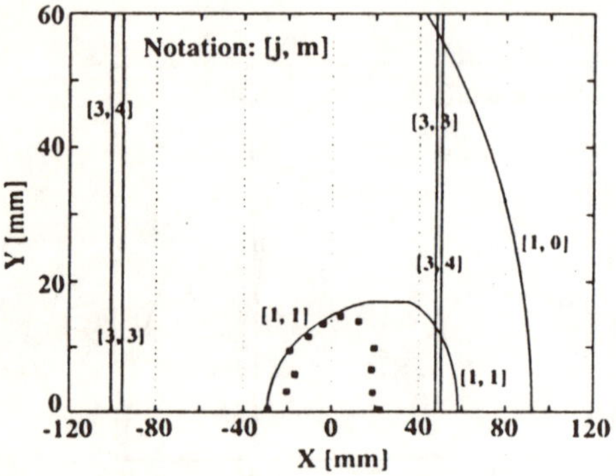

Fig. 5. Same as Fig. 2 for TBA.

large to conclude that the stability is solely determined by the resonance $\nu_x = 1$. In fact, the horizontal phase space trajectories obtained from the tracking indicate the excitation of 7th-order resonance close to the stability limit (Fig. 6). Bearing in mind the feedback of the analysis to the previous optimization using the harmonics, we shall take the view that the observed phenomenon is a combined effect of several different resonance driving terms of the lowest order (Eq. 1), influencing simultaneously the motion.

Since one can no longer make use of the rotating frame to eliminate the time dependence of the Hamiltonian, we shall rely on a numerical approach: We shall analyze the motion by tracking a particle in a "fictitious" sextupole field composed only of certain harmonics. .Provided that the observed motion is reproduced in this way, one

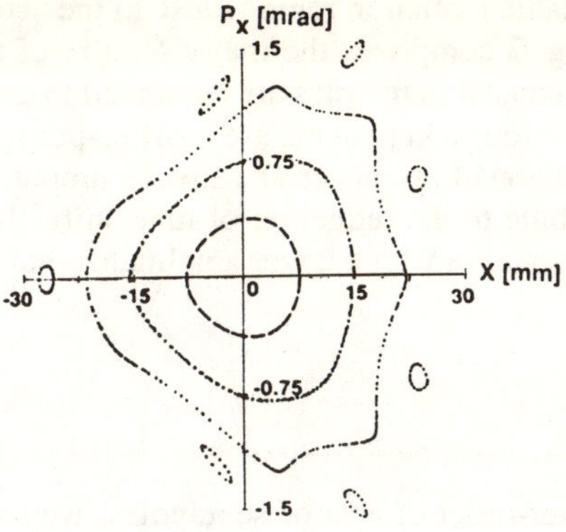

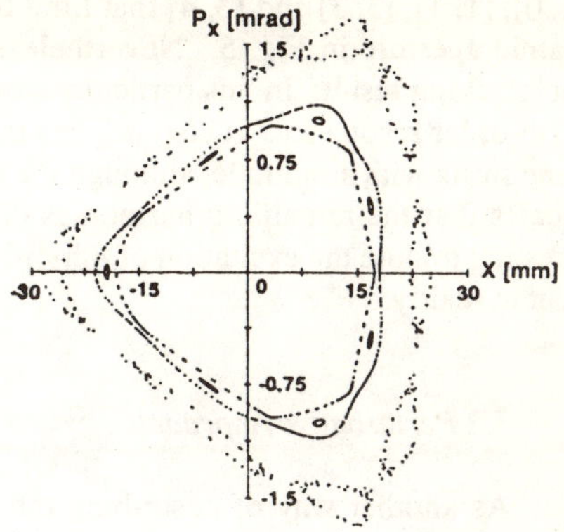

Fig. 6. (Upper Left) Horizontal phase space trajectories of TBA close to a 7th-order resonance, obtained from the tracking.

Fig. 7. (Upper Right) Horizontal phase space trajectories of TBA, obtained from the fictitious tracking.

has identified the harmonics that determine the dynamics. To this end, we transcript the Hamiltonian of Eq. 1 in terms of

$$u = \sqrt{2I_u}\cos\phi_u, \quad p_u = -\sqrt{2I_u}\sin\phi_u \quad (u = x, y) \tag{2}$$

as

$$H(x, p_x, y, p_y; \theta) = \frac{p_x^2}{2\mu_x} + \frac{p_y^2}{2\mu_y} + V(x, p_x, y, p_y; \theta), \tag{3}$$

where $\mu_u = \nu_u^{-1}$. The sextupole part of the potential consists of ten different cubic terms of u and p_u, each having a time-dependent coefficient which is expressed as an infinite sum over the products of harmonics A_{jm} etc. and the trigonometric functions.

Out of the infinite sum over m, our aim is to retain only those interested ones. This requires us to carry out the numerical integration at every point in the ring, since harmonic expanded fields are, by themselves, distributed everywhere. To guarantee the symplecticity of the integration, we adopt the technique of Ref. 6. The motions obtained by including a single harmonic are found to agree with the analytical ones, which serve as a check of the computation.

We show in Fig. 7, the phase space trajectories in TBA, calculated by the fictitious tracking. In the computation we naively included only four components [j, m]

= [1, 0], [1, 1], [3, 3] and [3, 4] that limit the stable motion in region close to the actual dynamic aperture in Fig. 5. Nevertheless, Fig. 7 comprises the major feature of the "real" tracking result: In this particular case, effects of harmonics are combined to drive the 7th-order resonance. It also implies that the four components are most responsible to tune shifts with amplitude, although the excitation of resonance at a smaller amplitude indicates that the remaining harmonics contribute to the reduction of tune shift. It is interesting to note the excitation of 8th-order resonance with a larger amplitude which is absent in reality.

3.2 *Perturbative Approach*

As another way of describing the higher-order effects of sextupoles, we shall take the perturbative approach. We shall employ the theory developed by Poincare and von Zeipel (PZ) in one dimension to analyze the horizontal motion.

The PZ theory defines a canonical transformation $(\phi, I) \rightarrow (\varphi, J)$ with a generating function[3]

$$S(J, \phi; \theta) \equiv J \cdot \phi + \varepsilon S_1 + \varepsilon^2 S_2 + \varepsilon^3 S_3 + \ldots, \tag{4}$$

where ε is a parameter representing the strength of the perturbation. The idea of the theory is to find successively the functional form of S_n (n=1,...) that eliminates the angle dependence of the new Hamiltonian $H'(\varphi, J; \theta) = H(\phi, I; \theta) + \partial S/\partial\theta$ up to $O(\varepsilon^n)$, so that the new action J becomes an invariant of motion up to that order.

In the present study, we have derived generating functions up to S_3, thus enabling us to study the perturbative effects of sextupoles up to 4th order in amplitudes. The steps taken are as follows: On the basis of harmonics predefined in a lower-order perturbation, we derive S_n in a harmonic expanded form as a solution of the differential equation. We then take the summation over the harmonics analytically to obtain a closed form of S_n, with which higher-order terms in the new Hamiltonian are evaluated. Next we expand them into harmonics again to define the higher-order counterparts of A_{jm} of the lowest order. Double integrals over angles θ and ϕ can be evaluated analytically in the thin lens treatment. In this way, higher-order harmonics, which are necessary for the resonance analysis, are obtained in closed forms which can be readily computed. Our formulation
is made also to enable a systematic evaluation of higher-order perturbations where one is involved with a large number of terms. Details will be found in Ref. 7.

The resulting Hamiltonian may be expressed as follows,

<u>(i) Up to 2nd order:</u>

$$H'(\varphi, J; \theta) = \nu J + J^2 \sum_{K} \sum_{M=-\infty}^{+\infty} b_{KM} \cos(K\varphi - M\theta) + O(\varepsilon^3), \qquad (5)$$

where $K = 0, 2, 4$ and 6

(ii) Up to 3rd order:

$$H'(\varphi, J; \theta) = \nu J + b_{00} J^2 + J^{5/2} \sum_{\Lambda} \sum_{M=-\infty}^{+\infty} d_{\Lambda M} \cos(\Lambda\varphi - M\theta) + O(\varepsilon^4), \qquad (6)$$

where $\Lambda = 1, 3, 5, 7$ and 9

(iii) Up to 4th order:

$$H'(\varphi, J; \theta) = \nu J + b_{00} J^2 + J^3 \sum_{\Omega} \sum_{M=-\infty}^{+\infty} e_{\Omega M} \cos(\Omega\varphi - M\theta) + O(\varepsilon^5), \qquad (7)$$

where $\Omega = 0, 2, 4, 6, 8, 10$ and 12 .

With above formulae, an attempt has been made to reproduce the resonances observed in the tracking. We apply the single resonance approximation to the higher order Hamiltonian to calculate the motion, which is transformed back to the original frame

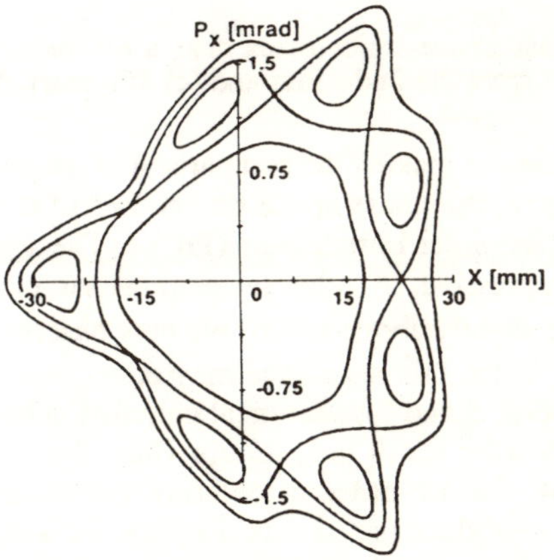

Fig. 8. Horizontal phase space trajectories of TBA, obtained from the 3rd-order PZ Hamiltonian.

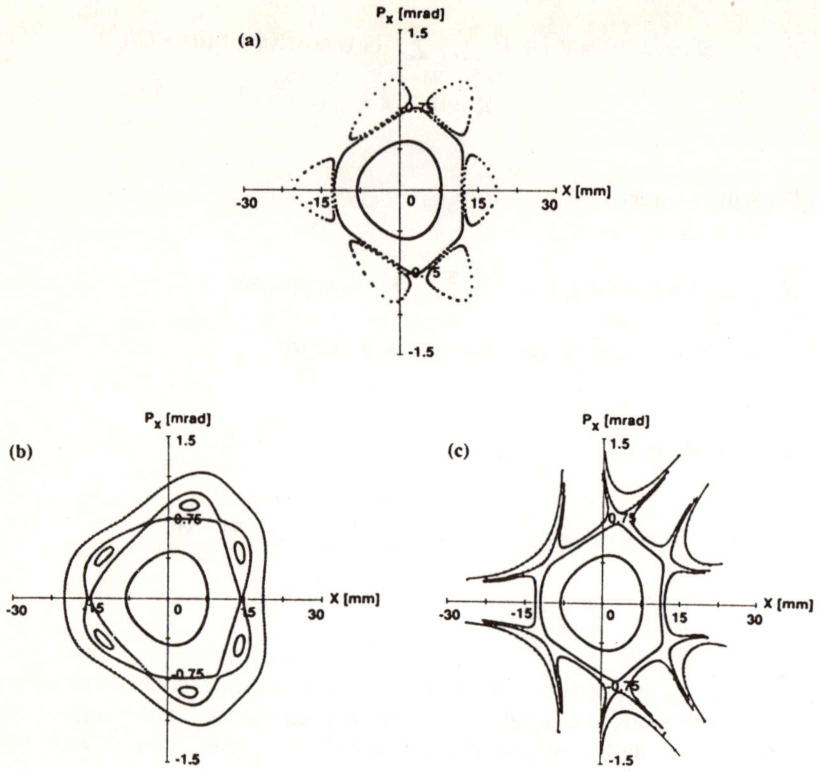

Fig. 9. Horizontal phase space trajectories near a 6th-order resonance of TBA. (a) Tracking. (b) Predicted from the 2nd-order and (c) 4th-order PZ Hamiltonian.

with the exact relations before making the comparison of the phase space structures.

We show in Fig. 8, the phase space trajectories of the 7th-order resonance of TBA obtained from the 3rd-order Hamiltonian (Eq. 6). The agreement with the tracking result (Fig. 6) is quite impressive. Another interesting example is found with this lattice by tuning ν_x to a value slightly above 1+1/6: We now observe a 6th-order resonance in the tracking (Fig. 9a). Since this resonance may either be driven as a 2nd- or a 4th-order perturbation (cf. Eqs. 5 and 7), analytical calculation has been performed in two way. Whereas the 2nd-order result fails to reproduce the phase space structure even topologically (Fig. 9b), the 4th-order calculation achieves fairly good agreement, although it still fails to reproduce the islands (Fig.9c). Finally, we refer to the fact that we have also found many cases with the low-emittance lattice in which the theory fails completely to reproduce the tracking results. Obviously, one possible interpretation is to assume the perturbations to be even higher than 4th order, but further studies may be necessary to examine the limitation of the theory.

References

1. E.A. Crosbie, *Internal Report* ANL-HEP-CP-87-21.
2. R. Nagaoka, H. Tanaka and M. Hara, *RIKEN Accel. Prog. Rep.* **21** (1987) 228.
3. A.J. Lichtenberg and M.A. Liebeman, *Applied Mathematical Science*, **38**, (Springer-Verlag, 1983), p. 71.
4. E.D. Courant, R.D. Ruth and W.T. Weng, *AIP Conf. Proc.* **127** (1985), p. 294
5. R. Nagaoka et al., *RIKEN Accel. Prog. Rep.* **22** (1988) 248.
6. R.D. Ruth, *AIP Conf. Proc.* **153** (1987), p. 150.
7. R. Nagaoka, K. Yoshida and M. Hara, *submitted to Nuclear Instruments & Methods*.

ON PERIODIC ORBITS OF NON-LINEAR DYNAMICAL SYSTEMS WITH MANY DEGREES OF FREEDOM

GAMAL M. MAHMOUD
Department of Mathematics, Faculty of Science, Assiut University,
Assiut, EGYPT

ABSTRACT

The purpose of this article is to extend our investigation to periodic orbits of non-linear dynamical systems with many degrees of freedom. The method of generalized averaging and the indicatrix method are developed. An example is studied to illustrate our techniques. The analytical expressions are in good agreement with numerical results computed by the indicatrix method for this example.

1. INTRODUCTION

In recent years, there has been an increased interest in non-linear dynamical systems of multiple degrees of freedom [1-3,6-9]. Periodic solutions (orbits) have been known to play a very important role in the study of these systems [4-7].

The aim here is to study periodic orbits of a class of nonlinear dynamical systems of the form

$$\ddot{x}_i + \omega_i^2 x_i + \lambda g_i (x, \dot{x}) \; P_i(\Omega_i t) = 0 , \tag{1}$$

$i=1,2,\dots,n$, $\omega_i > 0$, where $x \equiv (x_1,x_2,\dots,x_n)$, $\dot{x} \equiv (\dot{x}_1,\dot{x}_2,\dots,\dot{x}_n)$, g_i is a nonlinear function of x and $\dot{x}$, P_i is a periodic function, of t, λ is a small parameter, and dotes represent as usual differentiation with respect to t.

Equation (1) arises in numerous physical applications (e.g. the displacements of colliding particle beams in high energy accelerators, vibrations of shells, betatorn oscillations, and pendulum with vibrating length) [6-11].

Certain limiting cases of equation (1) are well known to mathematicians, physicists and engineers and have been treated at length in the literature (see for example [3, 6-8, 11]).

Section 2 presents the development of the generalized averaging method [6]. Using this technique we derive approximate expressions for the amplitude and phase of the periodic components $x_1(t)$, $x_2(t)$, ..., $x_n(t)$ of the solutions of equation (1).

Section 3 contains the extension of the indicatrix method [6]. This numerical approach is used to study the existance of the periodic orbits of equation (1).

Finally, in section 4, the case n=3 is studied to illustrate our techniques. Good agreement is found between theoretical predictions and numerical results computed by the method of section 3.

Currently, we are in the process of studing the stability properties of the periodic solutions of equation (1), $n \geq 2$. This is considerably more complicated than the case n=1 [6], and will appear in future publications.

2. THEORETICAL DEVELOPMENT

In this section we develop a generalized averaging method [6] to study periodic solutions of systems governed by the equation of motion

$$\ddot{x}_i + \omega_i^2 x_i + \lambda g_i(x,\dot{x}) P_i(\Omega_i t) = 0 , \qquad i=1,2,\ldots,n \quad . \tag{2}$$

The method starts from the so-called generating solutions

$$x_i = a_i \cos(\omega_i t + \phi_i), \quad \dot{x}_i = -a_i \omega_i \sin(\omega_i t + \phi_i), \quad i=1,\ldots,n \tag{3}$$

which satisfy equation (2) to zero order ($\lambda=0$). In order to solve equation (2), it is assumed that the constants of integration a_i and ϕ_i depend on time, so that in (3) $a_i \to a_i(t)$, $\phi_i \to \phi_i(t)$. Then inserting (3) into (2) we get a system of equations which we solve for $\dot{a}_i(t)$, $\dot{\phi}_i(t)$ to obtain (For more details see [7])

$$\frac{d}{dt}\begin{bmatrix} a(t) \\ \\ \phi(t) \end{bmatrix} = \lambda \begin{bmatrix} R(v) \\ \\ S(v) \end{bmatrix} \tag{4a}$$

where

$$\left.\begin{aligned} v &= (a(t),\ \dot{a}(t),\ \phi(t),\ \dot{\phi}(t),t) \ , \\ a(t) &= [a_1(t),\ a_2(t),\ldots,a_n(t)]^t,\ \ \phi(t) = [\phi_1(t),\phi_2(t),\ldots,\phi_n(t)]^t, \\ R &= [\frac{1}{\omega_1} g_1(v)P_1(\Omega_1 t)\sin(\omega_1 t+\phi_1),\ldots,\ \frac{1}{\omega_n} g_n(v)P_n(\Omega_n t)\sin(\omega_n t+\phi_n)]^t, \\ S &= [\frac{1}{a_1\omega_1} g_1(v)P_1(\Omega_1 t)\cos(\omega_1 t+\phi_1)\ldots,\ \frac{1}{a_n\omega_n} g_n(v)P_n(\Omega_n t)\cos(\omega_n t+\phi_n)]^t \end{aligned}\right\} \tag{4b}$$

(with $[\ldots]^t$ denoting transpose).

Equation (4a) with (4b), can now be solved by an extension of the technique of generalized averaging as follows :

Casting the original equation (4) in the form

$$\frac{dy}{dt} = \lambda\, f(y,t) \quad , \qquad y = [a(t),\ \phi(t)]^t , \tag{5a}$$

and let

$$\left.\begin{aligned} f(y,t) &= \bar{f}(u) + \tilde{f}(u,t) , \\ u \equiv [a_1(t_1), a_2(t_2), \ldots, a_n(t_n), \phi_1(t_1), \ldots, \phi_n(t_n)]^t & \\ &= [A_1, A_2, \ldots , A_n, \Phi_1, \ldots, \Phi_n]^t \end{aligned}\right\} \quad , \tag{5b}$$

where the vector function $\bar{f}(u)$ and $\tilde{f}(u,t)$ contain the terms of smallest frequency (plus constants) and the terms of higher frequency respectively. We associate with equation (4) the reduced system

$$\frac{du}{dt} = \lambda\, \bar{F}(u) \tag{6a}$$

where

$$y = u + \lambda\, \tilde{G}(u,t) \quad . \tag{6b}$$

Expand $\bar{F}(u)$ and $\tilde{G}(u,t)$ in powers of λ as follows

$$\bar{F}(u) = \bar{F}_1(u) + \lambda\, \bar{F}_2(u) + \ldots \tag{7a}$$

$$\tilde{G}(u,t) = \tilde{G}_1(u,t) + \lambda\, \tilde{G}_2(u,t) + \ldots \tag{7b}$$

and substitute from equation (7) into equation (6), using equation (5) to obtain, upon equating like powers of λ

$$\bar{F}_1(u) = \overline{f(u,t)}, \qquad \bar{F}_2(u) = \overline{\frac{\partial \tilde{f}}{\partial u} \int \tilde{f}(u,t)\, dt}, \ldots , \tag{8a}$$

$$\tilde{G}_1(u,t) = \int \tilde{f}(u,t)\, dt , \ \ldots \quad . \tag{8b}$$

From equation (6b), the approximate formulas for the amplitude and phase of the periodic components $x_1, \ldots, x_n$, of the solutions of equation (1) are written as

$$\left.\begin{aligned} a_i(t) &= A_i + \lambda \tilde{G}_{1a_i}(u,t) + O(\lambda^2) \\ \phi_i(t) &= \Phi_i + \lambda G_{1\phi_i}(u,t) + O(\lambda^2) \end{aligned}\right\} \quad i=1,2,\ldots,n \qquad \textbf{(9)}$$

with

$$\tilde{G}_k(u,t) \equiv [\tilde{G}_{ka_1}, \tilde{G}_{ka_2}, \ldots, \tilde{G}_{ka_n}, \tilde{G}_{k\phi_1}, \ldots, \tilde{G}_{k\phi_n}]^t, \qquad k=1,2,\ldots,n \, .$$

3. NUMERICAL DEVELOPMENT

In this section, we present the extension of the indicatrix method [6] in order to be applied to nonlinear systems having the form :

$$\ddot{x}_1 + \omega_i^2 x_i + \lambda g_i(x,\dot{x}) P_i(\Omega_i t) = 0, \qquad i=1,2,\ldots, n \, . \qquad \textbf{(10)}$$

Taking in equation (10) $\lambda \neq 0$, the sets of points $[x_1(t_1,t_1+T_1), \dot{x}_1(t_1,t_1+T_1)),\ldots, (x_n(t_n,t_n+T_n), \dot{x}_n(t_n,t_n+T_n))]$ with periods $T_1 = \frac{2\pi}{\omega_1}, \ldots, T_n = \frac{2\pi}{\omega_n}$, form closed curves parameterized by $t_i \in [0, \frac{2\pi}{\Omega_i}]$ in the spaces $(x,\dot{x}_1), \ldots, (x_n,\dot{x}_n)$ respectively, which may or may not pass by the specified initial conditions. These closed curves are called the indicatricies.

By varying slightly $\omega_1^2, \omega_2^2, \ldots, \omega_n^2$, and λ, we obtain the periodic components $x_1,x_2,\ldots,x_n$ when the indicatricies in the spaces $(x_1,\dot{x}_1), \ldots, (x_n,\dot{x}_n)$ are observed to pass by the prescribed initial conditions at some values of $t_1, t_2, \ldots, t_n$.

In this way we obtained the numerical values of $\omega_1^2, \omega_2^2,\ldots,\omega_n^2,\lambda$ at which periodic components $x_1,x_2,\ldots,x_n$ of the solutions of equation (10) with periods $T_1,T_2,\ldots,T_n$, respectively pass by specified initial conditions.

4. AN EXAMPLE

To demonstrat the techniques of sections 2,3, they can be applied to the following system :

$$\ddot{x}_1 + \omega_1^2 x_1 + \lambda(x_1x_2 + x_1^2) \cos t = 0 \quad (11)$$

$$\ddot{x}_2 + \omega_2^2 x_2 + \lambda(x_1^2 + x_2x_3) \cos 2t = 0 \quad (12)$$

$$\ddot{x}_3 + \omega_3^2 x_3 + \lambda(x_2^2 + x_3^2) \cos 3t = 0 \quad . \quad (13)$$

The main methods of this paper are not affected by the precise form of $g_i(x,\dot{x})$ and $P_i(\Omega_i t)$. (cf. equation (1)).

Performing the calculations in (4), one obtains the following for equations (11), (12), (13) :

$$\frac{d}{dt}\begin{bmatrix} a(t) \\ \phi(t) \end{bmatrix} = \lambda \begin{bmatrix} R(v) \\ S(v) \end{bmatrix} \quad (14a)$$

where

$v = (a(t),\ \dot{a}(t),\ \phi(t),\ \phi(t), t)$, $a(t) = [a_1(t), a_2(t), a_3(t)]^t$,

$\phi(t)= [\phi_1(t), \phi_2(t), \phi_3(t)]^t$, $R = [R_1,R_2,R_3]^t$, $S = [S_1,S_2,S_3]^t$,

R_i , S_i (i=1,2,3) are sums of trigonometric functions with arguments

$$\psi_{j,k} \equiv (\omega_j \pm k)\ t + \phi_j \quad , \quad j,k = 1,2,3, \quad , \quad (14b)$$

$$R_1 = \frac{1}{8\omega_1}[a_1a_2 \sin(2\psi_{1,0} \pm \psi_{2,0}) + a_1^2 \sin \psi_{1,1} + a_1^2 \sin(2\psi_{1,0} + \psi_{1,1})],$$

$$R_2 = \frac{1}{8\omega_2}[a_1^2(2 \sin\psi_{2,2} + \sin(\psi_{2,2}\pm 2\psi_{1,0}))+a_2a_3(\sin\psi_{3,2}+\sin(2\psi_{2,1}\pm\psi_{3,0})],$$

$$R_3 = \frac{1}{8\omega_3}[(2a_2^2 + a_3^2) \sin \psi_{3,3} + a_3^2 \sin 3\psi_{3,1} + a_2^2 \sin (\psi_{3,1} + 2\psi_{2,1})],$$

$$S_1 = \frac{1}{8\omega_1}[2a_2 \cos\psi_{2,1} + a_2\cos(2\psi_{1,0}\pm\psi_{2,1}) + 3a_1\cos\psi_{1,1}+a_1\cos(2\psi_{1,0}+\psi_{1,1})],$$

$$S_2 = \frac{1}{8\omega_2 a_2}[2a_1^2\cos\psi_{2,2}+a_1^2\cos(2\psi_{1,0}\pm\psi_{2,2})+a_2a_3(2\cos\psi_{3,2}+\cos(\psi_{3,0}\pm 2\psi_{2,1}))],$$

$$S_3 = \frac{1}{8\omega_3 a_3}[(3a_3^2+2a_2^2) \cos\psi_{3,3} + a_3^2 \cos 3\psi_{3,1} + a_2^2 \cos(\psi_{3,1} + 2\psi_{2,1})] \;. \qquad \text{....} \quad \textbf{(14c)}$$

From equation (14) small frequency components occur at

$$\omega_1 \cong \frac{1}{3}, 1, \quad \omega_2 \cong 1,2, \quad \omega_3 \cong 1,2,3, \quad \omega_3 \pm 2\omega_2 \cong 2,3, \quad 2\omega_1 \pm \omega_2 \cong 1,2. \qquad \text{....} \quad \textbf{(15)}$$

Choosing

$$\omega_1 \cong \frac{1}{3}, \quad \omega_2 \cong 2 \quad \text{and} \quad \omega_3 \cong 3 \quad . \qquad \textbf{(16)}$$

For this choice (other values of ω_i can be similarly treated) the reduced system (6a) becomes

$$\frac{du}{dt} = \lambda \bar{F}_1(u) + O(\lambda^2) \;, \qquad \textbf{(17a)}$$

where

$$u = [A_1, A_2, A_3, \Phi_1, \Phi_2, \Phi_3]^t \;, \quad \text{and} \qquad \textbf{(17b)}$$

$$\bar{F}_1(u) = [\bar{F}_{11}, \bar{F}_{12}, \bar{F}_{13}, \bar{F}_{14}, \bar{F}_{15}, \bar{F}_{16}]^t \quad \text{with}$$

$$\bar{F}_{11} = \frac{A_1^2}{8\omega_1} \sin(2\Psi_{1,0}+\Psi_{1,1}), \quad \bar{F}_{12} = \frac{A_1^2}{4\omega_2} \sin\Psi_{2,2}, \quad \bar{F}_{13} = \frac{1}{8\omega_3}(2A_2^2+A_3^2)\sin\Psi_{3,3}$$

$$\bar{F}_{14} = \frac{A_1}{8\omega_1} \cos(2\Psi_{1,0}+\Psi_{1,1}), \quad \bar{F}_{15} = \frac{A_1^2}{4\omega_2 A_2} \cos\Psi_{2,2}, \quad \bar{F}_{16} = \frac{1}{8A_3\omega_3}(3A_3^2+2A_2^2)\cos\Psi_{3,3}$$

and

$$\Psi_{j,k} = (\omega_j - k)t + \Phi_j \;. \qquad \textbf{(17c)}$$

To satisfy the conditions $\frac{dA_i}{dt} = 0$, $i=1,2,3$, we make the corresponding contributions of $\bar{F}_1(u)$ vanish by setting

$$\left.\begin{aligned} 2\Psi_{1,0} + \Psi_{1,1} &= (3\omega_1 - 1)t + 3\,\Phi_1 = \ell_1\pi, \\ \Psi_{2,2} &= (\omega_2 - 2)t + \Phi_2 = \ell_2\pi \quad, \\ \Psi_{3,3} &= (\omega_3 - 3)t + \Phi_3 = \ell_3\pi \end{aligned}\right\} \ell_1,\ell_2,\ell_3 = 0,1,2,.. \tag{18}$$

From equation (18) one obtains

$$\omega_1 = \frac{1}{3} - \frac{d\Phi_1}{dt}\,, \qquad \omega_2 = 2 - \frac{d\Phi_2}{dt}\,, \qquad \omega_3 = 3 - \frac{d\Phi_3}{dt}\,, \tag{19}$$

From equations (17), (19), we get the following expressions for ω_i^2 to first order in λ:

$$\omega_1^2 = \frac{1}{9} - \frac{1}{4}\lambda\, A_1(-1)^{\ell_1} + O(\lambda^2) \quad, \tag{20a}$$

$$\omega_2^2 = 4 - \frac{1}{2}\lambda\,\frac{A_1^2}{A_2}\,(-1)^{\ell_2} + O(\lambda^2) \quad, \tag{20b}$$

$$\omega_3^2 = 9 - \frac{1}{4A_3}\lambda(-1)^{\ell_3}\,(3A_3^2 + 2A_2^2) + O(\lambda^2) \quad. \tag{20c}$$

For different values of ℓ_1, ℓ_2, $\ell_3 = 0,1$, we pick different branches in (ω_1^2, λ), (ω_2^2, λ), (ω_3^2, λ) planes, along which periodic solutions of the system (11-13) can be found, with initial conditions $x_1(t_1) = A_1$, $x_2(t_2) = A_2$, $x_3(t_3) = A_3$, and $\dot{x}_1(t_1) = \dot{x}_2(t_2) = \dot{x}_3(t_3) = 0$. These solutions have period $x_1(t)$ component with period $T_1 = 6\pi$, periodic $x_2(t)$ with period $T_2 = \pi$ and periodic component $x_3(t)$ with period $T_3 = \frac{2\pi}{3}$. Their overall period, therefore, is $T = T_1 = 6T_2 = 9T_3 = 6\pi$.

To verify the validity of the analytical results we apply the numerical technique of section 3. By varying slightly ω_1^2, ω_2^2, ω_3^2 and λ, we obtain the periodic components x_1, x_2, x_3 when the indicatricies are observed to pass by our initial conditions (e.g.

$A_1 = A_2 = A_3 = 0.5)$ for different values of t_1, t_2, t_3, see Figure 1. As expected from the analytical approach we obtained 6π-periodic solutions with $x_1(t)$, $x_2(t)$, $x_3(t)$ components oscillating with periods 6π, π, $\frac{2}{3}\pi$ $(\omega_1^2 \cong \frac{1}{9}$, $\omega_2^2 \cong 4$, $\omega_3^2 \cong 9)$ respectively. Finally, expressions (20) are compared with the results of this technique and very good agreement is found for $\lambda \leq 0.1$, see Table 1.

Table 1

λ	Analytical Results (Eq. 20)			Numerical Results		
	ω_1^2	ω_2^2	ω_3^2	ω_1^2	ω_2^2	ω_3^2
0.001	0.111236 0.110986	4.00025 3.99975	9.000625 8.999375	0.111236 0.110986	4.00036 3.99965	9.000693 8.999447
0.005	0.111736 0.110486	4.00125 3.99875	9.003125 8.996875	0.111736 0.110486	4.00183 3.99825	9.003250 8.996895
0.010	0.112361 0.109861	4.00250 3.99750	9.006250 8.993750	0.112360 0.109851	4.00370 3.99655	9.00635 8.99377
0.050	0.117361 0.104861	4.01250 3.98750	9.031250 8.968750	0.117400 0.104662	4.01790 3.983395	9.03050 8.968573
0.100	0.123610 0.098610	4.02500 3.97500	9.062500 8.937500	0.123600 0.097700	4.03520 3.96850	9.06247 8.93702

for $\ell_1 = \ell_2 = \ell_3 = 1$ $(t_1 = \pi,\ t_2 = \frac{\pi}{2},\ t_3 = 1.09956)$

and

$\ell_1 = \ell_2 = \ell_3 = 0$ $(t_1 = t_2 = t_3 = 0)$.

REFERENCES

[1] J. Kevorkian, Perturbation Techniques for Oscillatory Systems with Slowly Varying Coefficients, SIAM REVIEW, Vol. 29, No.3, (1987).

[2] C.-C. Chi and R.M. Rosenberg, On Damped Nonlinear Dynamics Systems with Many Degrees of Freedom, Int. J. Non-Linear Mechanics, Vol. 20, No. 5/6, pp. 371-384 (1985).

[3] D.E. Gilsinn, A High Order Generalized Method of Averaging, SIAM J. APPL. MATH. Vol. 42, No. 1, (1982).

[4] J. Guckenheimer and P. Holmes, Nonlinear Oscillations, Dynamical Systems and Bifurcations of Vector Fields, Springer (1983).

[5] M. Lieberman and A. Lichtenberg, Regular and Stochastic Motion, Springer (1983).

[6] G.M. Mahmoud and T. Bountis, Synchronized Periodic Solutions of a Class of Periodically Driven Nonlinear Oscillators, J. Appl. Mech. 55, pp. 721-727 (1988).

[7] T. Bountis and G.M. Mahmoud, Synchronized Periodic Orbits in Beam-Beam Interaction Models of One and Two Spatial Dimensions, Particle Accelerators, Vol. 22, pp. 129-147 (1987).

[8] H. Sprysl, Internal Resonance of Non-Linear Autonomous Vibrating Systems with Two Degrees of Freedom, J. of Sound and Vibration 112 (1), 63-67 (1987).

[9] P.R. Sethna, Coupling in Certain Classes of Weakly Nonlinear Vibrating Systems, Int. Symposium on Nonlinear Diff. Eqns. and Nonlinear Mech., Edited by J. Lasalle and S. Lefschetz, Academic Press, N.Y (1963).

[10] R. Cohen and I. Porat, Coupled Torsional and Transverse Vibration of Unbalanced Roter, J. Appl. Mech. Vol. 52, pp. 701-705 , (1985).

[11] A.N. Nayfeh and D.T. Mook, Nonlinear Oscillations, New York, Wiley (1979).

FIGURE CAPTION

Figure 1 : Indicatrix plots of Eq. (10) for $\lambda = 0.001$,

$x_1(t_1) = x_2(t_2) = x_3(t_3) = 0.5$, $t_1 \varepsilon [0,2\pi]$, $t_2 \varepsilon [0,\pi]$ $t_3 \varepsilon [0, \frac{2}{3}\pi]$ and

(a) $\omega_1^2 = 0.110986$: (0.5, 0) at $t_1 = 0$,

(b) $\omega_2^2 = 3.99965$: (0.5, 0) at $t_2 = 0$,

(c) $\omega_3^2 = 8.999447$: (0.5, 0) at $t_3 = 0$,

(d) $\omega_1^2 = 0.111236$: (0.5, 0) at $t_1 = \pi$,

(e) $\omega_2^2 = 4.00036$: (0.5, 0) at $t_2 = \frac{\pi}{2}$, and

(f) $\omega_3^2 = 9.000693$: (0.5, 0) at $t_3 = 1.09956$.

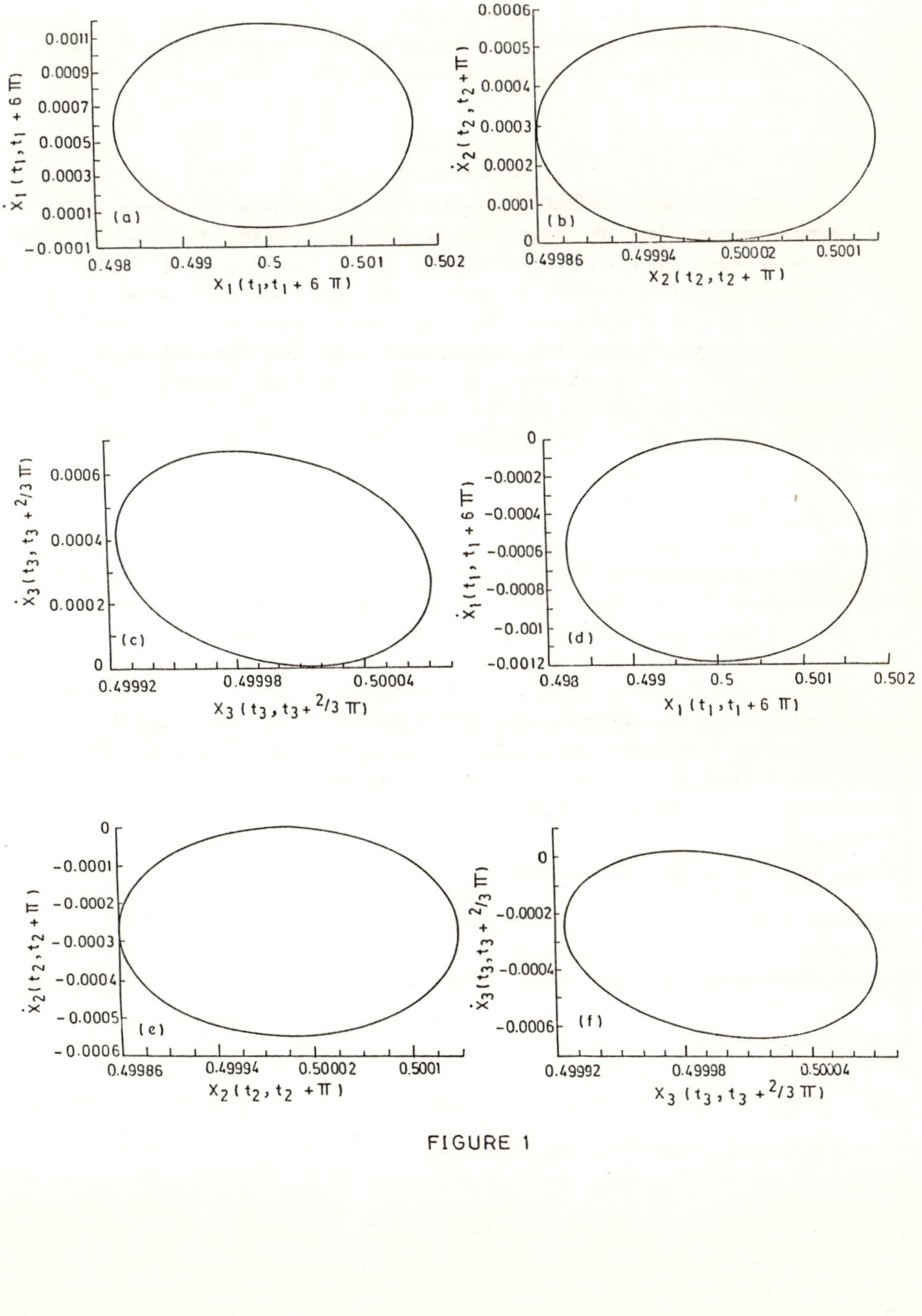
(a)
0.0011
0.0009
0.0007
0.0005
0.0003
0.0001
-0.0001
0.498
0.499
0.5
0.501
0.502
$\dot{X}_1(t_1, t_1 + 6\pi)$
$X_1(t_1, t_1 + 6\pi)$
(b)
0.0006
0.0005
0.0004
0.0003
0.0002
0.0001
0
0.49986
0.49994
0.50002
0.5001
$\dot{X}_2(t_2, t_2 + \pi)$
$X_2(t_2, t_2 + \pi)$
(c)
0.0006
0.0004
0.0002
0
0.49992
0.49998
0.50004
$\dot{X}_3(t_3, t_3 + 2/3\,\pi)$
$X_3(t_3, t_3 + 2/3\,\pi)$
(d)
0
-0.0002
-0.0004
-0.0006
-0.0008
-0.001
-0.0012
0.498
0.499
0.5
0.501
0.502
$\dot{X}_1(t_1, t_1 + 6\pi)$
$X_1(t_1, t_1 + 6\pi)$
(e)
0
-0.0001
-0.0002
-0.0003
-0.0004
-0.0005
-0.0006
0.49986
0.49994
0.50002
0.5001
$\dot{X}_2(t_2, t_2 + \pi)$
$X_2(t_2, t_2 + \pi)$
(f)
0
-0.0002
-0.0004
-0.0006
0.49992
0.49998
0.50004
$\dot{X}_3(t_3, t_3 + 2/3\,\pi)$
$X_3(t_3, t_3 + 2/3\,\pi)$

FIGURE 1

A FUNCTIONAL EQUATION ARISING FROM NORMAL FORM THEORY

XIE RUIFENG†

Dipartmento di Fisica, Università di Bologna, Via Irnerio, 46
I-40126 Bologna ,Italy

ABSTRACT

We discuss a functional equation arising in the Nekhoroshev estimate for symplectic maps. Some numerical results show that the preimages of the solution function on the unit disk give a good approximation to the fractal formed by the singular points of the solution. The first two steps of analytic continuation are rigorously proved. It is also shown that the fractal has a quadratic cusp at the origin and that the density of the singular points is exponentially small near it. These results are in agreement with the analysis of the leading singularities for the quadratic map made by Turchetti in his contribution.

1. Introduction

The theory of normal forms is a basic method of investigation for differential equations or discrete time dynamical systens. The method has its origin in Poincaré's thesis and consists in the reduction of the original system to a simpler form by means of a formal or analytic transformation of variables[1,2]. Usually the existence of a formal solution is easily established, but the power series giving the normalizing transformation are generally divergent[2]. However, these series can provide very useful dynamical infoarmation since they tipically verify suitable asymptotic estimates. Moreover, one can sometimes prove that these series are Borel summable ane reconstruct an analytic function which solves the normal form problem.

In this paper, we discuss the solution of a functional equation first derived by Turchetti[3] and arising from normal theory. In section 2 we show the existence and the basic properties of the solution. In section 3 we discuss the analytic continuation and describe the singularity structure: first we outline the proof of the first two steps of analytic continuation by the inverse of h. Then we numerically show that the fractal formed by the singular points has a quadratic cusp near the origin and that the density of the singular points is about $2^{-1/x}$, where x is the distance from the origin itself. In section 4, a numerical method is introduced to solve the functional

† Supported by ICSC World Laboratory

equation approximately. The numerical solution has been used in analyzing the continuation of the analyticity domain.

2. Rigorous Results on the Functional Equation

In the proof of Nekhoroshev stability estimate for symplectic maps[1,4] one needs to give an explicit bound to the sequence recursively defined by

$$\begin{aligned} g_1 &= g_2 = 1 \\ g_n &= \sum_{s=2}^{n-1} g_s \sum_{k_1+\ldots+k_s=n\,,\,k_i\geq 1} g_{k_1}\cdots g_{k_s}\ ,\ n\geq 3 \end{aligned} \tag{2.1}$$

Each g_n, together with the contribution due to the small divisors, gives a bound to the norm of the homogenous polynomial truncation of the normal form.

Theorem 2.1 *For the number sequence defined by (2.1) we have the estimate*

$$(n-1)! \leq g_n \leq 4^{n-2}(n-1)! \ \text{ for all } n \geq 3\ . \tag{2.2}$$

Proof. First of all note that all the g_n's are positive integers. Let $n \geq 3$. From the definition (2.1) we have

$$\begin{aligned} g_n &= (n-1)g_{n-1}g_2g_1^{n-1} + \sum_{s=2}^{n-2} g_s \sum_{k_1+\ldots+k_s=n} g_{k_1}\cdots g_{k_s} \\ &\geq (n-1)g_{n-1} \end{aligned}$$

Since $g_2 = 1$ one immediately derives the l.h.s of (2.2).

To prove the upper bound we first show by induction that

$$g_k \leq 4\left(k-\frac{5}{2}\right) g_{k-1}\ , \tag{2.3}$$

for all $k \geq 3$.

Assume that (2.3) is verified for $3 \le k \le n$. Then

$$g_{n+1} = \sum_{s=2}^{n} g_s \sum_{k_1+\ldots+k_s=n+1} g_{k_1}\cdots g_{k_s} = \sum_{s=2}^{n} g_s g_1 \sum_{k_1+\ldots+k_{s-1}=n} g_{k_1}\cdots g_{k_{s-1}}$$

$$+ \sum_{s=2}^{n} g_s \sum_{k_1=2}^{n-s+2} g_{k_1} \sum_{k_2+\ldots+k_s=n+1-k_1} g_{k_2}\cdots g_{k_s}$$

$$= \sum_{s=1}^{n-1} g_{s+1} \sum_{k_1+\ldots+k_s=n} g_{k_1}\cdots g_{k_s}$$

$$+ \sum_{s=2}^{n-1} g_s \sum_{k_1=2}^{n-s+2} g_{k_1} \sum_{k_2+\ldots+k_s=n+1-k_1} g_{k_2}\cdots g_{k_s} + g_n$$

$$= \sum_{s=2}^{n-1} g_{s+1} \sum_{k_1+\ldots+k_s=n} g_{k_1}\cdots g_{k_s}$$

$$+ \sum_{s=2}^{n-1} g_s \sum_{k_1=2}^{n-s+2} g_{k_1} \sum_{k_2+\ldots+k_s=n+1-k_1} g_{k_2}\cdots g_{k_s} + 2g_n$$

Our induction hypotesis implies that for all $2 \le s \le n-1$

$$g_{s+1} \le 4\left(s - \frac{3}{2}\right) g_s$$

and we get

$$g_{n+1} \le \sum_{s=2}^{n-1} 4\left(s-\frac{3}{2}\right) g_s \sum_{k_1+\ldots+k_s=n} g_{k_1}\cdots g_{k_s}$$

$$+ \sum_{s=2}^{n-1} 4g_s\left(n-s-\frac{1}{2}\right) \sum_{k_1=2}^{n-s+2} g_{k_1-1} \sum_{k_2+\ldots+k_s=n+1-k_1} g_{k_2}\cdots g_{k_s} + 2g_n$$

It follows that

$$g_{n+1} \le \sum_{s=2}^{n-1} 4\left(s-\frac{3}{2}\right) g_s \sum_{k_1+\ldots+k_s=n} g_{k_1}\cdots g_{k_s}$$

$$+ \sum_{s=2}^{n-1} 4g_s\left(n-s-\frac{1}{2}\right) \sum_{k_1+\ldots+k_s=n} g_{k_1}\cdots g_{k_s} + 2g_n$$

$$= 4(n-2)g_n + 2g_n = 4\left(n+1-\frac{5}{2}\right) g_n$$

which yields (2.3) for $k = n+1$. Hence (2.3) holds for all $k \geq 3$ and the upper bound in (2.2) is now easy to prove.

We now formally write the generating function

$$g(z) := \sum_{n=1}^{\infty} g_n z^n \tag{2.4}$$

of the numerical sequence (2.1). It is easy to check that $g(z)$ is a formal power series solution of the functional equation

$$3g(z) = g(g(z)) + z^2 + 2z\ , \tag{2.5}$$

with the additional constraints $g(0) = 0$, $2g'(0) = g''(0) = 2$.

To study (2.5) it is convenient to transform to a form symmetric about 0. We choose the following transformation

$$\begin{aligned} w &= z + 1 \\ h(w) &= g(z) + 1 \end{aligned}$$

Then (2.5) takes the form

$$3h(w) = h(h(w)) + w^2 + 1 \tag{2.6}$$

It is obvious that (2.5) and (2.6) are equivalent. For (2.6) one has the following theorem[5]

Theorem 2.2 *There exists a unique solution to (2.6) analytic in the unit disk* $\mathbf{D} := \{w \in \mathbf{C} \mid |w| < 1\}$ *and such that* $h \in C^\infty(\overline{\mathbf{D}})$ *and* $h(\mathbf{D}) \subseteq \mathbf{D}$, $h(\overline{\mathbf{D}}) \subseteq \overline{\mathbf{D}}$.

Moreover the solution h of (2.6) has the following properties:

1. h is real on $[-1,1]$;
2. h is symmetric about $w = 0$: $h(w) = h(-w)$;
3. h is not decreasing on $[0,1]$;
4. $h([-1,1]) \subset [-1,1]$;
5. For all $w \in \overline{\mathbf{D}}$ one has $\overline{h(w)} = h(\overline{w})$;

6. h maps $\{\text{ Re } w \geq 0\} \cap \{\text{ Im } w \geq 0\} \cap \overline{\mathbf{D}}$ into itself;

7. $|\text{ Im } h(w)| \leq |\text{ Im } w||\text{ Re } w|$.

Further analysis of h shows that it has some properties similar to the function $(1+w^2)/2$. Let us consider the sequence

$$\alpha_0 = 0$$
$$\alpha_{n+1} = \frac{1+\alpha_n^2}{2}, \; n \geq 0$$

and the domains

$$\Omega_n := \{w \in \mathbf{C} \mid |w - \alpha_n| \leq 1 - \alpha_n\}$$

Clearly $\Omega_n \subset \overline{\mathbf{D}}$ and moreover we have

Theorem 2.3 *If $h(w)$ is the solution of (2.6) then $h(\Omega_n) \subset \Omega_{n+1}$ and hence $h^{\circ n}(w) \to 1$ as $n \to \infty$ for all $w \in \mathbf{D}$.*

Proof. First we take $n = 0$. We know that $h(\Omega_0) \subset \Omega_0$ since $\Omega_0 = \overline{\mathbf{D}}$. Let

$$M_1 := \max_{z \in \Omega_0} |h(z) - \alpha_1|. \tag{2.7}$$

From the functional equation (2.6) we have

$$h(z) - \alpha_1 = \frac{1+z^2}{3} - \frac{1}{2} + \frac{1}{3}h(h(z)) = \frac{2}{3}\left(\frac{1+z^2}{2} - \frac{1}{2}\right) + \frac{1}{3}\left(h(h(z)) - \frac{1}{2}\right).$$

Thus for all $z \in \Omega_0$

$$|h(z) - \alpha_1| \leq \frac{2}{3}\frac{1}{2} + \frac{1}{3}M_1.$$

That is

$$M_1 \leq \frac{1}{3} + \frac{1}{3}M_1$$

which obviously implies

$$M_1 \leq \frac{1}{2} = 1 - \alpha_1.$$

Therefore we have proved (2.7) for $n = 0$. We now proceed by induction. Assume (2.7) holds for all $n \leq k$, i.e.

$$h(z) \in \Omega_{k+1}, \; \textit{for all } z \in \Omega_k,$$

and define

$$M_{k+2} := \max_{z\in\Omega_{k+1}} |h(z) - \alpha_{k+2}|.$$

Since $z \in \Omega_{k+1} \subset \Omega_k$ we have $h(z) \in \Omega_{k+1}$. Thus

$$\begin{aligned} h(z) - \alpha_{k+2} &= \frac{2}{3}\left(\frac{1+z^2}{2} - \alpha_{k+2}\right) + \frac{1}{3}(h(h(z)) - \alpha_{k+2}) \\ &= \frac{2}{3}\left(\frac{z^2 - \alpha_{k+1}^2}{2}\right) + \frac{1}{3}(h(h(z)) - \alpha_{k+2}) \end{aligned}$$

and

$$\begin{aligned} |h(z) - \alpha_{k+2}| &\le \frac{2}{3}\frac{|z+\alpha_{k+1}||z-\alpha_{k+1}|}{2} + \frac{1}{3}M_{k+2} \\ &\le \frac{2}{3}\frac{(1+\alpha_{k+1})(1-\alpha_{k+1})}{2} + \frac{1}{3}M_{k+2}. \end{aligned}$$

Therefore

$$M_{k+2} \le \frac{1-\alpha_{k+1}^2}{3} + \frac{1}{3}M_{k+2}$$

i.e.

$$M_{k+2} \le \frac{1-\alpha_{k+1}^2}{2} = 1 - \alpha_{k+2}.$$

Q.E.D.

Remark 1: From Theorem 2.2 we know that there exists a solution to the equation (2.1) which is analytic defined in the disk $|z+1| < 1$. This means that there exists analytic solution to the functional equation (2.1) in a domain near the origin in z–plane.

Remark 2: The "physical region" is the positive real axis $\mathbf{R}_+$, since z has the interpretation of the action r (radial distance from the fixed point at the origin of the sympletic map). So we want to extend the analyticity domain as close as possible to $\mathbf{R}_+$.

3. Analytic Continuation and Singularity Structure

By (2.2) we know $g_n \ge (n-1)!$. Thus $z = 0$ is a singular point of g, i.e. h is singular at $w = 1$. From (2.6) it is easy to check that h is analytic at a point $h(w)$ if h is analytic at w. This means that h is singular at $h^{-1}(w)$ if h is singular at w. By this property of h, from the knowledge that $w = 1$ is a singular point of h, we can get an infinite sequence of singular points by computing its preimages. To do this, we solve $h^{-1}(w)$ from the equation (2.6)

$$h^{-1}(w) = \pm\sqrt{3w - 1 - h(w)} \tag{3.1}$$

and we can compute the singular points recursively by

$$y_{-1} = 1 \,,\; y_0 = -1 \,;$$
$$y_n = \pm\sqrt{3y_{n-1} - 1 - y_{n-2}}.$$

As shown in Fig. 1, the set $\{y_n\}_{n\in\mathbf{N}}$ forms a dense subset of the boundary of a domain Ω in the complex plane. Therefore, the analytic continuation of h from the unit disk $\mathbf{D}$ can not get out of Ω.

Since h is a contracting map as stated in Theorem 2.3, we consider the inverse map h^{-1} of h in order to extend its analyticity domain.

To prove the first step of the analytic continuation, i.e. the extension of the analiticity domain of h obtained by one application of h^{-1} to the unit disk, we need the following properties of h:

i) $|\mathrm{Im}\, h(w)| \le \frac{1}{3}|\mathrm{Im} w|$, for all $w \in \mathbf{D}$;

ii) $|h'(w)| \le 1$ for all $w \in \mathbf{D}$. These properties can be easily proved by means of the approximations scheme used in the proof[5] of Theorem 2.2.

Let

$$\hat{\mathbf{D}} = \mathbf{D} \setminus [0,1]$$

and define the functions

$$\varphi_1^+(w) = \sqrt{3w - 1 - h(w)}\,, \quad \text{for all } w \in \hat{\mathbf{D}}\,;$$
$$\varphi_1^-(w) = -\sqrt{3w - 1 - h(w)}\,, \quad \text{for all } w \in \hat{\mathbf{D}}\,.$$

From **i)** , it is easy to check that $3w - 1 - h(w) = 0$ if and only if $w = h(0) \in [0,1]$. Thus the functions φ_1^+ and φ_1^- are well defined in $\hat{\mathbf{D}}$. Let

$$\mathbf{D}_1^+ = \varphi_1^+(\hat{\mathbf{D}})\,;$$
$$\mathbf{D}_1^- = \varphi_1^-(\hat{\mathbf{D}})\,.$$

Because of **i)** , φ_1^+ is a injective map on $\hat{\mathbf{D}}$. Let ψ_1^+ be the inverse of φ_1^+. By **ii)** , ψ_1^+ is analytic in the domain $\mathbf{D}_1^+$. Similarly, there is an analytic inverse function ψ_1^- of φ_1^-. Define the function

$$h_1(w) = \begin{cases} \varphi_1^+(w) & \text{if } w \in \mathbf{D}_1^+ \\ \varphi_1^-(w) & \text{if } w \in \mathbf{D}_1^- \end{cases}$$

and let $\mathbf{D}_1 = \mathrm{Int}(\overline{\mathbf{D}_1^+ \cup \mathbf{D}_1^-})$. One can check that

$$\mathbf{D} \subset \mathbf{D}_1$$

and

$$h_1(w) = h(w), \text{ for all } w \in \mathbf{D} \setminus [-1,1].$$

Hence h_1 can be defined on $\mathbf{D}_1$ and h_1 is the analytic continuation of h. Similarly, one can get the analytic continuation of h_1 by (3.1).

In general, let

$$\mathbf{D}_n = h^{-1}(\mathbf{D}_{n-1}).$$

The numerical results show that $\mathbf{D}_n$ approximates Ω well when n is large (see Fig. 2 where $n = 1$ and Fig. 3 where $n = 7$). However, the analytic continuations of h to the domains $\mathbf{D}_n$ when $n \geq 3$ are not rigorously proved yet.

Now we turn to discuss the singularity of h at $w = 1$, i.e. the singularity of g at $z = 0$. The numerical results show that the fractal formed by the singular points has a quadratic cusp (Fig. 4). From the datas in the Table 2, we know that $\text{Im } y_n \sim 3.7(\text{Re } (y_n - 1))^2$.

Moreover, the density of singularities near $y = 1$ is exponentially small. To see this we first introduce some notation. Let us define

$$A_{-1} = \{1\}\,;$$
$$A_n = h^{-1}(A_{n-1})\,.$$

and

$$x_{-1} = 1\,,\ x_0 = -1\,;$$
$$x_n = \sqrt{3x_{n-1} - 1 - x_{n-2}}\,.$$

Note that the sequence $\{x_n\}_{n\in\mathbf{N}}$ differs from the sequence $\{y_n\}_{n\in\mathbf{N}}$ since only the positive square root is considered. Set

$$S_{n,k} = \{y \in A_n\,,\ |y-1| < |x_k - 1|\}$$

Let $N_{n,k}$ denote the number of the points in $S_{n,k}$. By the numerical results in Table 1, we have

$$N_{n,(k+1)} \sim \frac{1}{2} N_{n,k}\,.$$

Hence

$$N_{n,k} \sim 2^{n-k}\,.$$

The datas in Table 2 show that

$$|x_k - 1| \sim \frac{1}{k}\,.$$

Let N_n be the number of the points in A_n. Then

$$\frac{N_{n,k}}{N_n} \sim 2^{-k} \sim 2^{-1/|x_k-1|} ,$$

and the density of singularities of g near $z = 0$ is about $2^{-1/x}$, where $x = \text{Re } z$.

This exponential weakness of the singularity at $z = 0$ is clearly in very good agreement with the effectiveness of normal forms and first integrals in the reproduction of phase plots and the determination of Nekhoroshev stability estimates.

4. Numerical Computation of the Solution

In this section we introduce a numerical method to solve the functional equation approximately and prove a convergence result for piecewise constant approximations of h (the solution of (2.6)) which have been used in the numerical computations.

Let us consider a partition of the unit disk $\mathbf{D}$ into small pieces $\mathbf{D} = \cup_{i=1}^{m} D_i$, and denote with d the maximum diameter of the D_i: $d := \max \text{diam } D_i$. We put

$$h^0(y) := \frac{y^2+1}{3} , \tag{4.1}$$

and let h_m^0 be a function defined on $\mathbf{D}$ such that h_m^0 is constant on each D_i and maps $\mathbf{D}$ into itself. For all $n \geq 0$ and $y \in \mathbf{D}$ we consider the sequence of functions $\{h_m^n\}_{n\in\mathbf{N}}$ defined as follows

$$h_m^{n+1}(y) := h_m^0(y) + \frac{h_m^n(h_m^n(y))}{3} . \tag{4.2}$$

Clearly each function h_m^n is constant on each set D_i and maps $\mathbf{D}$ into itself. We define

$$\varepsilon := \max_{y\in\overline{\mathbf{D}}} |h_m^0(y) - h(y)| , \tag{4.3}$$

$$\alpha_n := \max_{y\in\overline{\mathbf{D}}} |h_m^n(y) - h(y)| . \tag{4.4}$$

Then one obviously has

$$\begin{aligned}|h_m^{n+1}(y) - h(y)| &\leq \varepsilon + \frac{1}{3}|h_m^n(h_m^n(y)) - h(h_m^n(y))| + \frac{1}{3}|h(h_m^n(y)) - h(h(y))| \\ &\leq \varepsilon + \frac{2}{3}\alpha_n , \end{aligned}\tag{4.5}$$

since the first derivative of h on $\overline{\mathbf{D}}$ is bounded by 1. Thus

$$\alpha_{n+1} \leq \varepsilon + \frac{2}{3}\alpha_n \leq 3\varepsilon + \left(\frac{2}{3}\right)^{n+1} \alpha_0 \ . \tag{4.6}$$

Clearly one can make the initial choice of h_m^0 in such a way that there exists a positive constant c such that

$$\varepsilon \leq cd \ , \tag{4.7}$$

and then if $n \to 0$ and $d \to 0$ one finds that

$$\max_{y \in \overline{\mathbf{D}}} |h_m^n(y) - h(y)| \leq cd + \left(\frac{2}{3}\right)^n \alpha_0 \to 0 \ . \tag{4.8}$$

Acknowledgements

I wish to thank G. Turchetti and S. Marmi for their help in preparing this paper and G. Servizi and M. Malavasi for their help in the numerical work.

REFERENCES

1. Arnol'd V I 1980 *Chapitres supplémentaires de la théorie des équations différentielles ordinaires* (Moscow: Mir)

2. Brjuno A D 1989 *Local Methods in Nonlinear Differential Equations* (Berlin, Heidelberg, New York: Springer Verlag)

3. G. Turchetti, *Nekhoroshev Stability Estimates for Symplectic Maps and Physical Applications* in *Number Theory and Physics* J.–M. Luck, P. Moussa and M. Waldschmidt editors, Springer Proceedings in Physics **47**, 223–234, (1990).

4. A. Bazzani, S. Marmi and G. Turchetti, *Nechoroshev estimate for isochronous non resonant symplectic maps.* Celestial Mechanics, 1990, in press

5. D. Bessis, S. Marmi and G. Turchetti *Rendiconti di Matematica*, **9**, 645-659, (1989)

6. G. Turchetti, this volume

7. A. Bazzani, this volume

Table 1.

n	k	$N_{n,k}$
16	1	17978
16	2	10436
16	3	6324
16	4	3424
16	5	1874
16	6	976
16	7	512
16	8	272
16	9	148
16	10	78
16	11	40
16	12	20
16	13	10
16	14	6
16	15	2
16	16	2
16	17	0

Table 2.

n	$\mathrm{Re}(x_n - 1)$	$n \times \mathrm{Re}(x_n - 1)$	$\mathrm{Im} x_n / (\mathrm{Re}(x_n - 1))^2$
1	0.000000D+00	0.000000D+00	0.000000D+00
101	0.996325D–02	0.100629D+01	–.376109D+01
201	0.497446D–02	0.999866D+00	–.373625D+01
301	0.331763D–02	0.998607D+00	–.372897D+01
401	0.248941D–02	0.998255D+00	–.372550D+01
501	0.199235D–02	0.998167D+00	–.372347D+01
601	0.166085D–02	0.998174D+00	–.372214D+01
701	0.142399D–02	0.998217D+00	–.372121D+01
801	0.124629D–02	0.998275D+00	–.372051D+01
901	0.110803D–02	0.998338D+00	–.371997D+01
1001	0.997402D–03	0.998399D+00	–.371954D+01
1101	0.906866D–03	0.998459D+00	–.371919D+01
1201	0.831403D–03	0.998515D+00	–.371890D+01
1301	0.767539D–03	0.998569D+00	–.371865D+01
1401	0.712790D–03	0.998618D+00	–.371844D+01

Fig. 1

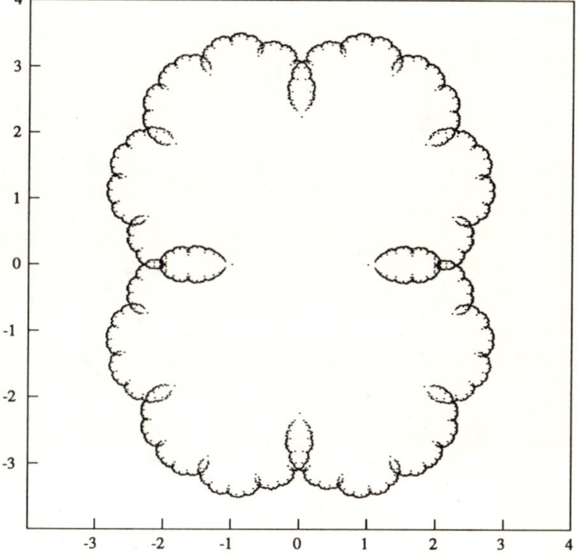

Fig. 2

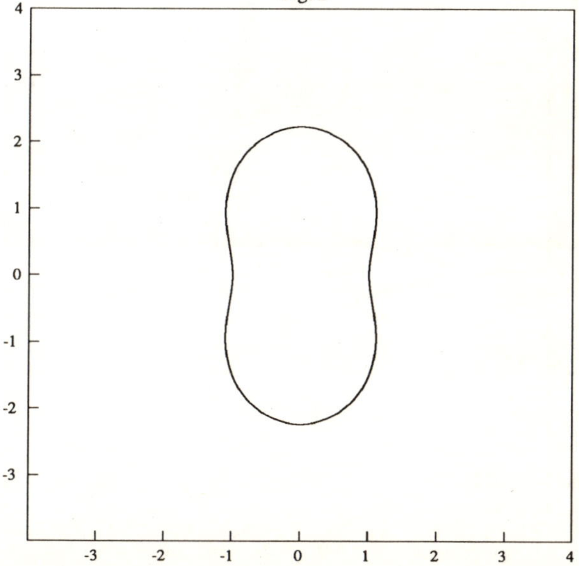

Fig. 3

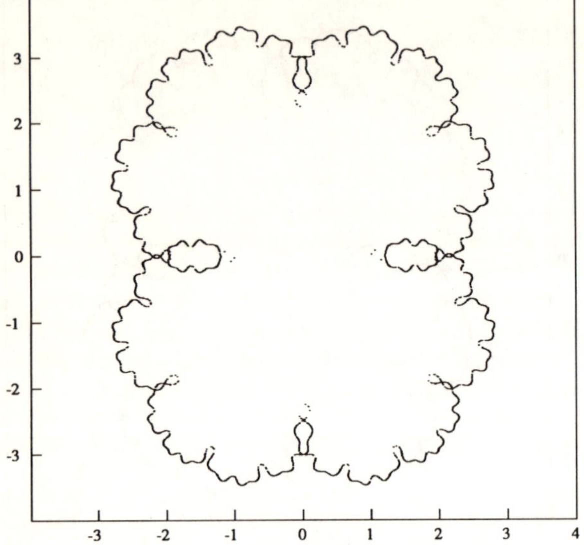
4
3
2
1
0
-1
-2
-3
-3
-2
-1
0
1
2
3
4

Fig. 4

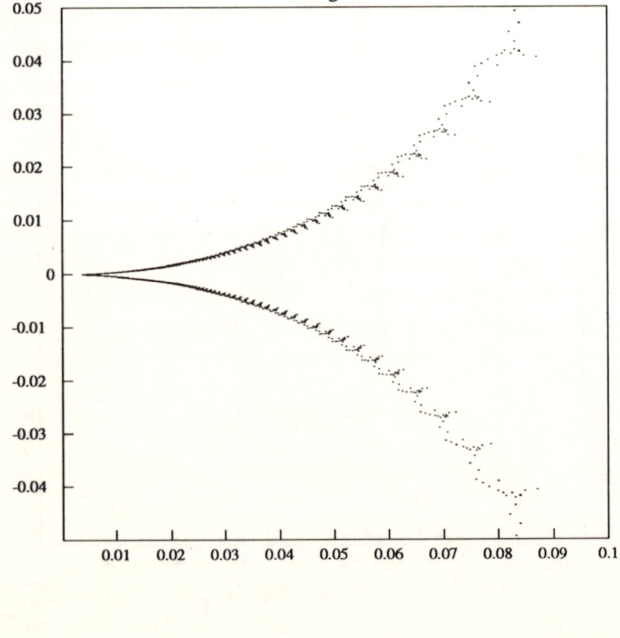
0.05
0.04
0.03
0.02
0.01
0
-0.01
-0.02
-0.03
-0.04
0.01
0.02
0.03
0.04
0.05
0.06
0.07
0.08
0.09
0.1

CRITICAL FUNCTIONS FOR AREA–PRESERVING MAPS

STEFANO MARMI †

Dipartimento di Matematica "U. Dini" - Università di Firenze
Viale Morgagni 67/A – 50134 Firenze – Italy

ABSTRACT

Given a near to integrable area–preserving twist map of the cylinder or of the plane, the critical function associates to each invariant circle of fixed rotation number ω the breakdown value $K(\omega)$ at which the perturbation destroys the invariant circle itself. In this note we study critical functions for complex area–preserving maps and for the standard map. We show that a purely arithmetic function, the Brjuno function, which only depends on the continued fraction expansion of the rotation number, gives their interpolation. This provides a good method for estimating the breakdown threshold for large classes of winding numbers, provided some normalization is fixed.

1. INTRODUCTION.

Most near to integrable Hamiltonian systems have a phase space divided into regions of regular and chaotic motion. These are separated by KAM invariant circles, on which the dynamics is conjugated to a rotation. The tori can be distinguished by the irrational rotation number ω of the motion on them. From the so–called "converse KAM theory"[1,2,3] one knows that if the perturbation's strength is too large there cannot be any torus with a given rotation number ω. Therefore, to each ω corresponds a critical value $K(\omega)$ of the perturbation's strength, the breakdown theshold, beyond which the torus labeled by ω is destroyed. For example if ω is rational one has $K(\omega) = 0$. The dependence of this critical parameter on the rotation number is the critical function $\omega \to K(\omega)$.

The problem of obtaining accurate estimates of the breakdown threshold for an invariant circle with given rotation number is still basically unsolved, since one lacks a rigorous and computationally effective method, and has been recently object of intense study[4,5,6,7].

For this reason the investigation – even if numerical – of critical functions is a very difficult one. On the other hand some basic facts are by now established: first of all critical functions have a complicated fractal structure[8,9], with some good

† Supported by ICSC World Laboratory

evidence of self–similarity. Moreover they seem to have some interesting scaling properties under the action of the modular group[10].

In this paper we report on some recent progress on the understanding of critical functions for area–preserving maps[11,12,13]. In particular we will show how one can interpolate critical functions by means of a universal function, the *Brjuno function*, which depends only on the continued fraction expansion of the rotation number[14,15,16,17,18]. For certain classes of quadratic irrationals, as the *noble numbers*, the Brjuno function can be exactly computed, since it verifies explicit functional relations under the action of modular transformations.

We will also describe a preliminary numerical result which suggests that the Brjuno function provides an interpolation of the standard map critical function. This may open the way to a new simple method for estimating the breakdown threshold of its invariant circles once a normalization (for example at the golden mean rotation number) is fixed.

The paper is organized as follows: in section 2 we will recall some basic facts on continued fractions and we define the Brjuno function. This is explicitely computed for a class of quadratic irrationals and for noble numbers. In section 3 we discuss two model problems, the Semi–Standard Map (SSM) and the Standard Map (SM). For the SSM, which is the complexification of the SM, one can prove that the convergence of the Brjuno function is a necessary and sufficient condition for the existence of invariant circles. We show how the Brjuno function effectively interpolates the critical function of the SSM and of the SM itself.

Some of these ideas[12] may be used to study the problem of dynamical aperture in particle accelerators, which can be modeled by area–preserving maps, as described by G. Turchetti in this volume. This is an interesting and challenging problem which however implies the understanding of the relation among *complex* and *real* dynamics which is, for the time being, an open mathematical problem.

2. Continued fractions and the Brjuno function.

Let $[a_0, a_1, \ldots]$ denote the coefficients of the continued fraction expansion of an irrational number $\omega \in \mathbf{R} \setminus \mathbf{Q}$, recursively determined by $a_j = [1/\omega_j]$ and $\omega_j = \{1/\omega_{j-1}\}$ for all $j \geq 1$, where $[\ ]$ and $\{\ \}$ respectively denote the integer part and the fractional part of a real number, $\omega_0 = \omega - [\omega]$ and $a_0 = [\omega]$. The partial fractions

$$\frac{p_k}{q_k} = a_0 + \cfrac{1}{a_1 + \cfrac{1}{a_2 + \cfrac{1}{\ddots + \cfrac{1}{a_k}}}} \tag{2.1}$$

are given by

$$\begin{aligned} p_k &= a_k p_{k-1} + p_{k-2} \\ q_k &= a_k q_{k-1} + q_{k-2} \end{aligned} \tag{2.2}$$

for all $k \geq 0$, with initial data $q_{-2} = p_{-1} = 1$, $q_{-1} = p_{-2} = 0$, and verify the inequalities

$$\frac{1}{q_k(q_k + q_{k+1})} < (-1)^k(\omega - \frac{p_k}{q_k}) < \frac{1}{q_k q_{k+1}} . \tag{2.3}$$

Siegel's diophantine condition on ω assumes the existence of two constants $\gamma > 0$ and $\mu \geq 2$ such that for all integers $p, q \in \mathbf{Z}$, $q \neq 0$ one has

$$\left|\omega - \frac{p}{q}\right| \geq \gamma |q|^{-\mu} , \tag{2.4}$$

and implies that successive denominators of the partial fractions of ω verify the estimate

$$\log q_{k+1} = \mathcal{O}(\log q_k) . \tag{2.5}$$

Brjuno[14,15] and Rüssmann[16,17] showed that for certain *complex* small divisors problems this condition can be weakened: it suffices to require that

$$\sum_{k=0}^{+\infty} \frac{\log q_{k+1}}{q_k} < +\infty . \tag{2.6}$$

Recently Jean–Christophe Yoccoz[18] has proved that the Brjuno condition Eq. 2.6 is also necessary for the Siegel problem, i.e. the problem of linearisation of germs of complex analytic diffeomorphisms in a neighborhood of an irrational indifferent fixed point, thus obtaining for the first time a necessary and sufficient condition for the existence of an analytic solution for a small divisors problem.

The set S of rotation numbers corresponding to linearizable germs is invariant under the action of the modular group PSL $(2, \mathbf{Z})$ on $\mathbf{R}$: let $T \in$ PSL $(2, \mathbf{Z})$. For all $x \in \mathbf{R}$

$$Tx := \frac{ax + b}{cx + d} , \tag{2.7}$$

where $a, b, c, d \in \mathbf{Z}$, $ad - bc = 1$.

Two irrational numbers ω and ω' are *equivalent* if and only if their continued fractions coincide apart from finitely many terms: $\omega = [a_0, \ldots, a_m, c_0, c_1, \ldots]$, $\omega' = [b_0, \ldots, b_n, c_0, c_1, \ldots]$. This condition is exactly equivalent to $\omega = T\omega'$ with $T \in$ PSL $(2, \mathbf{Z})$.

In his proof Yoccoz introduces a modified continued fraction expansion and defines an arithmetic function $B : \mathbf{R} \setminus \mathbf{Q} \to \mathbf{R}^+ \cup \{\infty\}$ which is even, $\mathbf{Z}$-periodic and finite if and only if evaluated at those irrational values of ω which verify Brjuno condition.

$B(\omega)$ is obtained from the modified continued fraction as follows: let $< >$ denote the nearest integer function and $\| \|$ the distance from the nearest integer of a real number, and let ω be irrational. For all $k \geq 1$ define

$$b_k = < \theta_{k-1}^{-1} > \\ \theta_k = \|\theta_{k-1}^{-1}\| \tag{2.8}$$

with initial data $b_0 = < \omega >$, $\theta_0 = \|\omega\|$. Moreover let $\beta_k = \prod_{i=0}^{k} \theta_k$ if $k \geq 0$, and $\beta_{-1} = 1$. $B(\omega)$ is defined by

$$B(\omega) := -\sum_{k=0}^{+\infty} \beta_{k-1} \log \theta_k \,. \tag{2.9}$$

The relation with the standard continued fraction expansion is due to the fact that can prove[18] that for all k there exists an integer n such that $\beta_k = (-1)^n(q_n\omega - p_n)$. The Brjuno condition Eq. 2.6 is equivalent to $B(\omega) < \infty$.

Clearly $B(\omega) = B(\omega + 1) = B(-\omega)$. Moreover, for all $\omega \in \mathbf{R} \setminus \mathbf{Q} \cap]0, 1/2]$, $B(\omega)$ verifies the functional equation

$$B(\omega) = -\log \omega + \omega B(\omega^{-1}) \,, \tag{2.10}$$

whereas if $\omega \in \mathbf{R} \setminus \mathbf{Q} \cap]1/2, 1]$ one has

$$B(\omega) = B(1-\omega) = -\log(1-\omega) + (1-\omega) B\left(\frac{1}{1-\omega}\right) \,. \tag{2.11}$$

From these relations one may read off immediately the behaviour of the Brjuno function under the action of the modular group PSL$(2, \mathbf{Z})$ which has generators $\omega \mapsto \omega + 1$ and $\omega \mapsto 1/\omega$.

The Brjuno function can be explicitely computed for some special classes of quadratic irrationals: the family $\omega_m = (\sqrt{m^2 + 4} - m)/2 = [m, m, \ldots]$, which are solutions of

$$m + \omega_m = \frac{1}{\omega_m} \tag{2.12}$$

and the so–called noble numbers, i.e. those irrationals which are obtained from the application of a modular transformation to the golden mean $\omega_1 = (\sqrt{5} - 1)/2 = [1, 1, \ldots]$. The noble numbers $\omega_{a_1, \ldots, a_n}$ have a continued fraction expansion $[a_1, \ldots, a_n, 1, 1, \ldots]$ which differs from the expansion of the golden mean only for finitely many terms, i.e.

$$\omega_{a_1, a_2, \ldots, a_n} = \cfrac{1}{a_1 + \cfrac{1}{a_2 + \cfrac{1}{\ddots + \cfrac{1}{a_n + \omega_1}}}} \,. \tag{2.13}$$

In the first case Eq. 2.12, by applying Eq. 2.10, one immediately has

$$b_m := B(\omega_m) = \frac{\log(\omega_m)}{\omega_m - 1}, \tag{2.14}$$

for all $m \geq 2$, whereas for the golden mean the result is obtained through Eq. 2.11, since $B(\omega_1) = B(1-\omega_1)$ and $\omega_1 + 2 = 1/(1-\omega_1)$

$$b_1 := B(\omega_1) = -\frac{\log(1-\omega_1)}{\omega_1} = 1.557234177\ldots . \tag{2.15}$$

We now consider the case of the noble numbers. Let us denote $b_{a_1,a_2,\ldots,a_n}$ the value of the Brjuno function $B(\omega_{a_1,a_2,\ldots,a_n})$. This can be computed in terms of b_1 by iterating Eq. 2.10 or Eq. 2.11 finitely many times. In fact, if $a_k \geq 2$ then $\omega_{a_k,a_{k+1},\ldots,a_n} \in]0,1/2[$ and by Eq. 2.10 one has

$$b_{a_k,a_{k+1},\ldots,a_n} = -\log \omega_{a_k,a_{k+1},\ldots,a_n} + \omega_{a_k,a_{k+1},\ldots,a_n} b_{a_{k+1},\ldots,a_n} . \tag{2.16}$$

If $a_k = 1$, so that $\omega_{a_k,a_{k+1},\ldots,a_n} \in]1/2,1[$, it suffices to observe that

$$\frac{1}{1-\omega_{a_k,a_{k+1},\ldots,a_n}} = a_k + a_{k+1} + \omega_{a_{k+2},\ldots,a_n}, \tag{2.17}$$

to find that, by Eq. 2.11

$$b_{a_k,a_{k+1},\ldots,a_n} = -\log(1-\omega_{a_k,a_{k+1},\ldots,a_n}) + (1-\omega_{a_k,a_{k+1},\ldots,a_n}) b_{a_{k+2},\ldots,a_n} . \tag{2.18}$$

3. The critical functions of the Semi–Standard Map and of the Standard Map.

The critical function $K = K(\omega)$ has been studied by various authors for complex area–preserving maps[19,8,20,9,11,12,13,22]. These maps have most of the relevant features of the real maps. However, the study of the perturbation series which parametrize the invariant circles is considerably semplified both algorithmically (since the recursive relation for the coefficients are very simple) and numerically (the series are absolutely convergent and therefore, on the contrary of what happens for real area–preserving maps as the standard map, the convergence of the series is not due to complicated cancellation mechanisms).

We will describe in this paragraph a model problem, the *Semi–Standard Map* (SSM)

$$\begin{aligned} y_{n+1} &= y_n + iKe^{ix_n} \\ x_{n+1} &= x_n + y_{n+1} \end{aligned} \tag{3.1}$$

where $(y_n, x_n) \in \mathbf{C} \times \mathbf{C}/2\pi\mathbf{Z}$, $K \in \mathbf{R}_+$, and we will show that the critical function of this map is interpolated by the Brjuno function. The SSM is a complexification of the well known *Standard Map* (SM)

$$\begin{aligned} y_{n+1} &= y_n + K \sin x_n \\ x_{n+1} &= x_n + y_{n+1} \end{aligned} \tag{3.2}$$

where $(y_n, x_n) \in \mathbf{R} \times \mathbf{R}/2\pi\mathbf{Z}$, $K \in \mathbf{R}_+$. Clearly Eq. 3.1 approximates Eq. 3.2 if $\mathrm{Im}\, x_n << 0$, so that e^{-ix_n} is negligible[21].

We will first state a theorem[13] which insures that the convergence of $B(\omega)$ is a necessary and sufficient condition for the existence of an invariant circle for the SSM. Then we will numerically give some evidence that $K(\omega) \approx \exp[-2B(\omega)]$ both for the SSM and the SM.

Let us suppose that $\omega \in \mathbf{R} \setminus \mathbf{Q}$ is a strongly irrational rotation number, so that there exists a corresponding invariant circle for $0 \leq K < K(\omega)$. Then the SSM is analytically conjugated to the rotation

$$z_{n+1} = e^{2\pi i\omega} z_n, \; z_n = K e^{i\phi_n} \tag{3.3}$$

where $\phi \in \mathbf{R}/2\pi\mathbf{Z}$ is an angle which parametrizes the invariant curve. This means that there exists a function $\Phi(z)$, analytic in the disk $\mathbf{D}_{K(\omega)}$, such that

$$\begin{aligned} \Phi(z_n) &= i(x_n - \phi_n) \\ y_n = x_n - x_{n-1} &= -i[\Phi(z_n) - \Phi(z_{n-1})] + 2\pi\omega \; . \end{aligned} \tag{3.4}$$

The conjugation function $\Phi(z)$ can be developed into a convergent power series $\Phi(z) = \sum_{n=1}^{+\infty} \Phi_n z^n$, and must solve the functional equation

$$(\Delta^2\Phi)(z) = -z \exp[\Phi(z)] \; , \tag{3.5}$$

where

$$(\Delta^2\Phi)(z) = \Phi(e^{2\pi i\omega} z) - 2\Phi(z) + \Phi(e^{-2\pi i\omega} z) \; . \tag{3.6}$$

To obtain the recurrence which defines the coefficients of $\Phi(z)$ we define

$$D_n := |2\sin(\pi n\omega)|^2 \; , \; n \geq 1 \tag{3.7}$$

which gives the small divisors sequence.

By equating powers in Eq. 3.5 one finds

$$\Phi_1 = \frac{1}{D_1}, \; \Phi_n = \frac{1}{D_n} \sum_{j=1}^{n-1} \frac{1}{j!} \sum_{k_1+\ldots+k_j=n-1} \Phi_{k_1} \ldots \Phi_{k_j} \; , n \geq 2 \, , \tag{3.8}$$

and the critical function $K(\omega)$ is obtained by Hadamard's formula: since all Φ_n are positive we can drop the modulus and

$$K(\omega) = [\limsup_{n\to\infty} \Phi_n^{1/n}]^{-1} . \tag{3.9}$$

The main rigorous result is the following theorem

Theorem. *A necessary and sufficient condition for the existence of an invariant circle for the SSM with rotation number ω is given by the Brjuno condition Eq. 2.6. Moreover*

$$\log K(\omega) \geq -\log \frac{21}{4} - 4 \sum_{k=0}^{+\infty} \frac{\log q_{k+1}}{q_k} . \tag{3.10}$$

This result has been generalized[22] to consider more general perturbations which are relevant in plasma physics.

It has been showed[11,13] how one is led to conjecture that

$$K(\omega) = \frac{e^{-2B(\omega)}}{C_S(\omega)} \tag{3.11}$$

where $C_S(\omega)$ is a continuous positive function defined on $\mathbf{R}$, periodic of period 1, symmetric about 1/2 and with strictly positive minimum on $[0, 1/2]$.

In figure 1 we have reported the critical function $K(\omega)$ of the SSM for 5000 random rotation numbers uniformly distributed $\omega \in [0, 1/2]$ computed by applying Hadamard's formula to Φ_{500}. The ratio $C_S(\omega) = e^{-2B(\omega)}/K(\omega)$ (figure 2) is clearly bounded and has angular points at all rationals. This supports the validity of conjecture Eq. 3.11.

For the SM one may use the perturbative approach to compute the critical function. Given a strongly irrational rotation number $\omega \in \mathbf{R} \setminus \mathbf{Q}$, so that there exists a corresponding invariant cricle for $0 \leq K < K(\omega)$, on this circle the SM is analytically conjugated to the irrational translation

$$T_\omega \ : t \mapsto t + 2\pi\omega . \tag{3.12}$$

This means[23] that there exists a function $u(t, K)$, analytic for $|K| < K(\omega)$ in some strip $|\operatorname{Im} t| \leq \delta(K)$, where $\delta(K) > 0$ if $|K| < K(\omega)$, such that

$$x = t + u(t, K) = t + \sum_{n=1}^{\infty} K^n \sum_{m=-n}^{n} u_{n,m} e^{imt} . \tag{3.13}$$

The conjugation function $u(t,K)$ must therefore solve the functional equation

$$u(t+2\pi\omega,K)-2u(t,K)+u(t-2\pi\omega,K)=K\sin(t+u(t,K))\;, \tag{3.14}$$

which gives an explicit recurrence for the coefficients $u_{n,m}$ of $u(t,K)$. Once these are known one may set

$$K(\omega):=\inf_{t\in[0,2\pi]}\left[\limsup_{n\to\infty}|u_n(t)|^{\frac{1}{n}}\right]^{-1}\;, \tag{3.15}$$

where $u_n(t)=\sum_{m=-n}^{n}u_{n,m}e^{imt}$. Another possibility is the method described in[24] and based on the exponential decay of the Fourier coefficients $u_{n,m}$ of $u(t,K)$ as $m\to\infty$. This is due to the analiticity of u in the strip $|\ \mathrm{Im}\ t|\le\delta(K)$.

If one may compute sufficiently fast high order expansions of $u(t,K)$ this would probably be the optimal scheme. The numerical implementation of this is presently carried out[25], but it requires some hard work in order to obtain an accurate and fast computer code (see, for example, references[24,5] for more details).

On the other hand one knows that converse KAM theory[1,2,3] gives surprisingly accurate upper bounds to the existence of invariant circles once a certain portion of phase space is considered. It is easy to prove that no rotational invariant circles of the SM exist if $|K|>4/3$, a value quite close to the conjectured[26] breakdown threshold of the golden circle $K_{cr}=0.97163\ldots$. Note that analytic proofs of KAM theory give lower bounds which are typically 2 or more orders of magnitude far from K_{cr}.

The criterion for non existence of a rotational invariant circle is the following: consider the tangent map to Eq. 3.2 which can be written in Newtonian form as

$$\delta x_{n+1}=(2+\varepsilon\cos x_n)\delta x_n-\delta x_{n-1}\;, \tag{3.16}$$

where $n\ge 1$ and δx_n is the δx component of a tangent orbit of Eq. 3.2. Let $\delta x_0=0$ and $\delta x_1=1$. If for some $n\ge 2$ one has

$$\delta x_n\le 0\;, \tag{3.17}$$

then the orbit $(x_n)_{n\in\mathbf{N}}$ does not lie on a rotational invariant circle.

It is evident that this criterion is very well adapted for numerical investigations, and that provides a method for deciding if an orbit does not lie on an invariant rotational circle from tracking data.

In figure 3 we have reported the ratio among the critical function $K(\omega)$ of the SM estimated by means of this criterion and the Brjuno function $\exp(-2B(\omega))$ analytically determined by Eq. 2.16 and Eq. 2.18 at 900 noble numbers distributed in the interval $[0,1/2]$. Their continued fraction expansion is given by Eq. 2.13, where $a_1\in\{2,\ldots,10\}$, a_2 and $a_3\in\{1,\ldots,10\}$. As one can see the Brjuno function seems to provide also for the SM a good interpolation of the critical function.

For $\omega_{2,1,1} = (3 - \sqrt{5})/2 = 1 - \omega_1$ the breakdown threshold is the same as for the golden mean ω_1, since the small divisors in the recurrence which determines $u(t, K)$ are the same. Therefore one has a normalization for the ratio $\exp\{-2B(\omega)\}/K(\omega)$ at this value $\omega = \omega_{2,1,1}$. In fact $K(\omega_{2,1,1}) = K_{cr} = .97163\ldots$, $\exp\{-2B(\omega_{2,1,1})\} = 0.044402107\ldots$ and

$$C := \frac{e^{-2B(\omega_2)}}{K_{cr}} = .0456983\ldots \,. \tag{3.18}$$

From figure 3 one realizes that at least in the interval $[1/10, 1/2]$ the ratio of the critical and the Brjuno function ranges from 0.02 to 0.05. Therefore one can immediately analytically obtain an estimate of the breakdown threshold of any invariant circle with rotation number in this interval and with, for example, a periodic continued fraction, simply by computing explicitely the value of the Brjuno function as shown in the previous section and dividing the result by C

$$K(\omega) \approx \frac{e^{-2B(\omega)}}{C} \approx 21.88 e^{-2B(\omega)} \,. \tag{3.19}$$

Note that the agreement of Eq. 3.19 with the numerical values of $K(\omega)$ may be considerably improved simply by interpolating figure 3 with a straight line. One may probably obtain further refinements by fractal interpolation[27] or modular smoothing[10].

In the region $0 < \omega < 1/10$ our numerical data do not allow to draw any conclusion. For these rotation numbers it is very difficult to compute the critical function since one is very close in phase space to the big island corresponding to the zero resonance (see figure 4, where we give a picture of the phase space of the standard map at $K = 0.2$). However, it would be very surprising if the interpolation by the Brjuno function were not valid here. In fact the basic nature of the problem is essentially number theoretical and we have stressed several times that small divisors and critical functions have important scaling properties under the action of the modular group which are perfectly reproduced by the Brjuno function. Note that applying a modular transformation to the golden mean one obtains a noble number which can be as close as wanted to 0.

Acknowledgments. The content of this paper summarizes present research activity which involves also other members of the Bologna group (Massimo Giovannozzi and Monica Malavasi). I am very grateful to them for their constant friendly collaboration and to Giorgio Turchetti for his encouragement. During the preparation of this work I have been partially supported by ICSC–WorldLab and by INFN – Sezione di Bologna.

References.

1. J. N. Mather, *Topology* **21** (1982) 457–467.
2. J. N. Mather, *Ergodic Th. and Dyn. Sys.* **4** (1984) 301–309.
3. R. S. Mackay and I. C. Percival, *Comm. Math. Phys.* **98** (1985) 469–512.
4. C. Liverani and G. Turchetti, *J. Stat. Phys.* **45** (1986) 1071–1086.
5. A. Celletti and L. Chierchia, *Comm. Math. Phys.* **118** (1988) 119–161.
6. S. Marmi, *J. Phys. A: Math. Gen.* **21** (1988) L961–L966.
7. R. De La Llave and D. Rana, *Bull. of the A.M.S.* **22** (1990) 85–90.
8. I. C. Percival, *Physica* **26D** (1982) 67–77.
9. I. C. Percival and F. Vivaldi, *Physica* **33D** (1988) 304–313.
10. N. Buric, I. C. Percival and F. Vivaldi, *Nonlinearity* **3** (1990) 21–37.
11. M. Malavasi and S. Marmi, *J. Phys. A: Math. Gen.* **22** (1989) L563–L569.
12. M. Giovannozzi and S. Marmi, *Rendiconti di Matematica* **9** (1989) 457–480.
13. S. Marmi, *Critical Functions for Complex Analytic Maps* J. Phys. A: Math. Gen., (1990), to appear.
14. A. D. Brjuno, *Trans. Moscow Math. Soc.* **25** (1971) 131–288.
15. A. D. Brjuno, *Trans. Moscow Math. Soc.* **26** (1972) 199–239.
16. H. Rüssmann, *J. Math. Mech.* **17** (1967) 523–532.
17. H. Rüssmann, *Nachr. Akad. Wiss. Göttingen Math.-Phys. Kl. 1* (1972) 1–20.
18. J. C. Yoccoz, *Théorème de Siegel, polynômes quadratiques et nombres de Brjuno. Preprint* Orsay, Paris, (1988).
19. J. M. Greene and I. C. Percival, *Physica* **3D** (1981) 530–548.
20. P. R. McGarr and I. C. Percival, *Physica* **14D** (1984) 49–66.
21. V. F. Lazutkin, I. G. Schachmannski and M. B. Tabanov, *Physica* **40D** (1989) 235–248.
22. S. Marmi, *Existence of analytic invariant circles for a generalized complex standard map* to appear in "Nonlinear Phenomena in Physics of Fluids and Plasmas", Maino G, Fronzoni L and Pettini M editors, World Scientific, Singapore, (1990).
23. D. Salamon and E. Zehnder, *Comm. Math. Helvetici* **64** (1989) 84–132.
24. G. Benettin, G. Turchetti and G. Zanetti, *Phys. Lett.* **105A** (1984) 436.
25. M. Malavasi and S. Marmi, work in progress.
26. J. M. Greene, *J. Math. Phys.* **20** (1979) 1183–1201.
27. M. Barnsley, *Fractals Everywhere*, (Academic Press, Boston 1988).

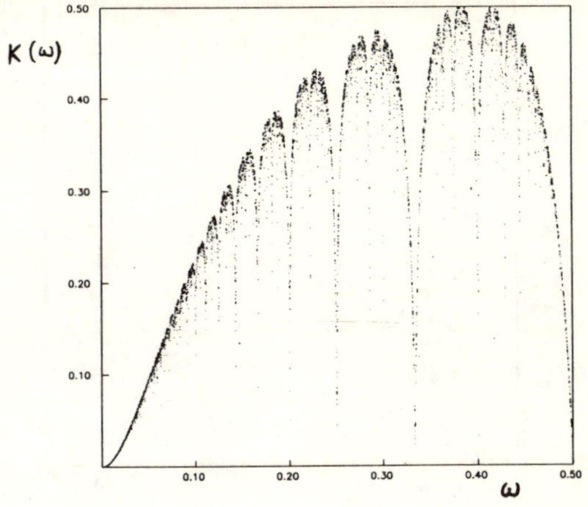

Figure 1. The critical function $K(\omega)$ of the SSM.

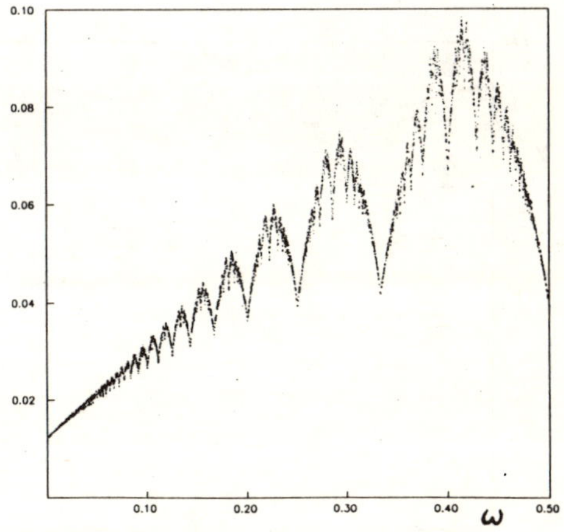

Figure 2. The ratio $C_S(\omega) = e^{-2B(\omega)}/K(\omega)$.

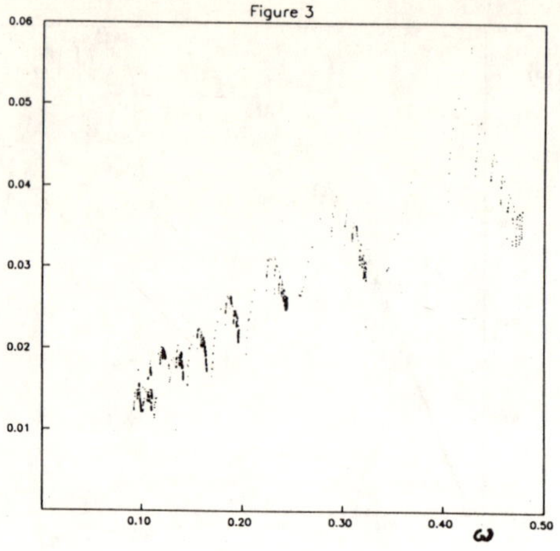

Figure 3. The ratio $e^{-2B(\omega)}/K(\omega)$ for the SM at 900 noble rotation numbers.

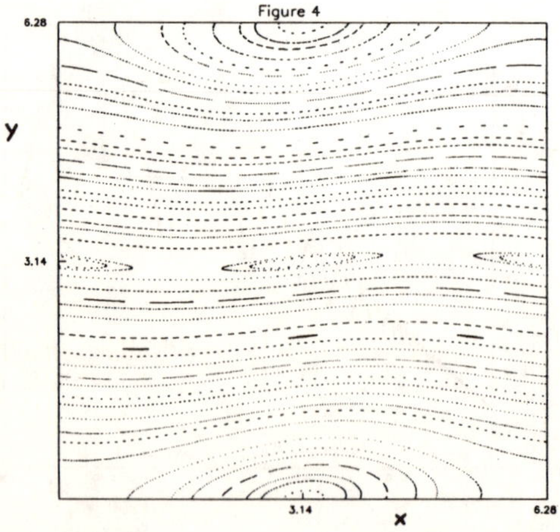

Figure 4. A phase space plot of the SM at $K = 0.2$.

CHAOS TRANSITION AND DIFFUSION IN SYMPLECTIC MAP: SOME NUMERICAL RESULTS

MONICA MALAVASI
Dipartimento di Fisica, Università di Bologna
INFN – Sezione di Bologna

ABSTRACT

We analyze the chaotic transition and the onset of diffusion in some symplectic maps of interest in physical system.
First the Generalized Standard Map is examined to investigate the dependence on the perturbation spectrum of the critical parameter and of the quasi-linear coefficient.
A model of modulated polynomial map is also briefly discussed and a definition for a distribution function is proposed.

1. Introduction

The understanding of the diffusion in the non linear Hamiltonian systems is of relevant interest in a variety of physical applications, including plasma confinement in nuclear fusion devices and particle loss from accelerators.
In the last decade these phenomena have been intensely studied in systems modelled by symplectic maps the diffusion rate is usually evaluated by the diffusion coefficient, defined as

$$D = \lim_{n \to \infty} \frac{\langle (r_0 - r_n)^2 \rangle_{r_0, \theta_0}}{2n} \tag{1}$$

where with $\langle \cdot \rangle$ we mean the average on the initial conditions [1].
In the case of two dimensional area–preserving maps we can distinguish three different regimes for the diffusion. In the first one we have no diffusion, because the phase space of an area–preserving map is divided by the KAM tori in invariant

regions that can not intersect each other; no Arnold diffusion can occur. So, as long as the perturbation is lower than a critical value, KAM curves acts as barriers to the orbits.
The KAM curves play an important role also when the perturbation exceeds the critical value, in fact in this case they appear to be partial barriers to the transport [2]. The study of the diffusion in this regime, in which the phase space is a mixture of regular and chaotic zones, is of particular relevance since it is concerned with most of the physical situations. However, this state is far from being completely understood: there is no analitical proof that the limit defining D exists and the numerical convergence is extremely slow because the trapping in the small islands. In this case D seems to be not a constant but a function of the action.
Finally when the perturbation strength is very large there is no correlation between the phase variables and it is possible to apply the quasi–linear approximation to evaluate D.
To distinguish these three regimes we must first of all determine the value of the critical parameter. In the first part of this paper we analyze its dependence on the perurbative spectra in a map that is a generalization of the well known Standard Map [3 4].
In the second part we treat the problem of the stability of the motion in a modulated quadratic map.
The presence of a slow modulation can affect the stability of a system in a very sensitive way [5 6 7]. This process can completely alter the character of the motion over long times and even yelds unbounded orbits. We have analyzed, in a specific model, the statistic of the system and to look for an estimate of the escape times , e.g. the value of t for which the motion is unbounded.
The system that we study is also a simplified model that could give an insight into the problem of diffusion due to ripples in the hadronic accelerators.

2. Dependence of the critical value from perturbation spectra

The perturbation spectra in most of the physical systems include many Fourier terms. In order to study the dependence of the transition to chaos on the spectra we introduce a generalization of the well known Standard Map [3], which is one of the most useful tools to study the main characteristics of a nonlinear system. This model can be used to describe the behaviour of the magnetic fields lines,in a magnetically confined plasma, when the equilibrium configuration is perturbed [8].
The map that we analyze is

$$\mathrm{M}:\begin{cases} r_{n+1} = & r_n + \varepsilon\dfrac{V(\theta_n)}{|V|} \\ \theta_{n+1} = & \theta_n + r_{n+1} \end{cases} \tag{2}$$

where M : $R \times T \to R \times T$ and $|\cdot|$ is the L_2–norm. In order to allow comparisons between different perturbation terms we have divided by the perturbative energy $\langle V \rangle$.
We study two different kinds of perturbative spectra decaying exponentially and linearly according to:

$$V(\theta) = \begin{cases} \sum_{k=1}^{\infty} e^{-\alpha(k-1)} \sin(k\theta) \\ \sum_{k=1}^{kmax} \frac{1}{k} \sin(k\theta) \end{cases} \tag{3}$$

We have computed numerically the critical perturbation value, ε_c, at which all the KAM curves break down by observing when the invariant curve whose rotation number is the golden mean, disappears [4 9].
It was proven, with different methods [3 10 11] that the stochastic transition for the Standard Map, in which $V(\theta) = \sin\theta$, occurs when

$$\frac{\varepsilon}{|V|} = \sqrt{2}\varepsilon_c \simeq 1 \tag{4}$$

The dependence on the spectra is very sensitive like one can observe in the Figures 1. The result of our analysis is that the critical value varies over more than an order of magnitude when we pass from an exponential decay to a linear decay of the spectrum. In fact in the case of the exponential decay with $\alpha = 0.1$ we obtain

$$\varepsilon_c \simeq 0.1 \tag{5}$$

and in the case of a linear decay with $kmax = 10$

$$\varepsilon_c \simeq 0.01 \tag{6}$$

Moreover we observe a decreasing of the critical value when we consider a greater number of frequencies in the linear case or a slower decay rate in the exponential case.
For the exponential decay we estimate the dependence of ε_c on the decay rate of the spectra.
To this end we have to make some consideration on the complexified model of the generalized Standard Map. The complex area–preserving maps are extremely interesting systems and they have a lot of common characteristics with the real area–preserving maps [12]. Moreover in the complex field we have more analytical tools to investigate the dynamics.
In the case of the complexified model of generalized Standard Map

$$\begin{cases} r_{n+1} = & r_n + \varepsilon \sum_{k \geq 0} e^{-\alpha k} e^{ik\theta_n} \\ \theta_{n+1} = & \theta_n + r_{n+1} \end{cases} \tag{7}$$

one can prove [12] a scaling law for the critical parameter.
Infact if one has two exponential spectra with different decay rates namely:

$$e^{-\alpha_1 k} \quad \alpha_1 > 0 \tag{8}$$

$$e^{-(\alpha_2)k} \quad \alpha_2 > 0 \tag{9}$$

the respéctive critical values,$\varepsilon_c^{(1)}$ and $\varepsilon_c^{(2)}$, satisfies the following law:

$$\varepsilon_c^{(2)} \simeq e^{-(\alpha_2-\alpha_1)}\varepsilon_c^{(1)} \tag{10}$$

V Numerical simulations show that this result is still valid in the real case.
When $\varepsilon \gg \varepsilon_c$ we can apply the quasilinear approximation to determine D.
In this approximation the phases are considered as completely uncorrelated. As a consequence we observe that the diffusion coefficient is almost independent from the spectral properties of the perturbating field [9].
Of course since the validity range of the quasilinear approximation is much above the chaotic transition threshold the diffusion coefficient in the weakly chaotic region remains the object of further investigation.

3. Analysis of the diffusion in the modulated quadratic map

Another interesting diffusion is related to the presence of a slow modulation in the system [5 6 7] leading to unbounded motion.
To study it we refer the one dimensional model of a FODO cell with a thin lens sextupoles where the quadrupole current is modulated by low frequency components. This is described by the equation

$$\ddot{x} + \omega^2(t)x = x^2\delta_p(t) \tag{11}$$

where

$$\omega(t) = \omega_0(1 + \varepsilon\sin(\Omega t)) \tag{12}$$

and

$$\delta_p(t) = \sum_{n\geq 0}\delta(t - n\pi) \tag{13}$$

Instead of solving this differential equation we analyzed the corresponding transfer matrix [13] which is computed up to the second order in ε and using the eikonal approximation ($\Omega/\omega_0 \rightarrow 0$) [14]. This is in agreement with our hypotesys of slow modulation and it is always justified in the physical systems.
The Poincaré Map of the system Eq.11

$$\begin{pmatrix} x_{n+1} \\ y_{n+1} \end{pmatrix} = M_n \begin{pmatrix} x_n \\ x_n^2 + y_n \end{pmatrix} \tag{14}$$

where

$$M_n = \begin{pmatrix} \sqrt{\beta_{n+1}} & 0 \\ 0 & \dfrac{1}{\sqrt{\beta_{n+1}}} \end{pmatrix} R\left[2\pi\omega_0(1+\varepsilon\sin(\Omega(n+\frac{1}{2}))\right] \begin{pmatrix} \dfrac{1}{\sqrt{\beta_n}} & 0 \\ 0 & \sqrt{\beta_n} \end{pmatrix} \quad (15)$$

and

$$\beta_n = \frac{1}{\omega_0}\left(1 - \frac{\varepsilon}{4}\sin(n\Omega)\right)^2 \quad (16)$$

In the phase plot, fig.2, the deformation due to the similarity transformation is removed by means of a suitable linear transformation. We note also from the pictures that when the modulation is turned on the characteristics of the motion drastically change.

A complete analysis of the diffusion can be done studying the statistical properties of the distribution function. To do that consider the previous map

$$\mathbf{x}_{n+1} = T^n(\mathbf{x}) \quad (17)$$

and take a uniform distribution of points in the disk $S_{r_0} : |\mathbf{x}| \leq r_0$. The Lebesgue measure μ is preserved by the map

$$\mu(S_r) = \mu(T^n(S_r)) \quad (18)$$

We normalize the measure as follows

$$\mu(S_r) \rightarrow \frac{\mu(S_r)}{\mu(S_{r_0})} \quad (19)$$

In this way it is possible to define the distribution function:

$$\mu(S_r \bigcap T^n(S_{r_0})) \equiv f(r, r_0, n) \quad (20)$$

If there exists an invariant curve γ which is the boundary of a domain D_γ letting

$$S_{r_1} \supset D_\gamma \supset S_{r_0} \quad (21)$$

we have

$$f(r, r_0, n) = 1 \quad \forall r \geq r_1 \quad (22)$$

and hence there is no diffusion.

When there is no modulation Eq(17) reduces to the quadratic Hénon map used in the study of the one dimensional betatron motion of a charged particle in a hadronic accelerator [15 16].

In this map the perturbative parameter is the distance from the origin and for $r < r_c$ the phase space is foliated in invariant curves. Therefore no diffusion can occur.

In our numerical analysis we use the invariant curves of the quadratic map to divide the phase space in different sectors and we study the evolution of a population of particles, initially uniformly distribuited on certain invariant curve, when the amplitude and the frequency of the modulation are varyed.
In this prelimanary work ω is kept fixed to the golden mean, $(\sqrt{5}-1)/2$, and we take a population of 1000 particles.
We observe that the diffusion depends essentially on the amplitude of the modulation and that there exists a threshold value depending on r, at which the particles begin to diffuse.
We also try determine whether the escape time satisfies the Nekhoroshev estimate. The Nekhoroshev theorem [17] establishes that given an integrable Hamiltonian with small perturbation, the variation of the actions can be bound for $0 \leq t \leq T_0 = \exp\left[b(\varepsilon_0/\varepsilon)^{(1/\alpha)}\right]$ where ε measures the intensity of the perturbation and b and α are costants. Recently the stability results known for the flows have been extended to the symplectic map [18].
In our model we define the escape time as the time at which half of the particles have $r > r_c$ and fixing $\Omega = 0.01$ we analyze the dependence of t_{esc} from ε.
The numerical simulation shows that

$$t_{esc} = A(r)e^{\frac{a(r)}{\varepsilon}} \tag{23}$$

as it is plotted in fig.3, in agreement with the Nekhoroshev estimate.
The problem of the modulated diffusion would require a deeper analysis formely investigating the moments of the distribution and the correlation functions [19] and later using a more complex model to understand the realistic physical situation.

Acknowledgements
The author would like to thank G.Turchetti,A.Bazzani,S.Marmi,E.Todesco and G.Servizi for the hepfull discussions.

References

1. A.J. Liechtenberg and M.A. Lieberman *Regular and Stochastic Motions*, Springer Verlag New York (1983).
2. R.S. Mackay, J.D. Meiss and I.C. Percival *Physica D* **29** (1987) 1.
3. B.V. Chirikov *Phy. Rep* **52** (1979) 263.
4. M. Malavasi and N. Reggiani (in preparation)
5. B.V. Chirikov, M.A. Lieberman, D.L. Shepelyansky and F.M. Vivaldi *Physica D* **14** (1985) 289.
6. F.M. Vivaldi *R. Modern Phys.* **56** (1984) 737.
7. A.L. Gerasimov, F.M. Izrailev and J.L. Tennyson (Preprint).
8. A. Bazzani, M. Malavasi, S. Siboni, C. Pellacani, S. Rambaldi and G. Turchetti *Nuovo Cimento* **B 103** (1989) 659.

9. M. Malavasi, A. Bazzani, N. Reggiani, S. Siboni and G. Turchetti in *Non linear phenomena in Physics of Fluids and Plasmas G. Maino L. Fronzoni M. Pettini Editors* World Scientific, Singapore (1990) in press.
10. J.M. Green *J. Math. Phys.* **20** (1986) 1183.
11. G. Servizi and G. Turchetti *Nuovo Cimento* **B 96** (1986) 121.
12. S. Marmi (in this volume and references therein)
13. E. Courant and H. Snyder *Ann. Phys.* **3** (1958) 1.
14. G. Turchetti (private comunication).
15. A. Bazzani, P. Mazzanti, G. Servizi and G. Turchetti *Nuovo Cimento* **B 102** (1988) 51.
16. A. Bazzani and G. Turchetti *CERN SPS report* **89–25 (SAMS)** (1989).
17. N.N. Nekhoroshev *Russ. Math Surv.* **32** (1977) 1.
18. A. Bazzani, S. Marmi and G. Turchetti *Cel. Mech.* (1990) (to appear.)
19. A. Bazzani, M. Malavasi and G. Turchetti (in preparation).

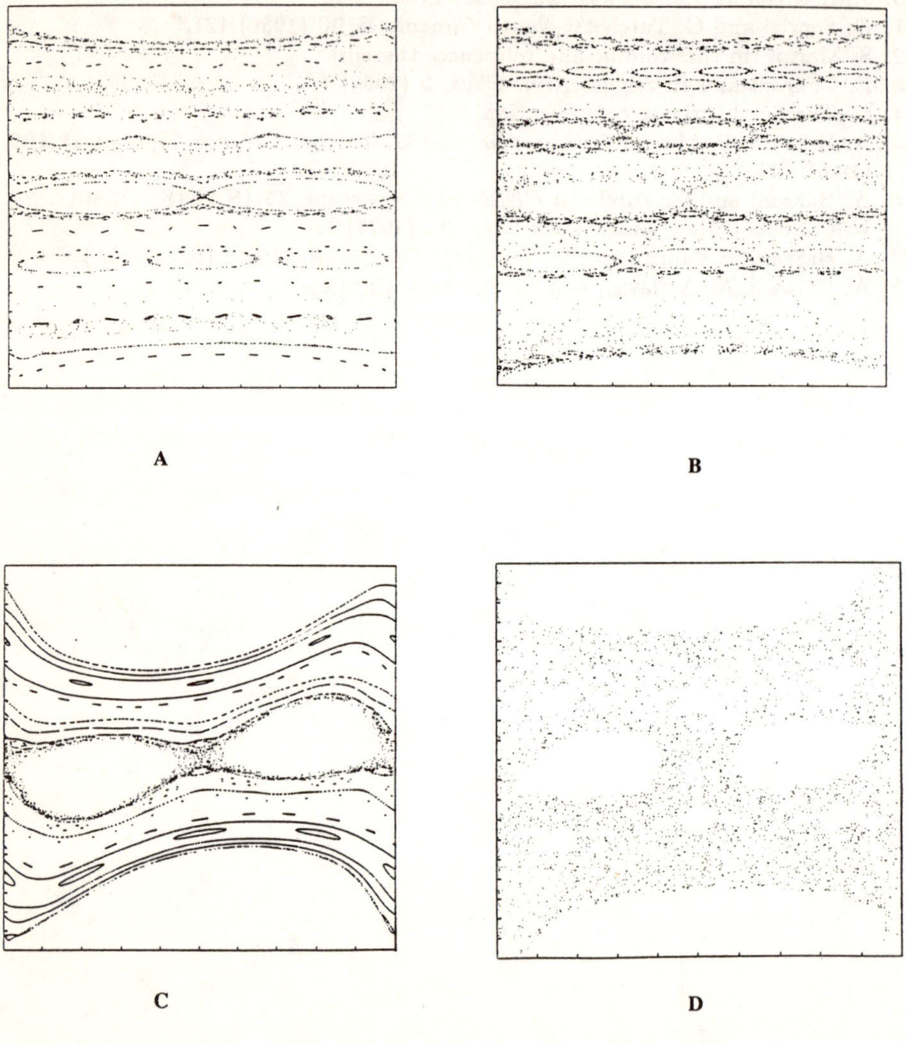

Fig 1. The phase plots of the Generalized Standard Map in the case of linear decaying for $\varepsilon = 0.05$ with A) kmax =10 , B) kmax =100 and in the case of exponential decaying for $\varepsilon = 0.5$ with C) $\alpha = 1$, D) $\alpha = 0.1$

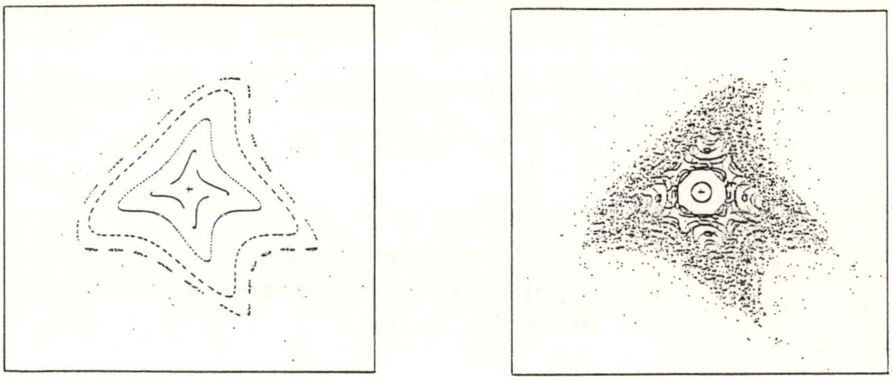

A B

Fig 2. The phase plot of the modulated quadratic map in the case of $\omega = 0.25$ for A) $\varepsilon = 0, \Omega = 0$ and B) $\varepsilon = 0.01, \Omega = 0.01$

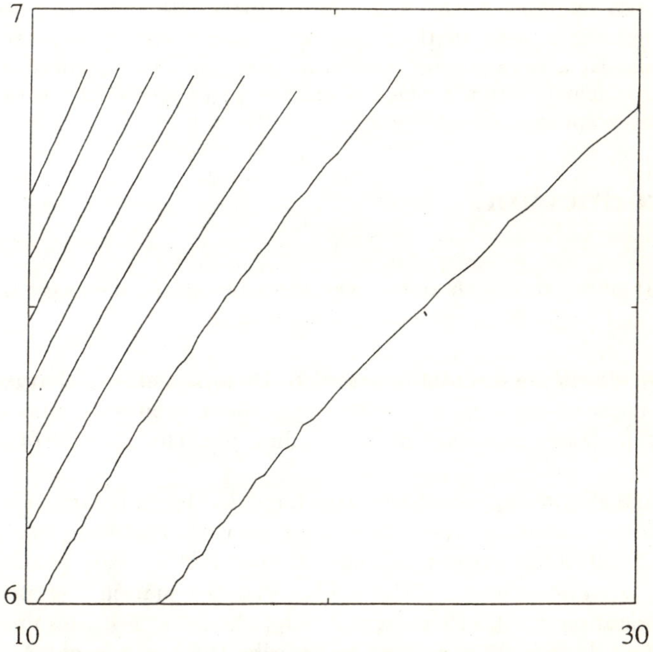

Fig 3.$\log t_{esc}$ as a function of $1/\varepsilon$ calculated for eight different initial populations situated at increasing r.

Bifurcation of the Longitudinal Equilibrium Distribution in Electron Storage Rings

S. Petracca
Dipartimento di Matematica e Applicazioni,
Universita' di Napoli, Via Mezzocannone 8,
80123 Napoli, Italy.

Abstract

The evolution of bunch length and energy spread under the action of localized wake forces can be described through a (nonlinear) mapping. The latter reveals the onset of period-one solution instability, the birth of period-two solutions, and the possible co- existence of period-one and period-two solutions, both stable. In this paper we present a preliminary comparison between the results for a simple step wake-potential, and those for a more general (nonlinear) one. The same qualitative behaviour is found, within a range of realistic parameter values, with the possible exception of bistability.

1 Introduction.

The study of bunch length and energy spread in electron storage rings [1] still reveals new and perhaps unexpected phenomena.

Collective effects are essentially described by wake fields: a charged particle traveling in an accelerator pipe interacts with its environment, thus creating a wake field. The latter acts back on the trailing particles thus perturbing their motion [1].

In the usual analytic approach the wake force, felt by each particle in a bunch, is assumed to be uniformly distributed along the ring. Following this approach, there exists a threshold current I_{th}, such that if $I < I_{th}$, the bunch profile is described by a static solution of the Fokker-Planck equation [2] (Potential Well Distortion Equation, [3–9]); otherwise if $I > I_{th}$ the latter becomes unstable and Mode Coupling Theory [10–13] is used to describe the bunch profile.

However, many sources of wake forces (cavities, bellows, etc.) are localized, especially for *large* machines, like LEP. The wake force is thus a time dependent function, where time means longitudinal position along the ring. Moreover, it should be easier to study localized sources first, and then extend the analysis to distributed sources. The opposite would be quite more difficult.

Numerical simulation codes are also available (Multi Particle Tracking (MPT) [14–17] codes) to analyze the collective effects.
In these codes, a time dipendent wake force is assumed, usually acting at one point in the ring. In this respect the usual analytic approach and MPT are different.

In an attempt to formulate an analytic model with time dependent localized wake force, Hirata [18] introduced a mapping approach describing the evolution of the distribution function from one turn to the next [18]. His main assumption that the distribution function, initially gaussian, remains gaussian even under the action of a wake force, represented by 'kicks', allows him to work in terms of the first and second order moments only. For the special case of a constant wake potential, he obtained an explicit analytic solution for period-one fixed points.

In [19] we thoroughly studied the stability of period-one fixed points of the moments mapping, for the simplest conceivable (step) wake potential. We obtained a number of new results, including the onset of instability of the period-one solution, the occurrence of period-two fixed solutions, as well as the possible coexistence of period-one and period-two solutions, both stable. All these results were found analitically, but were also favourably checked numerically, using MPT codes.

The aim of this paper is to extend the above analysis of the mapping equation to a more general nonlinear wake potential, to gain further insight into the possible dependence of the new results found in [19] on the assumption of a step wake potential.

The paper is organized as follows. In Sect. 2 we introduce the moments mapping approach for a general wake, and then specialize to the nonlinear and the constant wake case. In Sect. 3, we present a preliminary comparison of the stability and bifurcation properties of the moments mapping, for the nonlinear and step wake potential. Finally, in Sect. 4, we collect conclusions and recommendations for future work.

2 The Moments Mapping.

In this section we summarize the mapping approach introduced in [18] and used in [19]. The longitudinal beam dynamics in electron storage rings can be described in terms of the single particle motion in synchrotron phase space.

Accordingly, we introduce [18] the following dimensionless variables:

$$x_1 = \frac{\omega_s \tau}{\alpha_c}, \qquad x_2 = \frac{\epsilon}{E_s}, \tag{1}$$

where ω_s is the synchrotron angular frequency, α_c is the momentum compaction factor $\tau = (z - z_s)/v$, $\epsilon = E - E_s$, E_s, z_s being the energy and position of the synchronous particle, E, z those of the generic particle, and v the particle velocity.

In this phase space $\{x_1,\ x_2\}$ the equations of motion are the following stochastic Langevin equations:

$$\begin{aligned} \frac{dx_1}{d\theta} &= \nu\, x_2 \\ \frac{dx_2}{d\theta} &= -\nu\, x_1 - \frac{\Phi(x_1)}{2\pi} - \xi\, x_2 + \sqrt{2\, D}\, \hat{r}(\theta). \end{aligned} \tag{2}$$

where θ is time measured in units of the revolution period T_0 of the synchronous particle, $\nu = (\omega_s\ T_0)/(2\pi)$ is the synchrotron tune, ξ is a damping constant, describing the energy loss due to radiation, $\hat{r}$ is a Gaussian random variable with $< \hat{r} >= 0$ and $< \hat{r}^2 >= 1$, $D = \xi\ \sigma_0^2$, σ_0 being the nominal energy spread, and the wake force is represented by the integral:

$$\Phi(x_1) = \frac{e\ Q_{tot}}{\sigma_0\ E_s} \int_0^{+\infty} W(u)\ \rho(x - u)\ du \tag{3}$$

e being the electron charge, Q_{tot} the total bunch charge, $\rho(x)$ the bunch charge density normalized to one and $W(x)$ the wake potential.

Integrating the above system from one turn to the next, we get [20] a nonlinear stochastic mapping in x_1 and x_2. This stochastic mapping is equivalent to an infinite hierarchy of deterministic mappings in the unknown statistical moments of the distribution function $\psi(\vec{x})$, viz.: $\bar{x}_i =< x_i >$, $\sigma_{ij} =< (x_i - \bar{x}_i)(x_j - \bar{x}_j) >$, etc., $< * >$ indicating an average over all particles.
The main assumption made in [18,19] is that the distribution function in phase space is always Gaussian even in the presence of the wake force:

$$\psi(x_1, x_2) = \frac{1}{2\pi\sqrt{det\ \sigma}} \exp\left[\frac{1}{2} \sum_{i,j=1}^{2} \sigma_{i,j}^{-1}(x_i - \bar{x}_i)(x_j - \bar{x}_j)\right]. \tag{4}$$

This assumption corresponds to truncating the hierarchy to order two. Accordingly, one obtains two (uncoupled) maps: one (linear, 2-dimensional) in the first order moments, the other (nonlinear, 3-dimensional) in the second order moments. Splitting the 2nd order moments mapping into three parts, allows to describe separately the effect of radiation, wake force and synchrotron oscillation, as follows:

Radiation:

$$\begin{aligned} \sigma'_{11} &= \sigma_{11} \\ \sigma'_{12} &= \Lambda\sigma_{12} \\ \sigma'_{22} &= \Lambda^2\sigma_{22} + (1-\Lambda^2) \end{aligned} \tag{5}$$

Wake force:

$$\begin{aligned} \sigma''_{11} &= \sigma'_{11} \\ \sigma''_{12} &= \Lambda\sigma'_{12} - < (x'_1 - \bar{x}'_1)\phi(x'_1) > \\ \sigma''_{22} &= \Lambda^2\sigma'_{22} - 2 < (x'_2 - \bar{x}'_2)\phi(x'_1) > + < \phi^2(x'_1) > - < \phi(x'_1) >^2 \end{aligned} \tag{6}$$

Synchrotron oscillation:

$$\sigma'''_{ij} = \sum_{h,k=1}^{2} U_{ih}\sigma''_{hk}U^t_{kj} \tag{7}$$

Here U is the rotation matrix:

$$U = \begin{pmatrix} \cos\mu & \sin\mu \\ -\sin\mu & \cos\mu \end{pmatrix}, \tag{8}$$

and the other parameters are $\mu = 2\pi\nu$, $\Lambda = \exp(-2/T)$, and $T = 1/(\pi\xi)$, which is the synchrotron damping time measured in units of the revolution period. According to our main assumption, the averages in (6) are computed as follows:

$$\begin{aligned} &< \phi(x_1) >= \delta \int_0^\infty du\, W(u) \exp(\frac{-u^2}{4\sigma_{11}}), \\ &< \phi(x_1)(x_1 - \bar{x}_1) >= \frac{\delta}{2} \int_0^\infty du\, u\, W(u) \exp(\frac{-u^2}{4\sigma_{11}}), \\ &< \phi(x_1)(x_2 - \bar{x}_2) >= \frac{\sigma_{12}}{\sigma_{11}} < (x_1 - \bar{x}_1)\phi(x_1) >, \\ &< \phi(x_1)^2 >= \frac{2\delta^2}{\sqrt{3}} \int_0^\infty du_1\, W(u_1) \int_0^\infty du_2\, W(u_2) \exp(\frac{u_1u_2 - u_1^2 - u_2^2}{3\sigma_{11}}). \end{aligned} \tag{9}$$

where: $\delta = (e\ Q_{tot})/(2\ E_0\sqrt{\pi\sigma_{11}})$.

2..1 Wake Potential.

In this paper we consider the following pretty general (nonlinear) wake potential:

$$W(x) = \begin{cases} W_0\,(1-\beta\,x)\,exp\{-x^2/2\,\gamma\} & ,\,x>0 \\ 0 & ,\,x\le 0, \end{cases} \tag{10}$$

which depends on three parameters: W_0, β and γ. Calculating the averages in (6), we get:

$$\begin{aligned} \sigma''_{11} &= \sigma'_{11}, \\ \sigma''_{12} &= \sigma'_{12} - \frac{F_0}{8\,C_1\sqrt{\pi\sigma'_{11}}}\left(1-\frac{\beta\sqrt{\pi}}{2\sqrt{C_1}}\right), \\ \sigma''_{22} &= \sigma'_{22} - \frac{F_0(\sigma'_{12}/\sigma'_{11})}{4\,C_1\sqrt{\pi\sigma'_{11}}}\left(1-\frac{\beta\sqrt{\pi}}{2\sqrt{C_1}}\right)+ \end{aligned} \tag{11}$$

$$-\left[\frac{F_0}{4\sqrt{C_1\sigma'_{11}}}(1-\frac{\beta}{\sqrt{\pi C_1}})\right]^2 + \frac{F_0^2}{2\pi\sigma'_{11}\sqrt{3}}\left\{-\frac{\beta\sqrt{C_2\pi}}{4\,C_2^2}+\frac{\beta^2}{4\,C_2^2}+\right.$$

$$-\frac{18\beta\sigma'^2_{11}\sqrt{C_2\pi}}{2\,C_3}\left[1-\frac{\beta\sqrt{\pi}}{\sqrt{C_3C_2}}+\frac{1}{\sqrt{1+C_3}}+\frac{\beta}{\sqrt{\pi C_2}}(\frac{tan^{-1}(\sqrt{C_3})}{\sqrt{C_3}}-\frac{1}{1+C_3})+\right.$$

$$\left.\left.+\,3\sigma'_{11}\sqrt{\pi}\left[\sqrt{\frac{\pi}{C_3}}-\frac{\beta}{2C_3\sqrt{C_2}}-\frac{tan^{-1}(\sqrt{C_3})}{\sqrt{\pi C_3}}-\frac{\beta}{2C_3\sqrt{C_2}\,(1+C_3)}\right]\right]\right\}$$

where: $C_1=(1/2\,\gamma)+(1/4\,\sigma'_{11})$, $C_2=(1/2\,\gamma)+(1/3\,\sigma'_{11})$, $C_3=36\,C_2^2\,\sigma'^2_{11}-1$, and $F_0=(e\,Q_{tot}\,W_0)/(\sigma_0\,E_0)$.
Therefore, for the nonlinear wake potential (10), the second order moments mappings (5), (11), (7), depend on five main parameters: ν, T, F_0, β, γ.

Letting $\beta=0$ and $\gamma\to+\infty$, in (10) we obtain the simple step wake potential, already studied in our previous paper [19]. More specifically we get:

$$W(x) = \begin{cases} W_0 & ,\,x>0 \\ 0 & ,\,x\le 0, \end{cases} \tag{12}$$

and the mapping (6) becomes:

$$\begin{aligned} \sigma''_{11} &= \sigma'_{11}, \\ \sigma''_{12} &= \sigma'_{12} - \frac{F_0\sqrt{\sigma'_{11}}}{2\sqrt{\pi}}, \\ \sigma''_{22} &= \sigma'_{22} - \frac{F_0\sigma'_{12}}{\sqrt{\pi\sigma'_{11}}}+\frac{F_0^2}{12}. \end{aligned} \tag{13}$$

In this case the second order moments mappings (5), (13), (7), depend only on three main parameters: ν, T, F_0.

3 Comparison of Results between Nonlinear and Step Wake.

In this section we present some preliminary numerical results of the nonlinear second order moments mapping (5), (7), (11) with nonlinear wake potential (10).

Unfortunately, for the nonlinear wake potential (10) it is not easy to obtain an explicit (analytic) solution of the mapping equations (5), (7), (11). Thus, all results presented here are numerical.

Due to the presence of five main parameters, it is quite cumbersome to make an exhaustive study of (5), (7), (11). We accordingly restricted our analysis to a suitable subset of the whole parameter space including only *realistic* parameter values.

The main finding is that in some range of parameters period-one solutions become unstable, and correspondingly period two solutions develop, occurring in alternating pairs:

$$\vec{\sigma_a} \xrightarrow{S} \vec{\sigma_b} \xrightarrow{S} \vec{\sigma_a} \tag{14}$$

where $\vec{\sigma} = \{\sigma_{11}, \sigma_{12}, \sigma_{22}\}$, and S denotes the product of the mapping (5), (7), (11).

As an example, see *Fig.* 1*a*, where $\beta = .1$, $\gamma = 15$, and $\nu = .2$. In this figure black dots correspond to unstable period-one solutions, and to the the existence of period-two fixed points. It is seen that the (F_0, T) plane can be divided into three regions, each corresponding to a different range of T. In the first one, corresponding to ($T \lesssim 3$), the period-one solution becomes definitely unstable above a certain threshold value of F_0. In the second region, i.e. ($3 \lesssim T \lesssim 25$), the period-one solution is unstable only in a finite range of F_0 values. Finally in the third region ($T \gtrsim 25$), the period-one solution is definitely stable for all values of F_0.

Changing the tune value by 10% from $\nu = .2$ to $\nu = .22$, while keeping the other parameters fixed, as in *Fig.* 1*b*, the bifurcation area becomes significantly larger, though leaving the overall qualitative scenario essentially unchanged.

These results should be compared and contrasted to those obtained in [19] for a step wake potential. The latter are representatively summarized in *Fig.s* 2 and 3 *a*, *b*, *c*.

Comparing *Fig.s* 1*a*, 1*b* and 2 one can recognize several analogies.

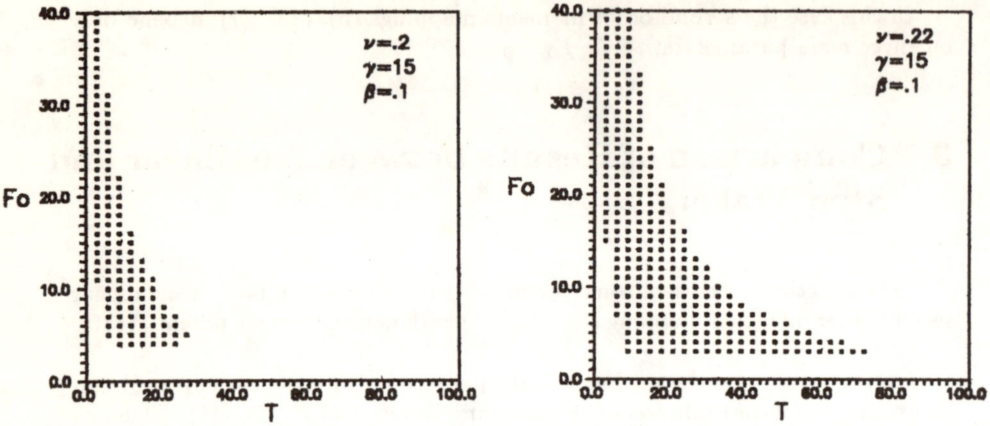

Fig. 1*a*, *b* – Bifurcation area, nonlinear wake-potential.

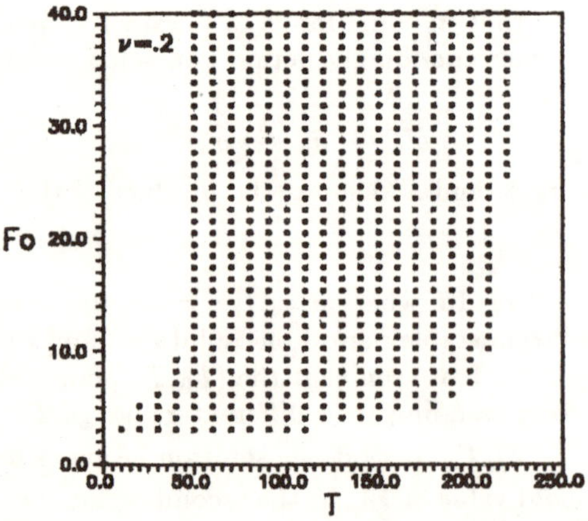

Fig. 2 – Bifurcation area, step wake potential.

From *Fig*.2, we can see that, for the step wake, depending on the value of T, the period-one solution can become definitely unstable above a certain F_0 value (e.g., at T=140), remain stable for all F_0 values (e.g., at T=250), or be unstable in a finite interval of F_0 values.

For this last case, we found that beyond the period-one solution instability interval, period-one and period two solutions, both stable, can co-exist. Such *bistable* regimes occur, e.g., within the interval from $T \sim 10$ to $T \sim 45$ in *Fig*.2.

This peculiar situation is further illustrated in *Fig.s* $3a, b, c$ where, respectively, σ_{11}, σ_{12} and σ_{22} are plotted as functions of F_0 , for $\nu = .2$, $T = 25$. The marks in *Fig.s* $3a, b, c$ have been obtained using an MPT simulation code [19], which, being based on an entirely different philosophy, indicates that our model (and in particular, our gaussian approximation (4)) is fairly accurate.

The graphs in *Fig.s* $3a, b, c$ should be compared to those represented in *Fig.s* $4a, b, c$, with $\nu = .2, T = 25$, $\beta = .1$ and $\gamma = 15$, for the nonlinear wake, where the bistable regime does *not* show up.

4 Conclusions.

From the preliminary results expounded above, we can draw the following tentative conclusions.

The occurrence of period-one solution instability and the corresponding birth of period two solutions, for realistic parameter values, probably does *not* depend qualitatively on the detailed shape of the wake potential, while the occurrence of bistable regimes possibly does.

Note that the parameter ranges covered in our exploration, viz.: $F_0 \in [0,\ 100]$, $T \in [0,\ 250]$ at collision, $\nu \in [0,\ .3]$, includes LEP [22] ($F_0 \sim$ a few units, $\nu \sim .1$, and $T \sim 170$).

It remains to investigate whether, by further increasing the beam intensity, for realistic parameter values, our model will show a cascaded-period-doubling route to chaos, or a back-bifurcation to the period-one solution, or won't undergo any further bifurcation at all. To the best of our knowledge, *no* general results are available for nonlinear maps with dimension three and co-dimension five, like the one at hand.

It could also be worth studying more general wake potentials (e.g., represented by a superposition of Gauss-Hermite functions), and hopefully more accurate descriptions of the phase-space distribution function (e.g., using Stratonovich expansion, as in [21]).

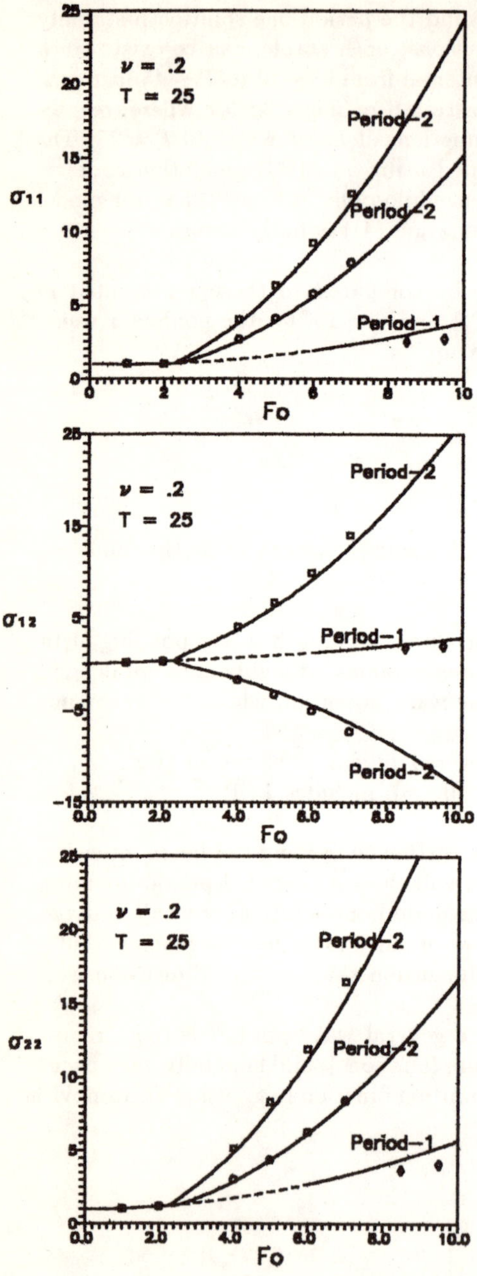

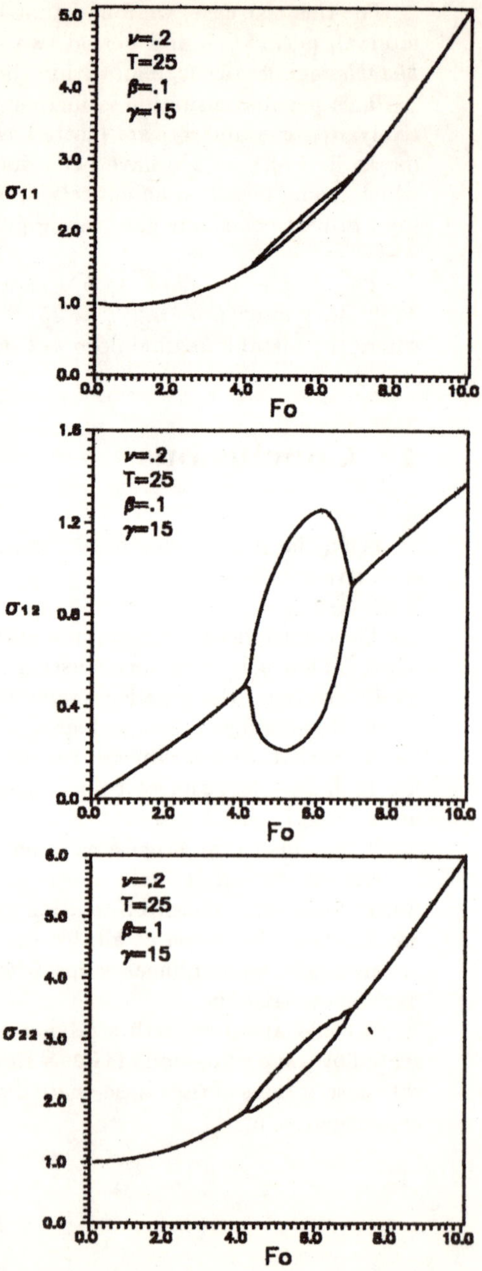

Left: *Fig.s* $3a, b, c$ – $\sigma_{i,j}$ vs. F_0, step wake; Right: *Fig.s* $4a, b, c$ – $\sigma_{i,j}$ vs. $F0$, nonlinear wake.

Acknowledgements.

The Author thanks L. Palumbo, W. Scandale, G. Turchetti and V. Vaccaro, the LEP Theory Group for their warm hospitality and help, and K. Hirata for his help and encouragement.

References.

1. A.W. Chao, SLAC rep.t PUB-2946 (1982).
2. E. Picasso, L. Radicati and F. Ruggiero, Proc. Joint US-CERN School on Particle Accelerators, S. Margherita di Pula, 1985, J. M. Jowett, M. Month and S. Turner, ed.s, Springer Verlag, 1986, p. 1.
3. A. Renieri, LNF-75/11R.
4. H. G. Hereward, PEP Note 53 (1973).
5. J. Haissinski, Nuovo Cimento B **18**, (1983), p. 72.
6. E. Keil, PEP Note 126 (1975).
7. A. W. Chao and J. Gareyte, Stanford Report PEP-224 (1976).
8. B. Zotter, IEEE Trans. Nucl. Sci., **NS-24**, (1977), p. 1393.
9. K. L. F. Bane and P. B. Wilson, SLAC-PUB 1909, PEP Note 241A (1977).
10. T. Suzuki, Particle Accel.,**14**, (1983), p. 91.
11. T. Suzuki, Y. Chin and K. Satoh, Particle Accel., **13**, (1983), p. 179.
12. G. Besnier, Nucl. Instrum. Methods, **164**, (1979), p. 235.
13. Y. Chin and K. Yokoya, Nucl. Instrum. Methods, **226**, (1984), p. 223.
14. T. Weiland, DESY 81-088 (1981).
15. D. Brandt, CERN/LEP Note 512 (1984).
16. V. Nys, CERN/LEP/TH/86-34 (1986).
17. P. B. Wilson, K. Bane and K. Satoh, IEEE Trans. Nucl. Sci., **NS-28**, (1981), p. 2525
18. K. Hirata, Particle Accel., **22**, (1987), p. 57.
19. K. Hirata, S. Petracca and F. Ruggiero, CERN/SL/AP Aug.1990, submitted for publication to Particle Accel., 1990.
20. S. Petracca, Ph. D. Thesis (unpublished).
21. K. Hirata, KH Note 2, CERN-LEP, Sept. 1988.
22. LEP Design Report, CERN/LEP/84-01.

CLASSICAL AND QUANTUM OPTICS FROM THE POINT OF VIEW OF ACCELERATOR PHYSICS

G. DATTOLI, C. MARI, M. RICHETTA[+] AND A. TORRE
ENEA — Dip. Sviluppo Tecnologie di Punta,
Centro Ricerche Energia Frascati,
C.P. 65 — 00044 Frascati, Rome, Italy

ABSTRACT

In this paper we show how the mathematical tools of charged beam transport theory can be applied to classical and quantum optics problems. We discuss the theory of optical beam transport in fibers, using Twiss parameters and Courant–Snyder invariants. The same quantities are then exploited to discuss the dynamical behaviour of quantum squeezed states. Finally we present generalized Twiss parameters and Courant–Snyder invariants for quadratic Hamiltonians with n degrees of freedom.

1. Introduction

The general theory of binormal differential equations has been formulated in a series of papers[1] to investigate the invariants associated to a differential equation of arbitrary order an with non constant coefficients. In particular it has been shown that, in the case of a second order equation, the invariant binorm is just the so called Courant–Snyder (C–S) invariant [2] of crucial importance in the theory of charged beam transport systems. More recently, however, the above quoted invariant has been shown to play a central role in other fields; as e.g. the propagation of electromagnetic waves in optical fibers or the dynamical evolution of the statistical properties of quantum states undergoing a time dependent harmonic perturbation[3]. In this note we will show that the C–S invariant is a powerful tool of analysis even in the case of coupled multimode time–dependent harmonic oscillators. The paper is therefore an attempt of presenting the relevance of a mathematical technique mainly adopted within the context of accelerator physics, to other areas of research.

The plane of the note is the following. In Sec. 2 we review the theory of binormal differential equations which is, in our opinion, the simplest mathematical framework to introduce the C–S invariant and its relevant generalizations. In Sec. 3 we discuss the propagation of optical beams in

---

[+]ENEA Guest.

a self–focusing fiber, using the language and the mathematical apparatus of accelerator physics. We define indeed, the transport "lattice" of the fiber, and introduce the relevant Twiss parameters and the associated emittance. In Sec. 4 it will be shown that, in the case of a time-dependent harmonic oscillator, the Twiss coefficients are linked to the quantum states uncertainties and to the squeezing parameters[4]. Furthermore a two photons hamiltonian is introduced to illustrate the applicability of the theoretical methods developed. In this context the motion of the C–S ellipse in the "phase–space" is analyzed and the evolution of the statistical properties of the quantum states is geometrically visualized. In Sec. 5 we will generalize the concept of C–S invariant to quadratic time dependent hamiltonians with n degrees of freedom. The relevant generalized Twiss parameters are shown to be $n \times n$ matrices and their equations of motion are discussed. The associated quantum problem is investigated and the Twiss parameters are again identified with the quantum uncertainties. Finally an example of two coupled oscillators hamiltonian, of interest both in quantum mechanics and in accelerator physics[5], is discussed and the relevance of the generalized Twiss parameters in the theory of charged beam transport is underlined.

2. The General Theory

Two differential equations of the type

$$\begin{aligned} \frac{\mathrm{d}}{\mathrm{d}\,x} \underline{y}_+ &= \hat{A}_+(x) \underline{y}_+(x) \\ \frac{\mathrm{d}}{\mathrm{d}\,x} \underline{y}_- &= \hat{A}_-(x) \underline{y}_-(x) \end{aligned} \tag{1}$$

where $\underline{y}_\pm$ is a n–column vector and $\hat{A}_\pm(x)$ is a $n \times n$ matrix, are said "binormal", whenever

$$\hat{A}_-(x) = -\hat{A}_+^+(x) \quad ^{(*)} \; . \tag{2}$$

The concept of binormality arises from the fact that the "scalar" product

$$\mathcal{I} = \underline{y}_-^+(x) \underline{y}_+(x) \tag{3}$$

is an invariant, namely a "constant" of motion as it is easy to check by taking the x–derivative of both sides of (3). We will see later that $\mathcal{I}$ can be understood as a generalization of the Courant–Snyder invariant of accelerator physics.

The link between $\underline{y}_+$ and $\underline{y}_-$ can be expressed in terms of an "evolution operator" defined by

$$\underline{y}_+(x) = \hat{n}(\quad x) \underline{y}_+(0) \quad ^{(**)} \; . \tag{4}$$

$^{(*)}\hat{A}^+$ denotes the transpose of $\hat{A}$.

Therefore, after setting $\hat{W}(x) = \hat{u}^{-1}(x)$, the following equality holds:

$$\underline{y}_-(x) = \hat{T}(x)\underline{y}_+(x) \tag{5}$$

where

$$\hat{T}(x) = \hat{W}^+(x)\hat{T}(0)\hat{W}(x) \tag{6}$$

is the so called Twiss matrix[6] which satisfies an evolution equation of the type:

$$\begin{aligned} \frac{\mathrm{d}\hat{T}}{\mathrm{d}x} &= -\left[\hat{A}_+^+(x)\hat{T}(x) + \hat{T}(x)\hat{A}_+(x)\right] \\ \hat{T}(0) &= \hat{1} \; . \end{aligned} \tag{7}$$

The $\hat{T}$–matrix is symmetric as clearly displayed by Eq. (7).

According to Eq. (5) the invariant $\mathcal{I}$ can be written in terms of $\underline{y}_+$ only and reads:

$$\mathcal{I} = \underline{y}_+^+(x)T(x)\underline{y}_+(x) \; . \tag{8}$$

These few elements represent the general framework in which a great deal of evolution problems may be stated.

3. Classical Optics: Beam Propagation in Optical Fibers

Beam propagation in optical fibers shares many features in common with the problem of charged beam transport in accelerator physics.

Starting from the Helmholtz equation[7]

$$\nabla^2 u + n^2(x,y,z)u = 0 \tag{9}$$

describing the wave propagation in a medium with index of refraction $n(x,y,z)$, we recover the geometric optics limit in a series of successive steps.

We assume

$$n(x,y,z) = n_0(x) - \frac{1}{2}n_{2x}(z)x^2 - \frac{1}{2}n_{2y}(z)y^2 \; . \tag{10}$$

Furthermore we seek for quasi–plane wave solutions of the form[8]

$$u(x,y,z) = \frac{1.}{\sqrt{n_0(z)}}\Phi(x,y,z)\exp\left\{-ik\int^z n_0(z')\mathrm{d}\,z'\right\} \tag{11}$$

(**) $\hat{u}(x) = \left\{\exp\int_0^x \hat{A}_+(x')\mathrm{d}x'\right\}_D$ where D denotes the Dyson "time ordering" whenever $[\hat{A}_+(x), \hat{A}_+(x')] \neq 0$.

where $\Phi(x, y, z)$ satisfies the following propagation equation in the paraxial approximation (i.e. after neglecting the $\frac{\partial^2}{\partial z^2}$ term)

$$\frac{i}{k}\frac{\partial}{\partial z}\Phi = \left\{\frac{1}{2k^2 n_0(z)}\nabla_\perp^2 - \frac{1}{2}\left[n_{2x}(z)x^2 + n_{2y}(z)y^2\right]\right\}\Phi \tag{12}$$

where $\nabla_\perp^2$ denotes the transverse Laplacian, namely

$$\nabla_\perp^2 = \frac{\partial^2}{\partial x^2} + \frac{\partial^2}{\partial y^2} \; . \tag{13}$$

Finally setting

$$\Phi(x, y, z) = \psi_x(x, z)\psi_y(y, z) \tag{14}$$

i.e. performing a variables separation, we find that $\psi_x(x, z)$ or $\psi_y(y, z)$ satisfy the following Schrödinger–type equation:

$$i\lambda\!\!\!^{-}\frac{\partial}{\partial z}\psi(\eta, z) = \frac{1}{2}\left[\frac{\lambda\!\!\!^{-2}}{n_0(z)}\frac{\partial^2}{\partial \eta^2} - n_2(z)\eta^2\right]\psi(\eta, z) \quad *$$

$$\left(\lambda\!\!\!^{-} = \frac{1}{k} \qquad \eta = x, y\right) \tag{15}$$

with a "time"–dependent hamiltonian.

In this framework $\lambda\!\!\!^{-}$ plays the same role of $\hbar$ in quantum mechanics and the "time" evolution of ψ is described by an harmonic oscillator quantum hamiltonian with "time" dependent mass and coupling.

The solutions of Eq. (15) can be found using Lie algebraic methods[9] in which the lowest dimensional matrix realization of the Lie differential operators is used to reduce the initial partial differential equation to an easily solvable ordinary differential equation written in a "spinorial" form. We introduce, for later convenience, the following differential operators:

$$\hat{k}_0(\eta) = \frac{1}{2}\left(\eta\frac{\partial}{\partial \eta} + \frac{1}{2}\right) \quad \hat{k}_+(\eta) = \frac{i}{2}\eta^2 \quad \hat{k}_-(\eta) = -\frac{i}{2}\frac{\partial^2}{\partial \eta^2} \tag{16}$$

which satisfy the relations of commutation

$$[\hat{k}_0, \hat{k}_\pm] = \pm\hat{k}_\pm \qquad\qquad [\hat{k}_+\hat{k}_-] = -2\hat{k}_0 \tag{17}$$

thus displaying a $SU(1,1)$ group structure which can be used to solve our "evolution" problem. The "evolution" operator $\hat{u}(z)$ satisfies the equation

$$i\lambda\!\!\!^{-}\frac{\mathrm{d}}{\mathrm{d}\,z}\hat{u}(z) = \hat{H}u(z)$$

$$\hat{u}(0) = \hat{\mathbf{1}} \; . \tag{18}$$

*We have omitted the subscript η for simplicity.

The above problem can be solved in the lowest dimensional matrix representation of the $SU(1,1)$ generators given by the following 2×2 matrices

$$\hat{k}_0 = \frac{1}{2}\begin{pmatrix} -1 & 0 \\ 0 & 1 \end{pmatrix} \qquad \hat{k}_+ = \begin{pmatrix} 0 & 0 \\ -1 & 0 \end{pmatrix} \qquad \hat{k}_- = \begin{pmatrix} 0 & 1 \\ 0 & 0 \end{pmatrix} . \tag{19}$$

Thus reducing Eq. (15) to the following "spinorial" form

$$\frac{d}{dz}\begin{pmatrix} \psi_1 \\ \psi_2 \end{pmatrix} = \begin{pmatrix} 0 & \frac{\lambda\!\!\!^-}{n_0(z)} \\ -\frac{n_2(z)}{\lambda\!\!\!^-} & 0 \end{pmatrix}\begin{pmatrix} \psi_1 \\ \psi_2 \end{pmatrix} . \tag{20}$$

The solution of the above equation can be written formally as

$$\begin{pmatrix} \psi_1(z) \\ \psi_2(z) \end{pmatrix} = \hat{u}(z)\begin{pmatrix} \psi_1(0) \\ \psi_2(0) \end{pmatrix} \tag{21}$$

where $\hat{u}$ is ordered a lá Wey–Norman[10] (W.N.)

$$\hat{u}(z) = \exp[2h(z)\hat{k}_0]\exp[g(z)\hat{k}_+]\exp[f(z)\hat{k}_-] \tag{22}$$

and its matrix image is

$$\hat{u}(z) = \underbrace{\begin{pmatrix} e^{-h} & 0 \\ 0 & e^{h} \end{pmatrix}}_{a}\underbrace{\begin{pmatrix} 1 & 0 \\ -g & 1 \end{pmatrix}}_{b}\underbrace{\begin{pmatrix} 1 & f \\ 0 & 1 \end{pmatrix}}_{c} = \begin{pmatrix} e^{-h} & fe^{-h} \\ -ge^{h} & (1-gf)e^{h} \end{pmatrix} \tag{23}$$

where f, g, h, namely the W.N. ordering functions, satisfy the following non–linear motion equations

$$\begin{aligned} f' &= \frac{\lambda\!\!\!^-}{n_0(z)}e^{2h} , && f(0) = 0 \\ g' &= -\frac{\lambda\!\!\!^-}{n_0(z)}g^2e^{2h} + \frac{n_2(z)}{\lambda\!\!\!^-}e^{-2h} , && g(0) = 0 \\ h' &= \frac{\lambda\!\!\!^-}{n_0(z)}ge^{2h} , && h(0) = 0 . \end{aligned} \tag{24}$$

The "evolution" operator in the differential representation is given by Eq. (22) in which $\hat{k}_0$, $\hat{k}_\pm$ are replaced by their differential form (16) with the same f, g, h.

At this point a very appealing result must be underlined: the geometric optics limit is recovered if we identify

$$\psi_1(z) \equiv r(z) , \quad \psi_2(z) = \frac{n_0(z)}{\lambda\!\!\!^-}\frac{d}{dz}r(z) \tag{25}$$

and $\psi_2(z)$ is the so–called reduced slope of the ray.
Therefore, from the motion equation (20), we get the well known ray propagation equation

$$\frac{\mathrm{d}}{\mathrm{d}\,z}\left(n_0(z)\frac{\mathrm{d}}{\mathrm{d}\,z}r(z)\right)+n_2(z)r(z)=0 \tag{26}$$

whose solutions are given by the relationship (21) with $\hat{u}(z)$ given as in (23): i.e. the most general ray–motion is understood in terms of propagation through a succession of the following optical elements[7]:

a) a beam expander $\begin{pmatrix} e^{-h} & 0 \\ 0 & e^{h} \end{pmatrix}$

b) a this tens $\begin{pmatrix} 1 & 0 \\ -g & 1 \end{pmatrix}$

c) a straight section $\begin{pmatrix} 1 & f \\ 0 & 1 \end{pmatrix}$ (27)

Furthermore the global motion is characterized by a dynamical invariant, namely an "optical" C–S invariant, given by

$$\mathcal{I}=\gamma\psi_1^2+2\alpha\psi_1\psi_2+\beta\psi_2^2 \tag{28}$$

with ψ_1 and ψ_2 as in Eq. (25).

In this case the Twiss parameters α, β, γ satisfy the following equations

$$\begin{aligned} \gamma' &= 2\frac{n_2(z)}{\chi}\alpha & \gamma(0)&=1 \\ \beta' &= -2\frac{\chi}{n_0(z)}\alpha & \beta(0)&=1 \\ \alpha' &= \frac{n_2(z)}{\chi}\beta-\frac{\chi}{n_0(z)}\gamma & \alpha(0)&=0 \end{aligned} \tag{29}$$

and can be linked to the beam–section or divergence in the usual way

$$\sigma=\sqrt{\gamma\mathcal{I}}\ ,\ \sigma'=\sqrt{\beta\mathcal{I}} \tag{30}$$

σ satisfies the well known envelope equation

$$\sigma''+\frac{n_0'}{n_0}\sigma'+\frac{n_2}{n_0\chi^2}\sigma-\frac{\mathcal{I}^2\chi^2}{n_0^2\sigma^3}=0\ . \tag{31}$$

From the above considerations it is evident that the problem of beam propagation in optical fibers can be discussed with the same language used in accelerator physics. Namely in this framework the Twiss parameters and the C–S invariant acquire a deep meaning in the evolution of global quantities related to the propagation of optical waves in media.

4. Quantum Optics

The classical problem of a general time dependent harmonic oscillator shares many features in common with the quantum analog; we briefly discuss some relevant properties.

Let us consider the following time–dependent, quadratic hamiltonian

$$H = \frac{a(t)}{2}q^2 + \frac{b(t)}{2}p^2 + c(t)qp \tag{32}$$

where a, b, c are arbitrary functions of time.

The equations of motion for p and q, written in a matrix form, read

$$\frac{\mathrm{d}}{\mathrm{d}t}\begin{pmatrix} q \\ p \end{pmatrix} = \begin{pmatrix} c(t) & b(t) \\ -a(t) & -c(t) \end{pmatrix}\begin{pmatrix} q \\ p \end{pmatrix} . \tag{33}$$

Using the analysis developed in the previous section we can introduce a Twiss matrix $T = \begin{pmatrix} \gamma\alpha \\ \alpha\beta \end{pmatrix}$ and find a C–S invariant in this case too. In fact from the Eq. (7) the Twiss coefficients satisfy the following evolution problem

$$\begin{aligned} \dot{\alpha} &= a(t)\beta - b(t)\gamma \qquad & \alpha(0) &= 0 \\ \dot{\beta} &= -2b(t)\alpha + 2c(t)\beta \qquad & \beta(0) &= 1 \\ \dot{\gamma} &= 2a(t)\alpha - 2c(t)\gamma \qquad & \gamma(0) &= 1 \end{aligned} \tag{34}$$

and the C–S invariant is[11]

$$\mathcal{I} = \gamma q^2 + 2\alpha qp + \beta p^2 \ . \tag{35}$$

$\mathcal{I}$ is a really invariant dynamical variable as it is easy to check by using the well known relationship

$$\frac{\mathrm{d}\mathcal{I}}{\mathrm{d}t} = \frac{\partial \mathcal{I}}{\partial t} + \{\mathcal{I}, H\}_{\mathrm{P.B.}} \tag{36}$$

where $\{\quad\}_{\mathrm{P.B.}}$ denotes the "classical" Poisson brackets of the dynamical variables considered. The meaning of $\mathcal{I}$ is well understood in terms of its analytical representation in the phase–space of the physical system: the motion is described by an ellipse of constant area which gives the time evolution of the Twiss coefficients in terms of the ellipse parameters (see Fig. 1).

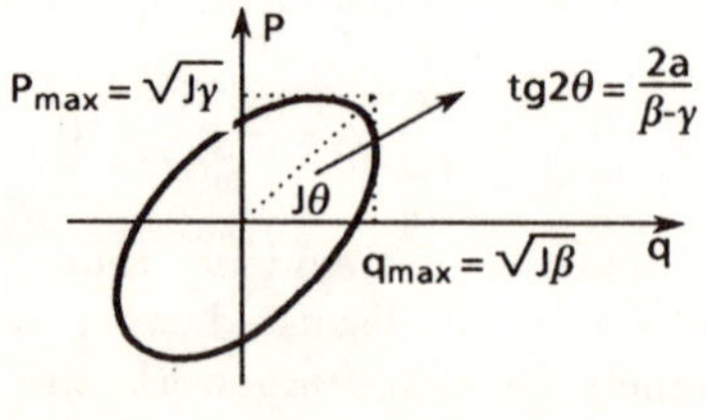

Fig. 1. Twiss coefficients and ellipse parameters.

This kind of analysis can be extended in a natural way to the quantum analog in which the classical hamiltonian (32) is replaced by the hermitian operator

$$\hat{H} = a(t)\frac{\hat{q}^2}{2} + b(t)\frac{\hat{p}^2}{2} + \frac{c(t)}{2}(\hat{q}\hat{p} + \hat{p}\hat{q}) \ . \tag{37}$$

The quantum version of the system (34) is obtained by introducing the $SU(1,1)$ generators ($\hbar = 1$):

$$\hat{k}_+ = -\frac{i}{2}\hat{q}^2 \ , \ \hat{k}_- = -\frac{i}{2}\hat{p}^2 \ , \ k_0 = \frac{i}{4}(\hat{q}\hat{p} + \hat{p}\hat{q}) \tag{38}$$

which satisfy the following commutation rules

$$[\hat{k}_+, \hat{k}_-] = -2\hat{k}_0 \ , \ [\hat{k}_0, \hat{k}_\pm] = \pm\hat{k}_\pm \tag{39}$$

and by direct substitution into Eq. (50), $\hat{H}$ becomes a linear combination of the above generators:

$$\hat{H} = ia(t)\hat{k}_+ + ib(t)\hat{k}_- - ic(t)\hat{k}_0 \ . \tag{40}$$

It is easy to show that with the help of the previous equations the time evolution of the $\hat{k}$'s can be cast in a familiar form:

$$\begin{aligned} \dot{\hat{k}}_+ &= -2b(t)\hat{k}_0 + 2c(t)\hat{k}_+ \\ \dot{\hat{k}}_- &= 2a(t)\hat{k}_0 - 2c(t)\hat{k}_- \\ \dot{\hat{k}}_0 &= a(t)\hat{k}_+ - b(t)\hat{k}_- \end{aligned} \tag{41}$$

i.e. the $SU(1,1)$ group generators obey, in the Heisenberg picture, the same evolution equations of the "classical" Twiss coefficients. In this framework the quantum analog of the C–S invariant is the well known Casimir invariant of the $SU(1,1)$ group:

$$\hat{C} = \hat{k}_0^2 - \hat{k}_1^2 - \hat{k}_2^2 \tag{42}$$

where $\hat{k}_1$, $\hat{k}_2$ are related to $\hat{k}_+$, $\hat{k}_-$ by the usual relationships

$$\hat{k}_1 = \frac{\hat{k}_+ + \hat{k}_-}{2} \qquad \hat{k}_2 = \frac{\hat{k}_+ - \hat{k}_-}{2i} \ . \tag{43}$$

But $\hat{C}$ is not the only invariant we can obtain: a linear one arises in a natural way and it is given by

$$\hat{\mathcal{I}} = \gamma\hat{k}_+ - 2\alpha\hat{k}_0 + \beta\hat{k}_- \tag{44}$$

as it can be shown from

$$\frac{\partial \hat{\mathcal{I}}}{\partial t} = -\frac{1}{i}[\hat{\mathcal{I}}, \hat{H}] \ . \tag{45}$$

Such invariants are very useful to understand the time evolution of the statistical properties of quantum states as we will see below. Namely, from the Ehrenfest's theorem the following identifications can be stated

$$\begin{aligned} \beta &= \langle \hat{q}^2 \rangle - \langle \hat{q} \rangle^2 \\ \gamma &= \langle \hat{p}^2 \rangle - \langle \hat{p} \rangle^2 \\ \alpha &= \langle \hat{q} \rangle \langle \hat{p} \rangle - \frac{1}{2}\langle \hat{q}\hat{p} + \hat{p}\hat{q} \rangle \end{aligned} \tag{46}$$

or in other words the elements of the Twiss matrix can be understood, in this quantum mechanical, framework as the components of the fluctuations tensor[12].

a) Two photons hamiltonian and squeezing

Let us consider a two photons hamiltonian of the parametric amplifier type

$$\hat{H} = \hbar\omega(a^+ a + 1/2) - i\hbar\frac{\Omega}{2}(a^{+^2} - a^2) \tag{47}$$

where a^+, a are the well known creation and annihilation operators for photons (of energy $\hbar\omega$) satisfying the usual commutation rules $[a, a^+] = 1$.

Introducing the hermitian quadrature operators

$$\hat{x}_c = \frac{1}{\sqrt{2}}(a + a^+) \qquad \hat{x}_s = \frac{1}{\sqrt{2}\, i}(a - a^+) \tag{48}$$

with the commutation rule $[\hat{x}_c, \hat{x}_s] = i$ we recast the hamiltonian (47) in the form

$$\hat{H} = \frac{\hbar\omega}{2}(\hat{x}_c^2 + \hat{x}_s^2) - \frac{\hbar\Omega}{2}(\hat{x}_c\hat{x}_s + \hat{x}_s\hat{x}_c) \tag{49}$$

or, if we use the $SU(1,1)$ realization (38) ($\hbar = 1$)

$$\hat{H} = i\omega(\hat{k}_+ + \hat{k}_-) + 2i\Omega\hat{k}_3 \ . \tag{50}$$

Within this framework it is immediately realized that

$$(\Delta\hat{x}_c)^2 = \frac{\beta}{2} \qquad (\Delta\hat{x}_s)^2 = \frac{\gamma}{2} \tag{51}$$

or that, in other words, β and γ can be viewed as the squeezing parameters. The equations specifying the entries of Twiss matrix, or in a quantum mechanical language, the squeezing parameters are therefore given by (see Eq. (41))

$$\begin{aligned} \dot{\beta} &= -2\omega\alpha - 2\Omega\beta \\ \dot{\gamma} &= 2\omega\alpha + 2\Omega\gamma \\ \dot{\alpha} &= \omega(\beta - \gamma) \ . \end{aligned} \tag{52}$$

At this point it is worth spending a few words on the practical use of the Twiss parameters and on the C–S geometrical characterization to specify the nature of a quantum state driven by a time–dependent harmonic oscillator hamiltonian.

1) Coherent Glauber States

From Fig. 2 and Eqs. (51), it is evident that minimum uncertainty states (M.U.S.) are characterized by a circle in the classical "phase–space" (x_c, x_s) and by $\alpha = 0$, $\beta = \gamma = 1$ which assure $\Delta p \cdot \Delta q = 1/2$ and $\Delta p = \Delta q = \frac{1}{\sqrt{2}}$ (see Fig. 2)

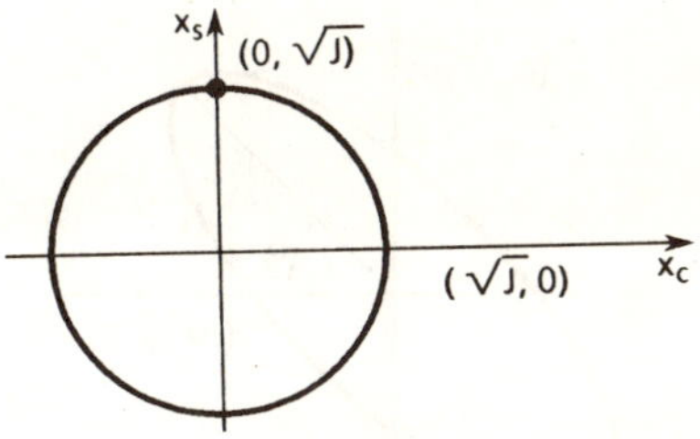

Fig. 2. Coherent Glauber state (M.U.S.)

2) Squeezed states: squeezed minimum uncertainty states (S.M.U.S.) are defined in the present context by $\alpha = 0$, $\beta \neq \gamma$ which assure $\Delta p \Delta q = 1/2$ but $\Delta q \neq \Delta p$ as shown in Fig. 3

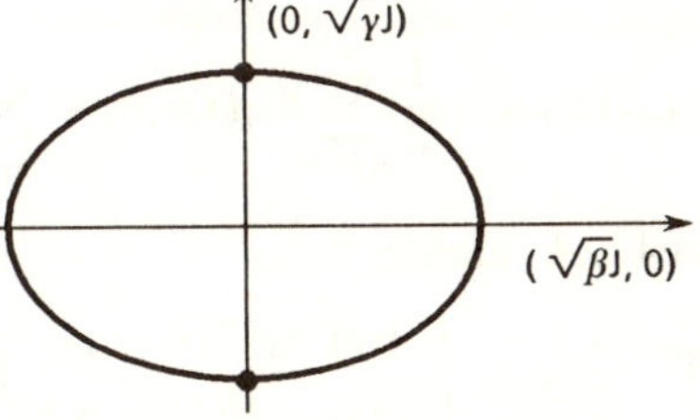

Fig. 3. A squeezed state (S.M.U.S.) is shown.

3) N.M.U.S.

The above achronim denotes "non minimum uncertainty states": the degree of deviation from a M.U.S. or a S.M.U.S. is just measured by α. These states are indeed specified by $\alpha \neq 0$ and $\beta\gamma = 1 + \alpha^2$. The orientation of the ellipse depends on the angle θ, defined in terms of the Twiss parameters by

$$\operatorname{tg} 2\theta = -\frac{2\alpha}{\beta - \gamma} \tag{53}$$

and when $\theta \neq 0$ the state does not belong to the M.U.S. or S.M.U.S. classes (see Fig. 4)

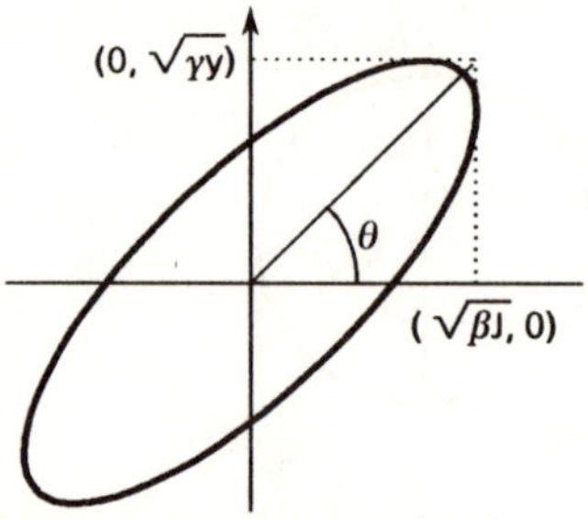

Fig. 4. Non minimum uncertainty state (N.M.U.S.).

For large β, γ and α the state is very noisy and its C–S representation looks like that given in Fig. 5

5. The Generalized C–S Invariant, Twiss Parameters and Quantum Optics

The general quadratic Hamiltonian, in the case of more than one degree of freedom, can be written as follows:

$$H = \frac{1}{2}\sum_{i,j=1}^{n} a_{ij}(t) q_i q_j + \frac{1}{2}\sum_{i,j=1}^{n} b_{ij}(t) p_i p_j + \sum_{i,j} c_{ij}(t) p_i q_j \tag{54}$$

which can be cast in the block matrix form

$$H = \frac{1}{2}(\underline{q}, \underline{p}) \begin{pmatrix} \hat{A}(t) & \hat{C}^T(t) \\ \hat{C}(t) & \hat{B}(t) \end{pmatrix} \begin{pmatrix} \underline{q} \\ \underline{p} \end{pmatrix} \tag{55}$$

where $\underline{q} = (q_1 \dots q_n)$, $\underline{p} = (p_1 \dots p_n)$ and $\hat{A}, \hat{B}, \hat{C}$ are $n \times n$ matrices with entries $a_{ij}(t)$, $b_{ij}(t)$, $c_{ij}(t) (i, j = 1 \dots n)$ respectively. The motion equations

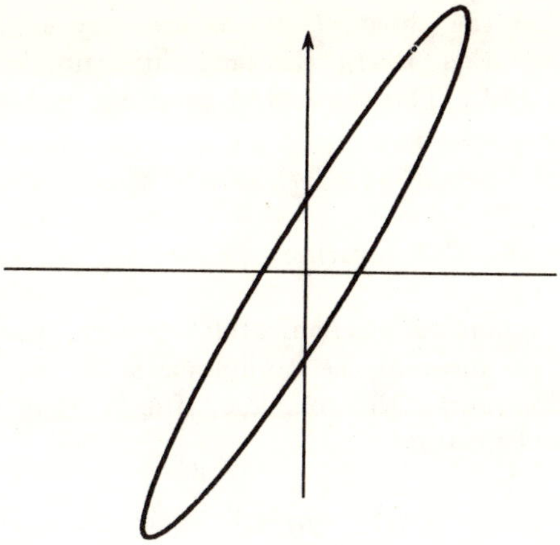

Fig. 5. A noisy state (large β, γ, α) is shown.

read therefore

$$\frac{\mathrm{d}}{\mathrm{d}t}\begin{pmatrix} \underline{q} \\ \underline{p} \end{pmatrix} \equiv \hat{S}\hat{H}\begin{pmatrix} \underline{q} \\ \underline{p} \end{pmatrix} = \begin{pmatrix} \hat{C}(t) & \hat{B}(t) \\ -\hat{A}(t) & -\hat{C}^{+}(t) \end{pmatrix}\begin{pmatrix} \underline{q} \\ \underline{p} \end{pmatrix} \tag{56}$$

where S is the generalized sympletic matrix

$$\hat{S} = \begin{pmatrix} 0 & \hat{\mathbf{1}}_n \\ -\hat{\mathbf{1}}_n & 0 \end{pmatrix} \tag{57}$$

written in a $n \times n$ block form.

The theory developed in the first part of this paper gives us the possibility of getting a dynamical invariant, namely a generalization of the C–S invariant of the one dimensional case. In fact, introducing the following generalized Twiss matrix[14]

$$\hat{T}(t) = \begin{pmatrix} \hat{\gamma} & \hat{\alpha} \\ \hat{\alpha}^T & \hat{\beta} \end{pmatrix} \tag{58}$$

where $\hat{\alpha}, \hat{\beta}, \hat{\gamma}$ are $n \times n$ time–dependent matrices (since by definition $\hat{T}$ is symmetric, $\hat{\beta}$ and $\hat{\gamma}$ must be symmetric matrices) they satisfy the set of equations listed below:

$$\begin{aligned} \dot{\hat{\beta}} &= -\hat{B}\hat{\alpha} + \hat{C}\hat{\beta} + \text{ transpose} \\ \dot{\hat{\gamma}} &= \hat{\alpha}\hat{A} - \hat{\gamma}\hat{C} + \text{ transpose} \\ \dot{\hat{\alpha}} &= -\hat{\gamma}\hat{B} + \hat{A}\hat{\beta} + [\hat{\alpha}, \hat{C}^{+}] \ . \end{aligned} \tag{59}$$

It is to be remarked that the above equations are very similar to those of the one–dimensional case (see Eqs. (34)): the only difference is due to the matrix structure of the system (59). The associated invariant is therefore

$$\mathcal{I} = \underline{q}^{+}\hat{\gamma}\underline{q} + 2\underline{q}^{+}\hat{\alpha}\underline{p} + \underline{p}^{+}\hat{\beta}\underline{p} \tag{60}$$

which is reminiscent of the C–S invariant derived in the previous section (see Eq. (35)).

Furthermore, in a quantum mechanical framework, the generalized Twiss coefficients preserve their meaning as components of the fluctations tensor: with the help of the Ehrenfest's theorem and after a little algebra we get the following interesting identification:

$$\begin{aligned} \beta_{ij} &= \langle \hat{q}_i \hat{q}_j \rangle - \langle \hat{q}_i \rangle \langle \hat{q}_j \rangle \\ \gamma_{ij} &= \langle \hat{p}_i \hat{p}_j \rangle - \langle \hat{p}_i \rangle \langle \hat{p}_j \rangle \\ \alpha_{ij} &= -\frac{1}{2} \langle \hat{q}_j \hat{p}_i + \hat{p}_i \hat{q}_j \rangle + \langle \hat{q}_j \rangle \langle \hat{p}_i \rangle \end{aligned} \tag{61}$$

where $\langle \quad \rangle$ denotes quantum mechanical averaging. Within the context of the present theory, squeezed states of the multimode electromagnetic field can be described: let us consider the following two photons multimode hamiltonian

$$\hat{H} = \sum_{j=1}^{n} \hbar\omega_j (\hat{a}_j^{+}\hat{a}_j + 1/2) + \sum_{j=1}^{n} \hbar\Omega_{ij} (\hat{a}_i^{+}\hat{a}_j + \hat{a}_j^{+}\hat{a}_i) + \sum_{j=1}^{n} \hbar\Gamma_{ij} (\hat{a}_i\hat{a}_j + \hat{a}_i^{+}\hat{a}_j^{+}) \tag{62}$$

where the operators $\hat{a}_i, \hat{a}_j^{+} \quad i,j = 1 \ldots n$ satisfy the usual commutation rules

$$[\hat{a}_i, \hat{a}_j] = 0 = [\hat{a}_i^{+}, a_j^{+}] \quad [\hat{a}_i, \hat{a}_j^{+}] = \delta_{ij} \qquad i,j = 1, \ldots, n \ . \tag{63}$$

If we introduce the quadrature operators

$$\hat{x}_{jc} = \frac{1}{\sqrt{2}}(\hat{a}_j + \hat{a}_j^{+}) \qquad \hat{x}_{jc} = \frac{1}{\sqrt{2}\ i}(\hat{a}_j - \hat{a}_j^{+}) \qquad j = 1, \ldots, n \tag{64}$$

the squeezing parameters are easily recovered

$$(\Delta\hat{x}_{jc})^2 = \frac{\beta_{jj}}{2} \quad (\Delta\hat{x}_{js})^2 = \frac{\gamma_{jj}}{2} \tag{65}$$

i.e. the evolution of the statistical properties of quantum states is described by the diagonal elements of the generalized Twiss matrix.

Concluding Remarks

Although, in the previous section, we have presented the meaning of the generalized Twiss coefficients in a quantum mechanical context, we will discuss in this last part the link between the diagonal Twiss elements and the physical global quantities (as the beam sections or divergences) which arise in a charged beam transport problem. As it is well known, the magnetic fields of an accelerator are designed in such a way that the horizontal and vertical motion of the charged particle are uncoupled. Small couplings, due to imperfections are however unavoidable. We limit our analysis to the case of $x-y$ coupling but the procedure is general.

Let us consider the following quadratic hamiltonian

$$H=\frac{1}{2}(p_x^2+k_x x^2)+\frac{1}{2}(p_y^2+k_y y^2)+L(yp_x-xp_y)+Qxy \tag{66}$$

where $k_{x,y}$ and Q are the strengths of the normal and skew quadrupoles respectively, L accounts for the solenoid. The equations of motion are

$$\frac{\mathrm{d}}{\mathrm{d}t}\begin{pmatrix}x\\y\\p_x\\p_y\end{pmatrix}=\begin{pmatrix}0&L&1&0\\-L&0&0&1\\-k_x&-Q&0&L\\-Q&-k_y&-L&0\end{pmatrix}\begin{pmatrix}x\\y\\p_x\\p_y\end{pmatrix}\equiv\hat{V}\begin{pmatrix}x\\y\\p_x\\p_y\end{pmatrix} \tag{67}$$

and the solutions can be found in terms of the evolution operator $\hat{u}(t)$ (see Eq. (4))

$$\hat{u}(t)=\begin{pmatrix}a_1&a_2&a_3&a_4\\b_1&b_2&b_3&b_4\\c_1&c_2&c_3&c_4\\d_1&d_2&d_3&d_4\end{pmatrix} \tag{68}$$

whose entries satisfy the equations:

$$\frac{\mathrm{d}}{\mathrm{d}t}\begin{pmatrix}a_\alpha\\b_\alpha\\c_\alpha\\d_\alpha\end{pmatrix}=\hat{V}(t)\begin{pmatrix}a_\alpha\\b_\alpha\\c_\alpha\\d_\alpha\end{pmatrix}\qquad \alpha=1,2,3,4\ . \tag{69}$$

The following parametrization of $\hat{u}(t)$ is exploited to introduce the generalized Twiss parameters

$$\begin{array}{ll}
\begin{cases}a_1=\sqrt{\beta_x}\ \cos\psi_x\cos\psi_{x,y}\\ a_2=\sqrt{\beta_x}\ \cos\Phi_x\sin\psi_{x,y}\end{cases} &
\begin{cases}a_3=\sqrt{\beta_x}\ \sin\psi_x\cos\psi_{x,y}\\ a_4=\sqrt{\beta_x}\ \sin\Phi_x\sin\psi_{x,y}\end{cases}\\
\begin{cases}b_1=\sqrt{\beta_y}\ \cos\psi_y\sin\psi_{y,x}\\ b_2=\sqrt{\beta_y}\ \cos\Phi_y\cos\psi_{y,x}\end{cases} &
\begin{cases}b_3=\sqrt{\beta_y}\ \sin\psi_y\sin\psi_{y,x}\\ b_4=\sqrt{\beta_y}\ \sin\Phi_y\cos\psi_{y,x}\end{cases}\\
\begin{cases}c_1=\sqrt{\gamma_x}\ \cos\chi_x\cos\chi_{x,y}\\ c_2=\sqrt{\gamma_x}\ \cos\tau_x\sin\chi_{x,y}\end{cases} &
\begin{cases}c_3=\sqrt{\gamma_x}\ \sin\chi_x\cos\chi_{x,y}\\ c_4=\sqrt{\gamma_x}\ \sin\tau_x\sin\chi_{x,y}\end{cases}\\
\begin{cases}d_1=\sqrt{\gamma_y}\ \cos\chi_y\sin\chi_{y,x}\\ d_2=\sqrt{\gamma_y}\ \cos\tau_y\cos\chi_{y,x}\end{cases} &
\begin{cases}d_3=\sqrt{\gamma_y}\ \sin\chi_y\sin\chi_{y,x}\\ d_4=\sqrt{\gamma_y}\ \sin\tau_y\cos\chi_{y,x}\end{cases}
\end{array} \tag{70}$$

in which the diagonal Twiss parameters $\gamma_{x,y}$ and $\beta_{x,y}$ explicitly appear. It is to be noted that we have parametrized a 4×4 sympletic matrix with sixteen not independent quantities; this number is reduced to ten once the sympleticity conditions are applied$^{(*)}$. Furthermore the initial condition for the "angles" in Eqs. (70) are derived from $\hat{u}(0) = \mathbf{1}$. If we assume that at $t = 0$

$$\begin{pmatrix} x_0 \\ y_0 \\ p_{x_0} \\ p_{y_0} \end{pmatrix} = \begin{pmatrix} \sqrt{\varepsilon_x}\ \cos\varphi_x \\ \sqrt{\varepsilon_y}\ \cos\varphi_y \\ \sqrt{\varepsilon_x}\ \sin\varphi_x \\ \sqrt{\varepsilon_y}\ \sin\varphi_y \end{pmatrix} \tag{71}$$

where $\varepsilon_{x,y}$ are the initial transverse emittances, the solutions of Eqs. (69) can be cast in the quasi–harmonic forms:

$$\begin{aligned} x(t) &= \sqrt{\varepsilon_x\beta_x}\ \cos\psi_{x,y}\cos(\psi_x - \varphi_x) + \sqrt{\varepsilon_y\beta_x}\ \sin\psi_{x,y}\cos(\Phi_x - \varphi_y) \\ y(t) &= \sqrt{\varepsilon_y\beta_y}\ \cos\psi_{y,x}\cos(\Phi_y - \varphi_y) + \sqrt{\varepsilon_x\beta_y}\ \sin\psi_{y,x}\cos(\psi_y - \varphi_x) \\ p_x(t) &= \sqrt{\varepsilon_x\gamma_x}\ \cos\chi_{x,y}\cos(\chi_x - \varphi_x) + \sqrt{\varepsilon_y\gamma_x}\ \sin\chi_{x,y}\cos(\tau_x - \varphi_y) \\ p_y(t) &= \sqrt{\varepsilon_y\gamma_y}\ \cos\chi_{y,x}\cos(\tau_y - \varphi_y) + \sqrt{\varepsilon_x\gamma_y}\ \sin\chi_{y,x}\cos(\chi_y - \varphi_x) \end{aligned} \tag{72}$$

which once averaged on the initial phases $\varphi_{x,y}$, yield for the r.m.s. section $\sigma_{x,y}$ and divergence $\sigma'_{x,y}$ of the beam:

$$\begin{aligned} \sigma_x &= \sqrt{\beta_x(\varepsilon_x\cos^2\psi_{x,y} + \varepsilon_y\sin^2\psi_{x,y})} \\ \sigma_y &= \sqrt{\beta_y(\varepsilon_y\cos^2\psi_{y,x} + \varepsilon_x\sin^2\psi_{y,x})} \\ \sigma'_x &= \sqrt{\gamma_x(\varepsilon_x\cos^2\chi_{x,y} + \varepsilon_y\sin^2\chi_{x,y})} \\ \sigma'_y &= \sqrt{\gamma_y(\varepsilon_y\cos^2\chi_{y,x} + \varepsilon_x\sin^2\chi_{y,x})}\ . \end{aligned} \tag{73}$$

As it is easy to see, the above relations are reminiscent of the one–dimensional case (see Eq. (30)).

$^{(*)}$The sympletic condition $u^+Su = S$ imposes, in the case of 4×4 matrices, six contraints, as it is easy to verify, which reduce to ten the number of independent parameters used in the relations (70).

References

1. D. Dattoli, V. Loreto, C. Mari, M. Richetta and A. Torre, *Biorthogonal states and ordinary differential equations*, part. (I–III), to be published in Il Nuovo Cimento B.
2. E. D. Courant and H. S. Snyder, Ann. Phys. **3** (1958), 1.
3. G. Dattoli, G. P. Gallerano, C. Mari, A. Torre and M. Richetta, *Twiss parameters and evolution of quantum harmonic oscillator states*, to be published in Il Nuovo Cimento.
4. D. F. Walls, Nature **306** (1983), 141.
5. L. C. Teng, *Frontiers of Particle Beams; observation, Diagnosis and Correction*, Proceedings Anacapri 1988, Springer–Verlag.
6. K. L. Brown and R. V. Servranckx, Nucl. Inst. A**258** (1987) 480.
7. A. Yariv, *Quantum Electronics*, 2nd edition (J. Wiley, New York, 1975).
8. G. Dattoli, S. Solimeno and A. Torre, Phys. Rev. A**35** (1987) 1668.
9. G. Dattoli, M. Richetta, G. Schettini and A. Torre, *Lie Algebraic Methods and solutions of linear partial differential equations* accepted for publication in J. Math Phys.
10. J. Wei and E. Norman, J. Math. Phys. **4** (1963) 575.
11. H. R. Lewis Jr, J. Math. Phys. **9** (1968) 1976; H. R. Lewis Jr. and W. B. Riesenfeld, J. Math. Phys. **10** (1969) 1458.
12. P. K. Aravind, J. Opt. Soc. Am. B**5** (1988) 1545.
13. M. Weissbluth, *Photon–Atom Interactions*, Academic Press (1989).
14. G. Dattoli, C. Mari, C. Ronsivalle and A. Torre, *Coupled transverse motion in charged beam transport dynamics and generalized Twiss coefficients* to be published in Il Nuovo Cimento B.

Conclusions of the panel on tracking

The discussion went on two main points: one turn map and Lyapunov exponent.

1) Tracking per element versus one turn map.

The symplectified one turn map appears as one good hope for investigating the long term behaviour ($\sim 10^8$) of particle trajectories in the future large hadron storage rings (SSC and LHC). Presently it is not possible to estimate the loss of accuracy (or its absence) due to the use of a one turn map instead of a tracking per element. It was stressed that the accuracy of the machine modelisation has to be seriously taken into account.

In this context it will be extremely valuable to compare, on the basis of a single turn tracking, the differences on the final coordinates observed when using different tracking per element and their related symplectified one turn map in the whole relevant machine aperture.

2) Lyapunov exponent.

Several methods for obtaining the Lyapunov exponent were discussed. Looking at the evolution in time of two close trajectories does not provide for all cases stable values: it depends on the distance chosen and on the time domain used.

It was proposed a long time ago to compute it with backwards tracking, which provided good results for the case of beam-beam interaction. It was not clear that it would work as well for the case of magnet errors. In order to overcome the above mentioned drawback a variational method was proposed, which can be implemented easily in the optics codes. This has been tried by the DESY people and was not considered as much more successful because of similar drawbacks.

It was agreed that the values of Lyapunov exponent should be provided with the tracking results in order to help understanding long term behaviour.

Conclusions of the panel on diffusion

The discussion tried to define the essential features of a model which describes the diffusion in a particle accelerator. Two mechanisms of diffusion were considered: the slow modulation in the current of the magnetic elements (modulation diffusion) and the stochastic noise in the magnetic lattice (stochastic diffusion).

The accelerator physicists agree in considering the modulation diffusion as the main present problem for the design of the future accelerators and many experiments on such a diffusion have been carried out on the existing accelerator machines. A well estabilshed theory for interpreting of the experimental results is far from being avaliable also if there is a strong evidence that a Nekhoroshev estimate holds for the beam stability time.

The non linear dynamicists pointed out the necessity of defining a simple model which would allow to obtain analytical results on the modulation diffusion and proposed the study of the dynamics of the Hénon map in $\mathbf{R}^2$ and $\mathbf{R}^4$ when the linear frequency is slowly modulated. In this case by using the normal forms theory it is possible to recover a Nekhoroshev estimate for the stability time.

A general agreement on these kinds of models was finally obtained.

On the other hand it seemed to be more difficult to define a model for the stochastic diffusion because of the different sources of the noise: imperfections of the magnets, electrons protons beam-beam effect, noises in the magnetic currents. The theoretical results on stochastic diffusion in dynamical systems were presented by prof. Y. Kifer.

It was pointed out that any model for the stochastic diffusion has to take into account the periodicity of the particle accelerators so that one has to consider periodic or almost periodic noises.

It was decided to continue the discussion on the stochastic diffusion in the next future on well defined mathematical models.

Concluding Remarks

Y. Kifer

The workshop was interesting but for me, as a mathematician, it could be much more interesting if I could derive from talks formulations of precise mathematical problems which one should try to solve. Unfortunately talks related to accelerators except for few ones were not at the state where mathematicians can enter the picture. There is some confusion in language, no commonly accepted physical models and no clear understanding what is more important and what is less. In order to construct a mathematical model to study accelerators, physicists should specify the most important features of the system and ignore less important ones. For instance, it is not important what is the degree of the polynomial describing the symplectic map which is a composition of kicks; the important thing is what are properties of such maps from the dynamical systems point of view. After constructing a simplified qualitative model which will explain roughly the processes in the system, a refined version of this model can be worked out on a computer.

List of Participants

ACIERNO E.
Dip. Informatica Salerno Univ.
Lancusi(SA)
ITALY

BAZZANI A.
Dip. Fisica Bologna Univ.
Via Irnerio, 46
40126 Bologna
ITALY

BERZ M. Lawrence Berkeley Laboraory
1 Cyclotron Road Berkeley
94720 CALIFORNIA
USA

BISCARI C.
INFN LNF Frascati
P.O. Box 13
00044 Frascati (ROMA)
ITALY

BOUNTIS T.
Univ. of Patras dep. of Mathematics
Faculty of sciences
261.10 Patras GREECE

CONTOPOULOS G.
Dep. of Astronomy Athens Univ.
Panepistemiopolis GR-157 83 Zografos
ATHENS
GREECE

DAINELLI A.
INFN LNL Legnaro (PD)
ITALY

DATTOLI G.
ENEA Frascati
TIB-FIS P.O. Box 65
00044 Frascati (ROMA)
ITALY

DOME G.
CERN, SL Div.
1211 Geneva 23
CH

DRAGT A.
Physics Dep. Univ. of Maryland
College Park, Maryland 20742
USA

FOREST E.
Lawrence Berkeley Laboratory
Bld 71, 1 Cyclotron Road, Berkeley
94720 CALIFORNIA
USA

GALLUCCIO F.
CERN, SL Div.
1211 Geneva 23
CH

GAREYTE J.
CERN, SL Div.
1211 Geneva 23
CH

GERASIMOV A.
Newman Lab. Cornell Univ.
Ithaca NY 14853
USA

GIORGILLI A.
Dip. di Matematica
Via Soldini 50
20133 MILANO
ITALY

GOLIN S.
Techn. Univ. Berlin
Fachbereich Math.. 17 Juni
136 Str. 1 Berlin 12
W. GERMANY

GUIDUCCI S.
INFN LNF Frascati
P.O. Box 13
00044 Frascati
ITALY

IAZZOURENE F.
Sincrotrone Trieste
Padriciano 99, 34012 Trieste
ITALY

IRWIN J.
SLAC Stanford
P.O. Box 4349 (Bin 68)
Stanford 94720 CA
USA

ISELIN F.C.
CERN, SL Div., AP Group
1211 Geneva 23
CH

KIFER Y.
The Hebrew Univ. of Gerusalem
Mathem. Inst. Hebrew Univ. 91904 Jerusalem
ISRAEL

MAHMOUD MOKHTAR G.
Math. Dept. Assiut Univ.
Assiut
EGYPT

MAIS H.
DESY Hamburg
Notkestr. 85, 2000 Hamburg 52
W. GERMANY

MALAVASI M.
Dip. Fisica Bologna Univ.
Via Irnerio, 46
40126 Bologna
ITALY

MARI C.
ENEA Frascati
ENEA TIB-FIS, P.O. Box 65
00044 Frascati
ITALY

MARMI S.
Dip. Fisica Bologna Univ.
Via Irnerio, 46
40126 Bologna
ITALY

MASULLO M.R.
INFN Napoli
Mostra d'Oltremare, Pad. 20
80125 Napoli
ITALY

NAGAOKA R.
Sincrotrone Trieste
Padriciano 99, 34012 Trieste
ITALY

PALUMBO L.
Dip. Energetica Roma Univ.
Via A. Scarpa 14
00161 Roma
ITALY

PETRACCA S.
Dip. Matematica Napoli Univ.
Via Mezzocannone 8
Napoli
ITALY

PIERRO V.
Dip. Fisica Teorica Salerno Univ.
Lancusi (SA)
ITALY

PINTO I.M.
Dip. Elettronica Salerno Univ.
84081 Baronissi (SA)
ITALY

PUSTERLA M.
Dip. Fisica Padova Univ.
Via Marzolo 8
35131 Padova
ITALY

REMIDDI E.
INFN Bologna
Via Irnerio 46
40126 Bologna
ITALY

RIPKEN G.
DESY Hamburg
Notkestr. 85, 2000 Hamburg 52
W. GERMANY

RITSON D.
SSC Lab. and SLAC
Stanford, CA 94305
USA

RUGGIERO A.G.
BNL Brookhaven
Upton NY 11973
USA

SCANDALE W.
CERN, SPS Div.
1211 Geneva 23
CH

SCHMIDT F.
CERN
1211 Geneva 23
CH

SERVIZI G.
Dip. Fisica Bologna Univ.
Via Irnerio, 46
40126 Bologna
ITALY

TALMAN R.
SSC Lab. Dallas
2550 Bekleymeade Av.
Dallas, TX 75237
USA

TENNYSON J.
LBL Berkeley
MS 71259, Berkeley CA 94720
USA

THIELHEIM K.O.
Inst. f. Reine u. Angew. Physik
Univ. Kiel, 2300 Kiel
W. GERMANY

TODESCO E.
Dip. Fisica Bologna Univ.
Via Irnerio, 46
40126 Bologna
ITALY

TURCHETTI G.
Dip. Fisica Bologna Univ.
Via Irnerio, 46
40126 Bologna
ITALY

VACCARO V.G.
Dip. Scienze Fis. Napoli Univ.
Mostra d'Oltremare, Pad.20
80125 Napoli
ITALY

VAZQUEZ L.
Fis. Dep. Comput. Univ. Madrid
Ciudad Universitaria
28040 Madrid
SPAIN

VERDIER A.
CERN, SL Div..
1211 Geneva 23
CH

VISNJIC V.
Fermilab
P.O. Box 500 Batavia, Chicago ILL
USA

VIVALDI F.
Math. School Queen's Mary College
Mile End Road London E1 4NS
U.K.

WARNOCK R.
SLAC Stanford
BIN 26, P.O. Box 4349,
Stanford CA 94309
USA

WILLEKE F.
DESY Hamburg
Notkestr. 85, 2000 Hamburg 52
W. GERMANY

XIE R.
Dip. Fisica Bologna Univ.
Via Irnerio, 46
40126 Bologna
ITALY

YAN Y.
SSC Lab., Dallas
2550 Beckleymade Ave., Dallas TX 75237
USA